Missile Design Guide

Missile Design Guide

Eugene L. Fleeman
Lilburn, GA

 AIAA EDUCATION SERIES

Joseph A. Schetz, Editor-in-Chief
Virginia Polytechnic Institute and State University
Blacksburg, Virginia

Published by the
American Institute of Aeronautics and Astronautics, Inc.
12700 Sunrise Valley Drive, Suite 200, Reston, Virginia 20191-5807

American Institute of Aeronautics and Astronautics, Inc., Reston, Virginia

1 2 3 4 5

Library of Congress Cataloging-in-Publication Data

Names: Fleeman, Eugene L., author.
Title: Missile design guide / Eugene L. Fleeman.
Description: Reston, Virginia : The American Institute of Aeronautics and
  Astronautics, Inc., [2022] | Series: AIAA education series | Includes
  bibliographical references and index.
Identifiers: LCCN 2022012417 (print) | LCCN 2022012418 (ebook) | ISBN
  9781624106187 (paperback) | ISBN 9781624106347 (ebook)
Subjects: LCSH: Tactical missiles—Design and construction. |
  Astronautics--Systems engineering.
Classification: LCC UG1310 .F579 2022 (print) | LCC UG1310 (ebook) | DDC
  623.4/519—dc23/eng/20220331
LC record available at https://lccn.loc.gov/2022012417
LC ebook record available at https://lccn.loc.gov/2022012418

Print ISBN: 978-1-62410-618-7
Ebook ISBN: 978-1-62410-634-7

Copyright © 2022 American Institute of Aeronautics and Astronautics, Inc. Printed in the United States of America. No part of this publication may be reproduced, distributed, or transmitted, in any form or by any means, or stored in a data-base or retrieval system, without prior written permission.

Data and information appearing in this book are for informational purposes only. AIAA is not responsible for any injury or damage resulting from use or reliance, nor does AIAA warrant that use or reliance will be free from privately owned rights.

AIAA EDUCATION SERIES

Editor-in-Chief
Joseph A. Schetz
Virginia Polytechnic Institute and State University

Editorial Board

Joao Luiz F. Azevedo
Comando-Geral De Tecnologia Aeroespacial

Marty Bradley
University of Southern California

James R. DeBonis
NASA Glenn Research Center

Kajal K. Gupta
NASA Dryden Flight Research Center

Rakesh K. Kapania
Virginia Polytechnic Institute and State University

Brian Landrum
University of Alabama, Huntsville

Michael Mohaghegh
The Boeing Company

Conrad F. Newberry
Naval Postgraduate School; California State Polytechnic University, Pomona

Brett Newman
Old Dominion University

Hanspeter Schaub
University of Colorado

David M. Van Wie
Johns Hopkins University

To Mary, Erika, Adrienne, Nick, Chris, Abby, and Ezra

CONTENTS

Preface — xiii

Chapter 1 Introduction — 1
1.1 Overview — 1
1.2 Missile Characteristics Comparison — 2
1.3 Conceptual Design and System Engineering Process — 4
1.4 System-of-Systems Comparison — 8
1.5 Examples of State-of-the-Art Missiles — 12
1.6 Examples of Alternatives in Establishing Mission Requirements — 13
1.7 Use of a Baseline Missile — 15

Chapter 2 Aerodynamics — 17
2.1 Introduction — 17
2.2 Missile Diameter Tradeoff — 19
2.3 Nose Fineness and Geometry Tradeoffs — 21
2.4 Body Drag Prediction — 22
2.5 Boattail Tradeoffs — 23
2.6 Body Normal Force and Lift-to-Drag Prediction — 24
2.7 Sign Convention of Forces, Moments, and Axes — 26
2.8 Static Stability and Body Aerodynamic Center Prediction — 27
2.9 Flare Stabilizer Tradeoffs — 28
2.10 Wings Versus No Wings — 29
2.11 Normal Force Prediction for Planar Surfaces — 32
2.12 Aerodynamic Center Location and Hinge Moment Prediction for Planar Surfaces — 33
2.13 Planar Surface Drag and Lift-to-Drag Prediction — 34
2.14 Surface Planform Geometry and Integration Alternatives — 36
2.15 Flight Control Alternatives — 37
2.16 Maneuver Law Alternatives — 47
2.17 Roll Angle and Control Surface Sign Convention — 49
2.18 Trim and Static Stability Considerations — 50
2.19 Stability and Control Conceptual Design Criteria — 52

Chapter 3 Propulsion — 55
3.1 Introduction — 55
3.2 Propulsion Alternatives Assessment — 56
3.3 Turbojet Flow Path, Components, and Nomenclature — 58
3.4 Turbojet Thrust Prediction — 63
3.5 Turbojet Specific Impulse Prediction — 64
3.6 Subsonic Turbojet Propulsion Efficiency — 65
3.7 Ramjet Flow Path, Components, and Nomenclature — 67
3.8 Ramjet Temperature and Specific Impulse Prediction — 68
3.9 Ramjet Thrust Prediction — 69
3.10 Ramjet Inlet Design Considerations — 70
3.11 Ramjet Combustor Design Considerations — 71

3.12	Ramjet Booster Integration	73
3.13	Ramjet Inlet Options	74
3.14	Supersonic Inlet/Airframe Integration	76
3.15	Fuel Alternatives	80
3.16	Solid Propellant Rocket Motor Flow Path, Components, and Nomenclature	80
3.17	Rocket Motor Performance Prediction	81
3.18	Rocket Motor Sizing Process	83
3.19	Solid Propellant Rocket Motor Production Alternatives	85
3.20	Solid Propellant Rocket Thrust Magnitude Control	86
3.21	Solid Propellant Alternatives	88
3.22	Solid Propellant Aging	89
3.23	Solid Propellant Rocket Combustion Stability	90
3.24	Rocket Motor Case and Nozzle Material Alternatives	92
3.25	Ducted Rocket Design Considerations	94

Chapter 4 Weight   97

4.1	Introduction	97
4.2	Missile Weight Prediction	99
4.3	Center-of-Gravity and Moment-of-Inertia Prediction	101
4.4	Missile Airframe Structure Manufacturing Processes	102
4.5	Missile Airframe Material Alternatives	107
4.6	Missile Structure/Insulation Trades	109
4.7	High Temperature Insulation Materials	110
4.8	Missile Aerodynamic Heating/Thermal Response Prediction	112
4.9	Localized Aerodynamic Heating and Thermal Stress	116
4.10	Missile Structure Design	117
4.11	Seeker Dome Alternatives	125
4.12	Missile Power Supply and Flight Control Actuators	127

Chapter 5 Flight Performance   129

5.1	Introduction	129
5.2	Missile Flight Performance Envelope	130
5.3	Equations of Motion Modeling	135
5.4	Driving Parameters for Missile Flight Performance	136
5.5	Steady-State Flight and Constant Bearing Intercept	137
5.6	Boost, Glide, Coast, Ballistic, and Divert Flight	139
5.7	Turn Performance	145

Chapter 6 Other Measures of Merit   149

6.1	Introduction	149
6.2	Robustness	150
6.3	Lethality	178
6.4	Accuracy	197
6.5	Carriage and Launch Observables	211
6.6	Missile Survivability and Safety	212
6.7	Reliability	227
6.8	Cost	230
6.9	Launch Platform/Fire Control Integration	237

Chapter 7 Sizing Examples and Sizing Tools   263

7.1	Introduction	263
7.2	Rocket Baseline Missile	264
7.3	Ramjet Baseline Missile	277
7.4	Turbojet Baseline Missile	286

7.5	Baseline Guided Bomb	301
7.6	Computer Aided Conceptual Design Sizing Tools	307
7.7	Soda Straw Rocket (DBF, Pareto, Uncertainty Analysis, HOQ, DOE)	309

Chapter 8 Development Process — 325
8.1 Missile Technology and System Development Process — 325
8.2 Examples of State-of-the-Art Advancement — 337
8.3 Enabling Technologies for Missiles — 340

Chapter 9 Lessons Learned — 341

Chapter 10 Summary — 353
10.1 Missile Design Guidelines — 353
10.2 Wrap Up — 355

Chapter 11 References, Bibliography — 357
11.1 References — 357
11.2 Bibliography — 358

Chapter 12 Appendices — 361

List of Figures — 429

Follow-up Communication — 455

Index — 457

Supplemental Materials — 467

PREFACE

Missile Design Guide was developed using material from the author's short courses on Missile Design, Development, and System Engineering. Emphasis is given to presenting the technical content through full color figures, photographs, videos, and graphics. It is oriented toward the needs of aerospace engineering students and professors, missile engineers, system engineers, system analysts, program managers, and others working in the areas of missile systems and missile technology development. Readers will gain an understanding of missile design, missile technologies, launch platform integration, targeting, fire control integration, missile system measures of merit, and the missile system development process. One objective is to provide a reference for the aerospace engineering curriculum of universities. Although the missile community is large, receives a significant amount of funding, and has many technical and system problems to address, currently there are only a few universities that offer courses in missile design. A second objective is to provide a quick reference for the missile community.

What You Will Learn

- Key drivers in the missile design, development, and system engineering process
- Conceptual design criteria for missiles
- Critical tradeoffs, methods, and technologies in aerodynamic, propulsion, structure, seeker, warhead, fuzing, and subsystems sizing to meet flight performance and other requirements
- Launch platform and fire control system integration
- Robustness, lethality, guidance, navigation and control, accuracy, observables, survivability, safety, reliability, and cost considerations
- Missile sizing examples
- Missile system and missile technology development process

A system-level, integrated method is provided for the missile configuration design, development, and analysis activities in addressing requirements such as cost, performance, and risk. Configuration sizing examples are presented for rocket-powered, ramjet-powered, and turbojet-powered baseline missiles as well as guided bombs. System engineering considerations include targeting, launch platform integration, and fire control system integration. Typical values of missile parameters and the characteristics of current missiles are discussed as well as the enabling subsystems and technologies for missiles and the current/projected state-of-the-art of missiles. Full color figures, photographs, and videos illustrate missile development activities and missile performance.

Missile Design Guide is a summary of information that I have collected during my 50+ years of experience in the design and development of missile systems. It distills the knowledge that I have gathered into a method for the design and system engineering of missiles. Generally used are simple, closed-form, analytical, physics-based equations. The equations provide insight into the primary driving parameters and are applicable to broad range of concepts and flight conditions. Closed-form analytical equations are a throwback to the way missile design was conducted fifty years ago. Although computers with modern numerical methods are more precise, they typically provide less insight into the underlying drivers. Also, the more sophisticated computer-based methods may be susceptible to modeling errors, software programming errors, and data input errors. The simple analytical methods of this textbook allow a quick "sanity comparison" with the more complex computer-based methods. Features of *Missile Design Guide* include example calculations of rocket-powered, ramjet-powered, and turbojet-powered baseline missiles as well as guided bombs, typical values of missile parameters, examples of the characteristics of current missiles, discussion of the enabling subsystems and technologies of missiles, and the current/projected state-of-the-art of missiles. When possible, the figures have self-standing content and parameters are defined on the figure, enhancing ease of use. A disadvantage of this approach is that the figures and equations may be cluttered and require a longer time to digest.

The organization of the material in this text is as follows:

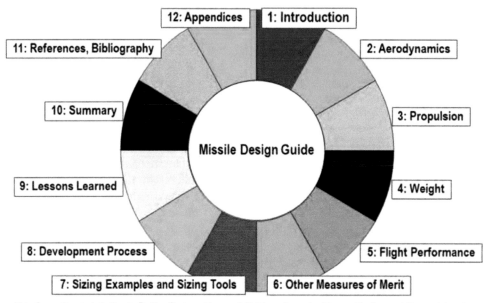

Note: Appendices include Chapter Problem Reviews, Homework Problems, Example of Request for Proposal, Nomenclature, Acronyms, Conversion Factors, Syllabus, Quizzes, Design Case Studies, TMD Spreadsheet, Soda Straw Rocket Science

- Chapter 1 includes an overview of the missile design and system engineering process, different missile characteristics, state-of-the art for missiles, missile configuration sizing parameters, examples of system-of-system integration, and Pareto sensitivity.
- Chapter 2 discusses aerodynamic design, aerodynamic system engineering considerations, flight control alternatives, maneuver law alternatives, and the aerodynamic technologies of low aspect ratio wing and wingless configurations.
- Rocket, ramjet, and turbojet propulsion conceptual design, system engineering, and technologies are included in Chapter 3.
- Chapter 4 addresses missile conceptual design weight prediction, system engineering weight, airframe manufacturing, aerodynamic heating prediction, and technologies for missile weight reduction.
- Conceptual design prediction methods and system engineering considerations for alternative flight trajectories, flight trajectory range, velocity, time-to-target, maneuverability, and off boresight are given in Chapter 5.
- Chapter 6 provides other measures of merit and launch platform integration/system engineering considerations. It includes conceptual design methods and technologies in the areas of robustness, seeker alternatives, warhead lethality, fuzing, guidance, miss distance, carriage and launch observables, other survivability considerations, safety, reliability, cost, firepower, store separation and carriage, fire control system integration, and storage/carriage environmental considerations.
- Conceptual design sizing examples and computer aided sizing tools are presented in Chapter 7. Sizing examples are for a rocket-powered missile, ramjet-powered missile, turbojet-powered missile, guided bomb, and a soda straw rocket. The soda straw rocket sizing example illustrates the use of Pareto analysis, house of quality, and Design of Experiment (DOE) methods to facilitate the sizing process. An example of an electronic spreadsheet sizing tool is also presented.
- Chapter 8 discusses the missile system and technology development process. It includes the typical funding and time frames for missile development, missile development histories and follow-on programs, tradeoff of missile cost, risk, and performance, missile test and integration facilities, and new technologies for missiles.
- Finally, some lessons learned from my career and a summary are presented in Chapters 9 and 10.

The appendices in the back of *Missile Design Guide* have problem reviews for each chapter, homework problems, an example of a request for proposal for a missile design study, list of nomenclature, a list of acronyms, a table for conversion of English to metric units, an example of a syllabus for a graduate course curriculum in Missile Design, Development, and System Engineering, quizzes, a listing of design case studies, a summary of a design spreadsheet, and a summary of a soda straw rocket design-build-fly competition.

References of the data and methods used and a bibliography of other reports and web sites that are related to missiles are provided in the back of the textbook.

Supplemental materials for this book are available for its website. Supplemental materials include the following:

- A presentation of the Missile Design and System Engineering short course that is the basis of this textbook. The slides are in full color. Embedded with the slides are over 100 videos illustrating missile design considerations, development testing, manufacturing, and technologies.
- A Tactical Missile Design spreadsheet in Microsoft Excel format. The original spreadsheet was developed by Georgia Tech graduate students Andrew Frits and Jack Zentner. The original spreadsheet models the configuration sizing methods of the earlier textbook *Tactical Missile Design*. It addresses the design of rocket-powered and ramjet-powered missiles. The updated spreadsheet was developed by Seth Spears and Randy Allen of Lone Star Aerospace. It includes sizing methods for turbojet-powered missiles and guided bombs.
- Missile design case studies. These were conducted by Georgia Tech graduate students.
- A presentation of Soda Straw Rocket Science projects. It is an aerospace engineering outreach program for students.

I would like to thank the faculty and students at the Georgia Institute of Technology for their support for the earlier *Tactical Missile Design* textbook that preceded this textbook. Special appreciation is expressed to Dr. Dimitri Mavris, Director of the Georgia Tech Aerospace Systems Design Laboratory.

I would like to also express my appreciation to the following persons who have supported my work in missiles: Bill Lamar, Charles Westbrook, and Don Hoak of the United States Air Force Research Laboratory, and Mike Yarymovych and Thad Sandford of the Boeing Company.

Finally, I would appreciate receiving your questions, comments and corrections, as well as any data, photographs, drawings, videos, examples, or references that you may offer. Please e-mail these to: GeneFleeman@msn.com or visit the web site: https://sites.google.com/site/eugenefleeman/home.

Eugene L. Fleeman
Lilburn, Georgia
September 2021

Chapter 1   Introduction

1.1 Overview

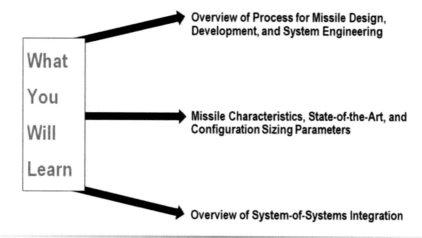

Fig. 1.1   What You Will Learn.

Area	Emphasis
Aero Configuration Sizing	●
Aero Stability & Control	●
Aero Flight Performance	●
Propulsion	●
Structure	●
Weight	●
System Engineering	●
Warhead and Fuzing	○
Seeker, Guidance, Navigation & Control	○
Miss Distance	○
Cost	○
Power Supply and Actuators	○
Survivability, Safety and Reliability	○
Electronics	-
Software	-

Primary Emphasis ●
Secondary Emphasis ○
Tertiary Emphasis -

Fig. 1.2   Emphasis of This Material Is on Physics-Based Conceptual Missile Design.

1.2 Missile Characteristics Comparison

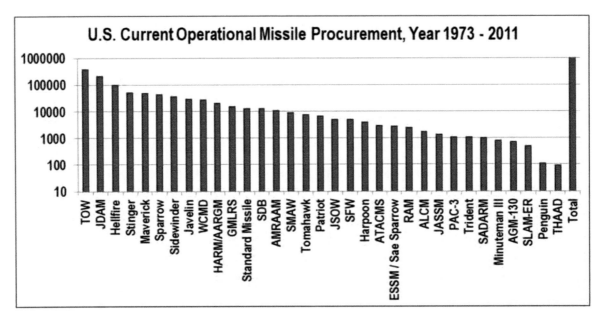

Driver	Example of Tactical Missile SOTA	Example of Strategic Missile SOTA	Tactical vs Strategic SOTA
Accuracy	SM3	GBI	○
Axial Acceleration	HARM	GBI	◒
Maneuverability	Archer	GBI	◒
Speed / Altitude	SM-3	GBI	–
Dynamic Pressure	PAC-3	Minuteman	○
Size / Weight	MHTK	ALCM	●
Quantity / Cost	GBU-31	ALCM	●
Observables	JASSM	AGM-129	●
Range	CALCM	Minuteman	–
# Targets per Use	Storm Shdw	Trident	–
Auto Target Acq	Brimstone	GBI	◒

Tact Strongly Driving ● Tact Driving ◒ Tact & Strat Comp ○ Strat Driving ∽ Strat Strongly Driving –

Fig. 1.3 Examples of State of the Art (SOTA) Drivers—Tactical Missiles: Cost, Strategic Missiles: Range.

Note: Top 23 US operational missile procurements are tactical missiles, with 99.6% of the US 1973-2011 total missile procurement (1,051,731 tactical missiles versus 3,898 strategic missiles (ALCM, Trident, Minuteman III)).

Source: Nicholas, T. and Rossi, R., "U.S. Missile Data Book, 2011," Data Search Associates, Nov 2010

Fig. 1.4 Tactical Missiles Are Produced in Larger Quantity Than Strategic Missiles.

Fig. 1.5 Missile Subsystems Are Analogous to a Human Boxer.

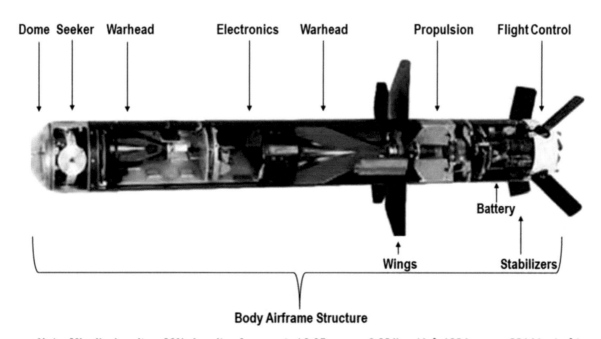

Note: Missile density ~ 60% density of concrete (0.05 versus 0.08 lbm / in³, 1384 versus 2214 kg / m³)

Fig. 1.6 Typical Missile Subsystems—Packaging Is Longitudinal, with High Density.

4 Missile Design Guide

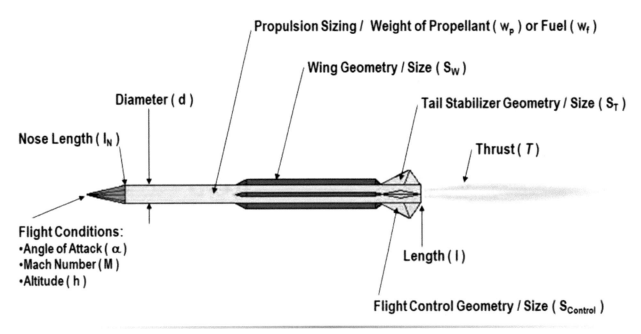

Fig. 1.7 Configuration Sizing Parameters Emphasized in This Material.

1.3 Conceptual Design and System Engineering Process

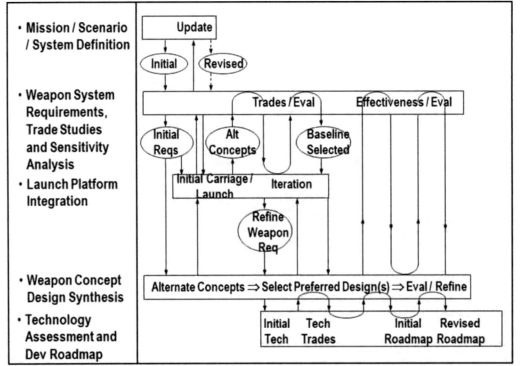

Note: Conceptual design requires fast cycle, ~ 3 to 9 months. Pareto may be used to determine driving parameters for each measure of merit. House of Quality may be used to translate customer requirements to engineering requirements. Design of Experiment may be used to efficiently evaluate and optimize the broad range of design solutions.

Fig. 1.8 Conceptual Design Should Be Unbiased, Creative, and Iterative—with Rapid Evaluations.

Example of Alternative Design Concepts	Example Constraints
Highest Performance Concept	Cost and Risk Threshold / Standard Requirements
Lowest Cost Concept	Performance and Risk Threshold / Standard Requirements
Lowest Risk Concept (e.g., Derivative Concept)	Performance and Cost Threshold / Standard Requirements
Best Value (Balanced Design) Concept (e.g., Best Combination of Performance, Cost, Risk) 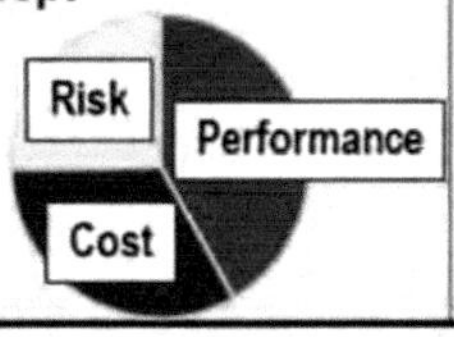	Performance, Cost, and Risk Standard / Goal Requirements

Example of Questions to be Answered before Picking a Preferred Concept / Approach
1. Preferred (best value) concept selected only after evaluating highest performance, lowest cost, and lowest risk concept alternatives
2. The preferred concept / approach strengths will lead to a successful program
3. The preferred concept / approach weaknesses are minor
4. Alternative concepts / approaches have more serious weaknesses

Conceptual design and system engineering studies are required to evaluate performance, cost, and risk of candidate concepts

Fig. 1.9 Conceptual Design and System Engineering Is Required to Explore Alternative Approaches.

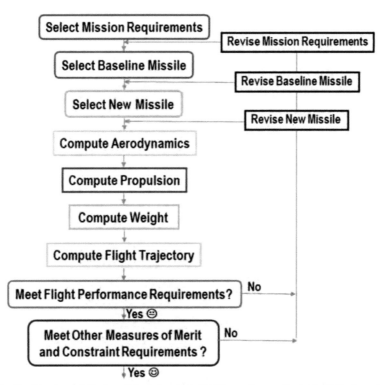

Note: House of Quality, Pareto, and Design of Experiments (DOE) may be used to evaluate the broad range of designs

Fig. 1.10 Missile Conceptual Design and System Engineering Requires Broad, Creative, Rapid, and Iterative Evaluations.

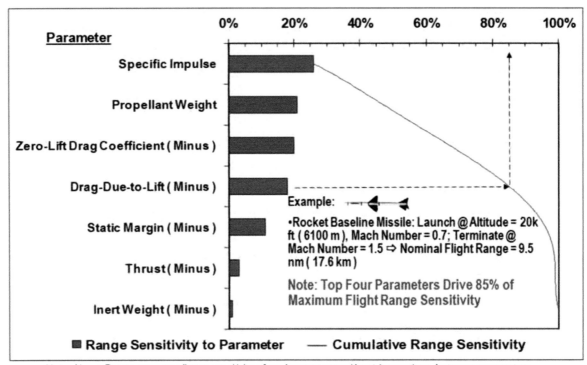

Fig. 1.11 Pareto Effect—Only a Few Parameters Are Drivers for Each Measure of Merit.

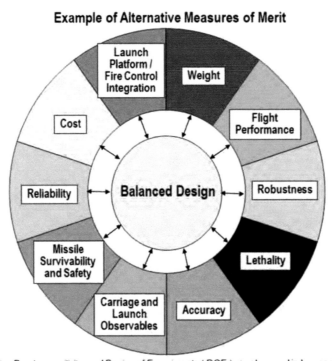

Fig. 1.12 A Balanced Missile Design Requires Harmonized Mission Requirements and Measures of Merit.

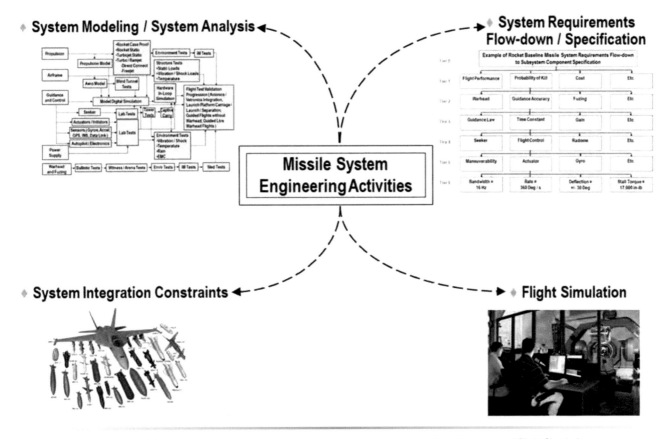

Fig. 1.13 System Engineering Includes Modeling, Analysis, Integration, Requirements, and Flight Simulation.

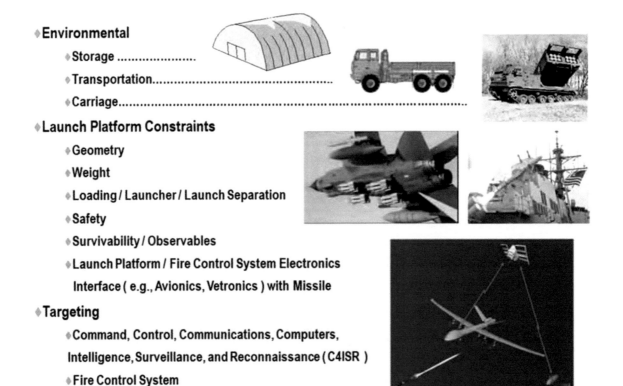

Fig. 1.14 Missile Design and System Engineering Require System Integration.

8 Missile Design Guide

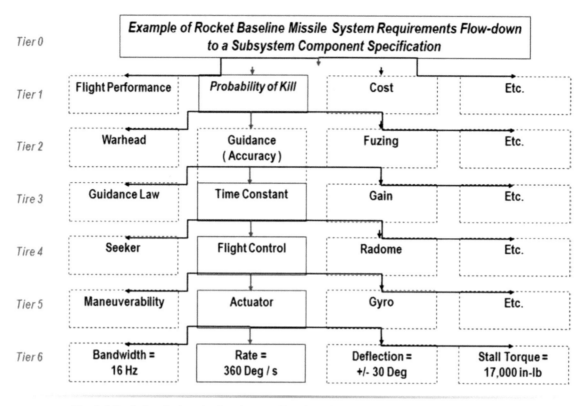

Fig. 1.15 System Engineering Requirements Flow-down Provides Subsystem Components Specifications.

1.4 System-of-Systems Comparison

Example: Typical US Carrier Strike Group, with Complementary Missile Launch Platforms / Load-out

- Air-to-Surface: JASSM, SLAM, Harpoon, JSOW, JDAM, Maverick, HARM, GBU-10, GBU-5, Penguin, Hellfire
- Air-to-Air: AMRAAM, Sparrow, Sidewinder
- Surface-to-Air: SM-3, SM-6, SM-2, Evolved Sea Sparrow, SeaRAM
- Surface-to-Surface: Tomahawk, Harpoon

Fig. 1.16a Missile Development Should Be Conducted in a System-of-Systems Context.
Note: See video supplement of typical US Carrier Strike Group, with Complementary Missile Launch Platforms/Load-out.

Example: Alternative Weapon Approaches for Short Range Ship Lethal Defense

Example Measures of Merit for Alternative Weapons for Short Range Ship Lethal Defense	Missile (RAM) 	Gun Projectiles (Phalanx) 	High Energy Laser 	Railgun Guided Projectile 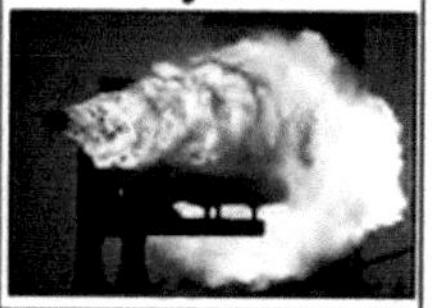
Cost and Weight per Engagement	○	◐	●	◐
Launcher Weight	○	○	–○	–
Ship Power Required	○	○	–	–
Time-to-Target	○	◐	●	●
Req Dwell on Target	○	–○	–	○
Max Range	○	–	–	○
Adverse Weather	○	○	–	○
Technology Maturity	○	○	–	–

● Superior ◐ Good ○ Average – Below Average

Note: Ship survivability alternatives include low observables, electronic countermeasures, decoys, and lethal defense weapons.

Fig. 1.16b Missile Development Should Be Conducted in a System-of-Systems Context (cont).

Example: US Missile Defense System

GMD Aegis THAAD PAC-3
Command, Control, Battle Mgmt, Comm
Missile Defense System

SRBM   MRBM   IRBM   ICBM
< 1000 km < 3000 km < 5500 km > 5500 km
< 9 min   < 19 min   < 26 min   > 26 min
Threats

Note:
——— Primary Mission
- - - - - Secondary Mission

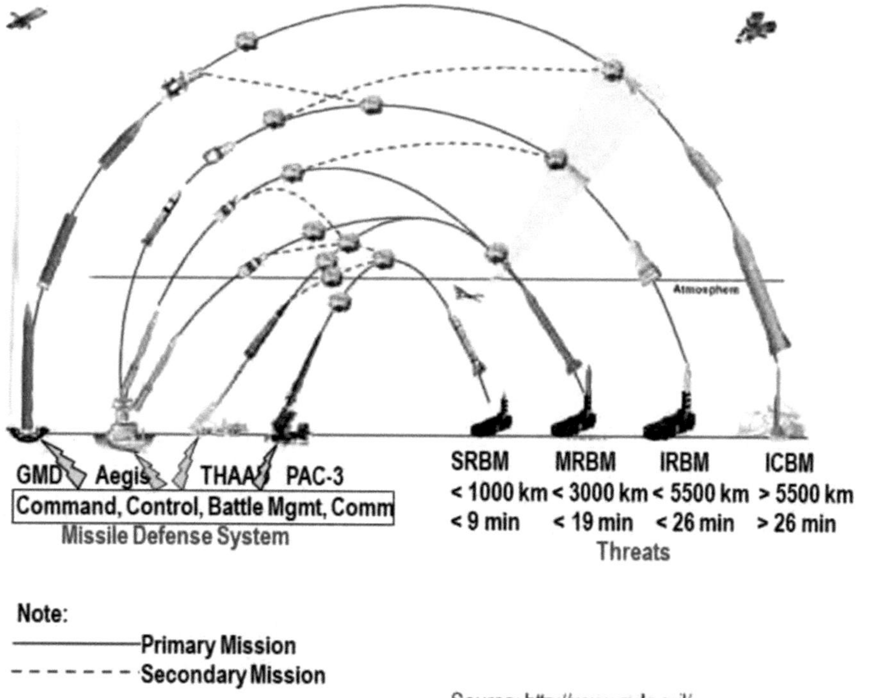

Source: http://www.mda.mil/

Fig. 1.16c Missile Development Should Be Conducted in a System-of-Systems Context (cont).
Note: See video supplement, US Missile Defense System.

Example: US Defense Support Program (DSP) and Space Based Infrared System (SBIRS) Ballistic Missile Launch Warning System

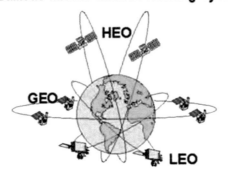

Fig. 1.16d Missile Development Should Be Conducted in a System-of-Systems Context (cont).
Note: See video supplement, DSP and SBIRS.

Example: US Missile Defense Discrimination Radars

Upgraded Early Warning Radar   Sea-Based X-Band Radar   AN/TPY-2 Forward Based Radar   Ship-Based SPY-1 Radar

Fig. 1.16e Missile Development Should Be Conducted in a System-of-Systems Context (cont).
Note: See video supplement, US Missile Defense Midcourse Discrimination Radars.

Example Circa Year 1999 Approach to System-of-Systems Integration for Precision Strike of Mobile Target: Manned Aircraft ⇒ Many Events, Many Personnel ⇒ Relatively Long Timeline

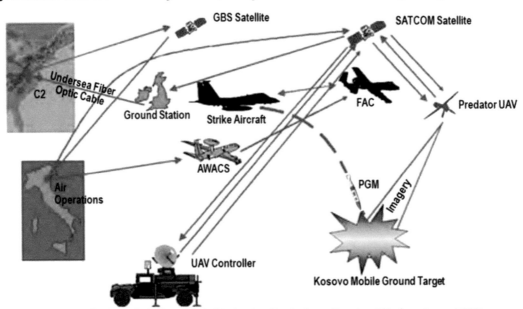

Ref: Alberts, D. S., Garstka, R. E., Hayes, D. A., Signori, D. A., *Understanding Information Age Warfare*, August 2001

Fig. 1.16f Missile Development Should Be Conducted in a System-of-Systems Context (cont).

Example of Circa Year 2002 System-of-Systems Integration for Precision Strike of Mobile Target:
UCAV ⇒ Fewer Events, Fewer Personnel ⇒ Shorter Timeline / Higher Lethality

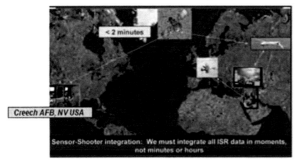

Fig. 1.16g Missile Development Should Be Conducted in a System-of-Systems Context (cont).
Note: See video supplement, Predator Unmanned Combat Air Vehicle (UCAV).

Example: Tactical Ballistic Missile Range Versus Launch Platform Load-out / Firepower

Tactical Ballistic Missile	Max Range	Missile Load-out on HIMARS Launcher
GMLRS (Diameter = 9 in, Weight = 675 lb)	~ 70 km ◒	6 ●
ATACMS (Diameter = 24 in, Weight = 3690 lb)	~ 300 km ●	1 -

HIMARS Carriage On C-130 Aircraft

Note: ● Superior   ◒ Good   - Low
Note: Precision Strike Missile Follow-on to ATACMS Is Treaty Range Limited (< 500 km)

Fig. 1.16h Missile Development Should Be Conducted in a System-of-Systems Context (cont).
Note: See video supplement, HIMARS with Guided MLRS and ATACMS.

1.5 Examples of State-of-the-Art Missiles

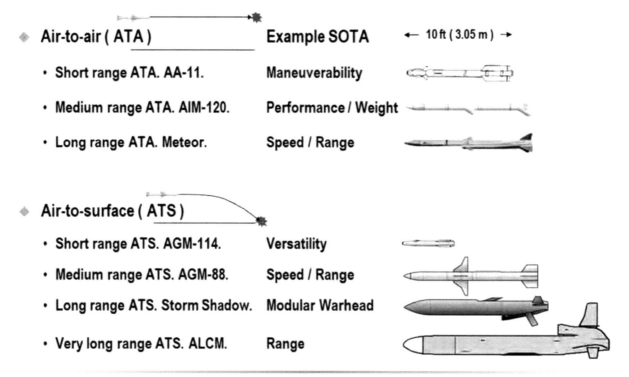

Fig. 1.17 Examples of Design Drivers/State-of-the-Art (SOTA) for Air-Launched Missile Missions, Types, and Attributes.

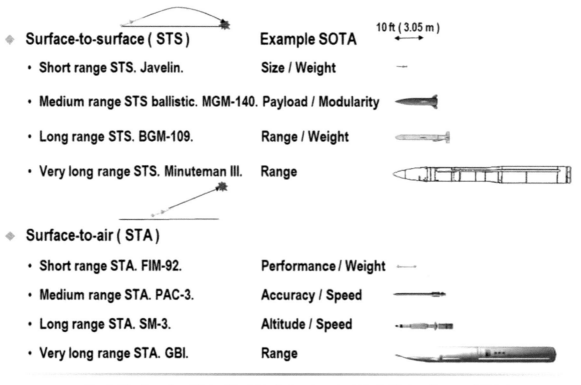

Fig. 1.18 Examples of State-of-the-Art for Surface-Launched Missile Missions, Types, and Attributes.

1.6 Examples of Alternatives in Establishing Mission Requirements

Alternatives for Precision Strike of Time Critical Targets	Measures of Merit		
	Cost per Shot	Number of Launch Platforms Required	Time Critical Target (TCT) Effectiveness
Future Systems			
• Standoff platforms / hypersonic missiles	○	●	◐
• Overhead loitering UCAVs / hypersonic missiles	◐	◐	●
• Overhead loitering UCAVs / lightweight PGMs	●	○	◐
Current Systems			
• Penetrating aircraft / subsonic PGMs	●	–	–
• Standoff platforms / subsonic missiles	○	●	–

Note: ● Superior   ◐ Good   ○ Average   – Poor

Note: Command, Control, Communication, Computers, Intelligence, Surveillance, and Reconnaissance (C4ISR) targeting state-of-the-art provides sensor-to-shooter / weapon connectivity time of less than 2 m and target location error (TLE) of less than 1 m for motion suspended target.

Note: Hypersonic missiles could be airbreathing (e.g., ramjet) or ballistic / semi-ballistic / lofted boost-glide rocket.

Fig. 1.19   Example of Assessment of Alternatives to Establish Requirements for Future Mission.

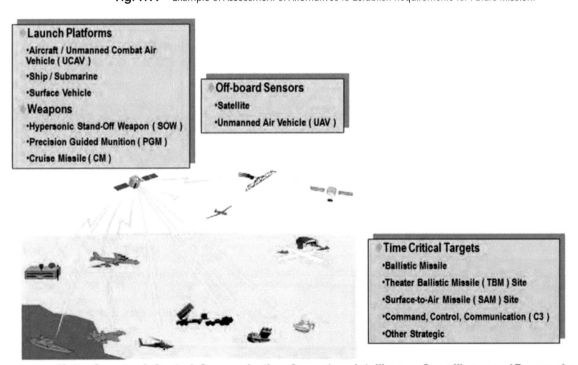

Note: Command, Control, Communication, Computers, Intelligence, Surveillance, and Reconnaissance (C4ISR) targeting state-of-the-art provides sensor-to-shooter / weapon connectivity time of less than 2 m and target location error (TLE) of less than 1 m for motion suspended target.

Fig. 1.20   C4ISR Satellites and UAVs Have High Effectiveness for Time Critical Target Cueing.
Note: See video supplement, Hermes 900 UAV ISR Sensors.

14   Missile Design Guide

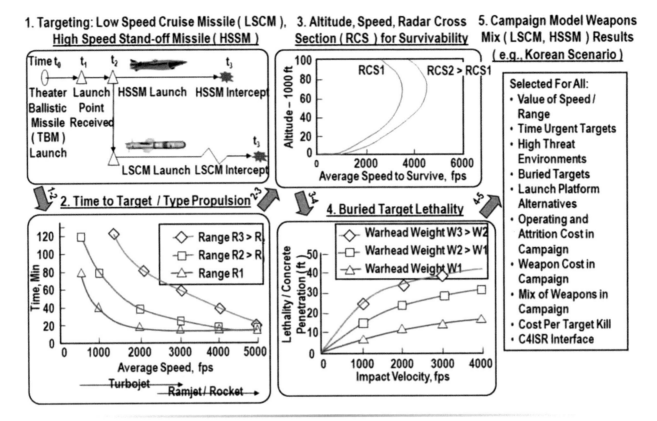

Fig. 1.21 Example of System-of-Systems Analysis to Develop Future Stand-off Missile Requirements.

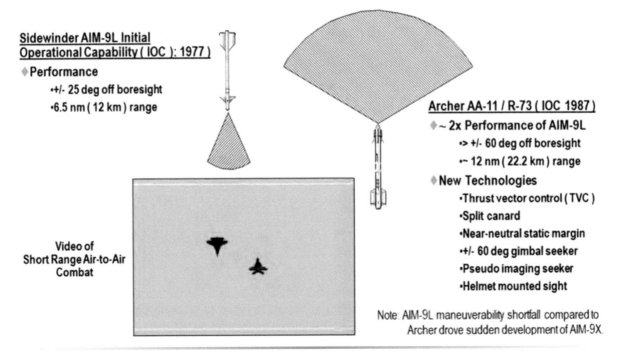

Fig. 1.22 Technological Surprise May Drive Immediate Mission Requirements.
Note: See video supplement, Short Range Air-to-Air Combat.

1.7 Use of a Baseline Missile

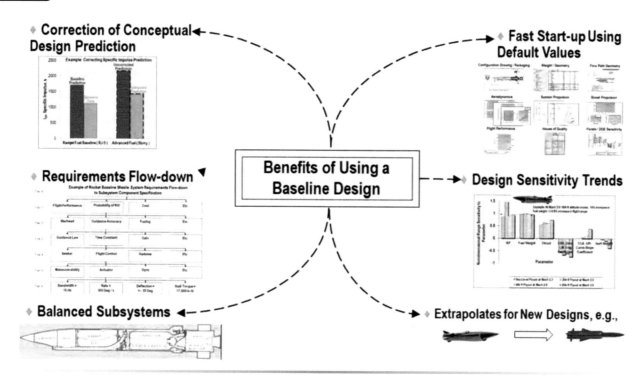

Fig. 1.23 Starting with a Baseline Design Expedites the Design Process.

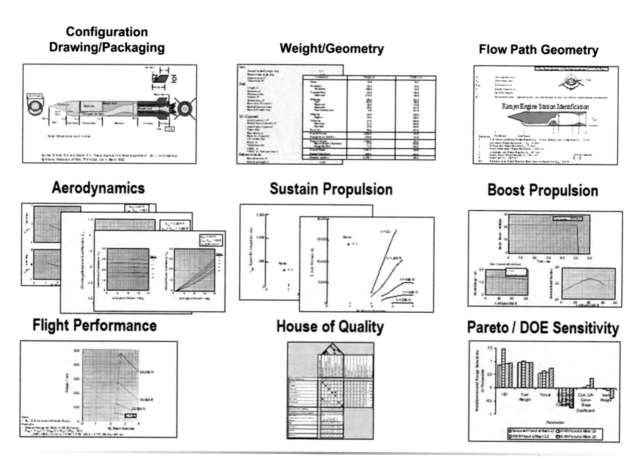

Fig. 1.24 Example of Missile Baseline Data—Chapter 7 Data of Ramjet Baseline Missile.

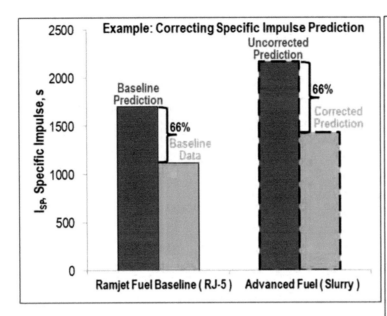

Example:

- Ramjet baseline RJ-5 hydrocarbon fuel:. Heating value H_f = 13.9 x 10^6 ft-lbf / lbm (30.7 x 10^6 J / kg)
- Advanced concept with slurry fuel (40% JP-10 / 60% boron carbide), heating value H_f = 18.5 x 10^6 ft-lbf / lbm (55.3 x 10^6 J / kg)
- Flight conditions: Mach 3.5 cruise, 60k ft (18.3 km) altitude, combustion temperature 4000 R (2222 K)
- Calculate specific impulse $(I_{SP})_{CD,C}$ for conceptual design, based on corrected baseline data
  - $(I_{SP})_{B,C}$ = 1120 s (based on test data)
  - $(I_{SP})_{B,U}$ = 1700 s (computed ideal w/o typical losses from inlet, combustor, nozzle)
  - $(I_{SP})_{CD,U}$ = 2258 s (ditto from above)
  - $(I_{SP})_{CD,C}$ = [$(I_{SP})_{B,C}$ / $(I_{SP})_{B,U}$] $(I_{SP})_{CD,U}$ = [(1120) / (1700)](2258) = 0.659 (2258) = 1488 s

Note:

$(I_{SP})_{CD,C}$ = Specific Impulse of conceptual design, corrected, $(I_{SP})_{B,C}$ = Specific Impulse of baseline, corrected (actual test data), $(I_{SP})_{B,U}$ = Specific Impulse of baseline, uncorrected (computed), $(I_{SP})_{CD,U}$ = Specific Impulse of conceptual design, uncorrected (computed)

Fig. 1.25 Baseline Missile Design Data Allows Correction of Conceptual Design Computed Parameters.

Chapter 2 Aerodynamics

2.1 Introduction

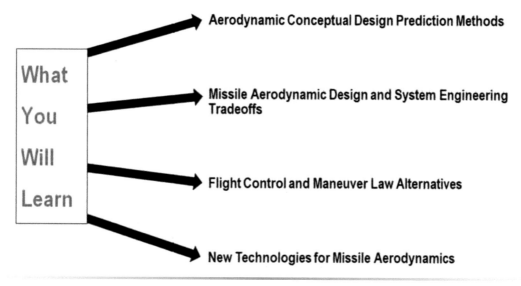

Fig. 2.1 Chapter 2 Aerodynamics—What You Will Learn.

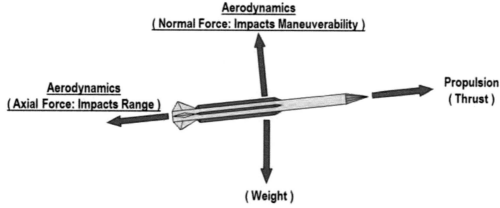

Fig. 2.2 Aerodynamic Forces Impact Missile Maneuverability and Flight Range.

18 Missile Design Guide

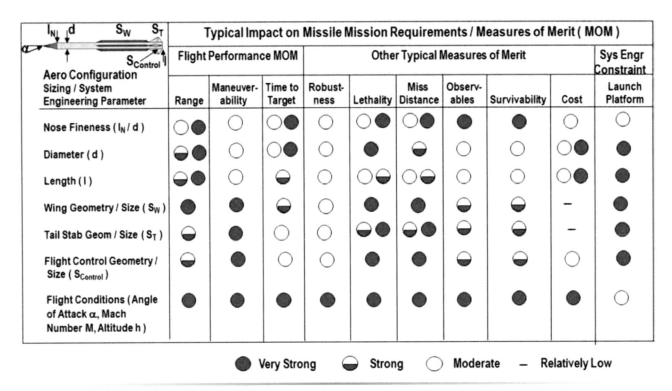

Fig. 2.3 Aero Configuration Sizing and System Engineering Has High Impact on Mission Requirements and Measures of Merit.

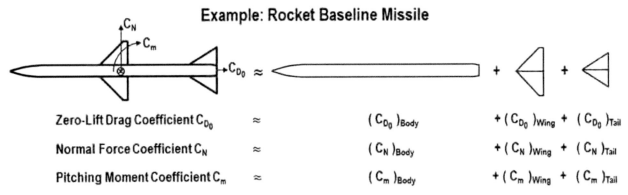

Note:
C_{D_0} = Zero-lift drag / (dynamic pressure x reference area), C_N = Normal force / (dynamic pressure x reference area), C_m = Pitching moment / (dynamic pressure x reference area x reference length).
Conceptual design prediction methods of this text assume independent aerodynamics of body, wing, and tail. These methods do not include aerodynamic interactions of
- Body-wing
- Body-tail
- Wing-body
- Wing-tail
- Wing-wing
- Tail-body
- Tail-wing
- Tail-tail

Fig. 2.4 Conceptual Design Aerodynamic Methods of This Text Are Based on Aero Configuration Buildup.

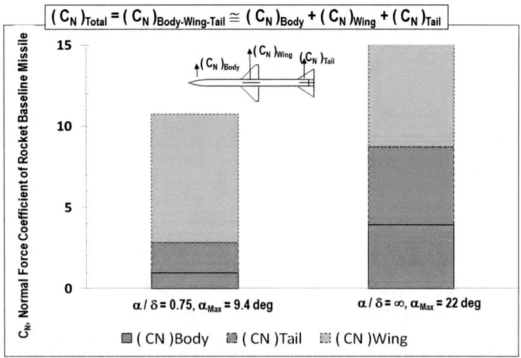

Fig. 2.5 Conceptual Design Total Aerodynamic Force May Be Estimated by Summing Individual Contributors.

2.2 Missile Diameter Tradeoff

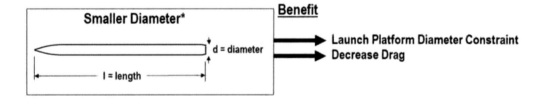

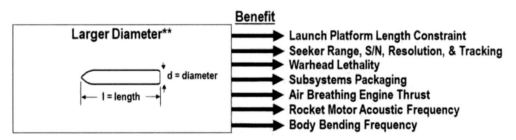

*Smaller diameter missile requires longer length to package subsystems
**Larger diameter missile can package subsystems in shorter length

Note: Typical body fineness ratio is $l/d > 5$ (Javelin $l/d = 8.5$) and $l/d < 25$ (AIM-120 $l/d = 20.5$)

Fig. 2.6 Missile Diameter Is a Tradeoff.
Note: S/N = signal-to-noise ratio.

20 Missile Design Guide

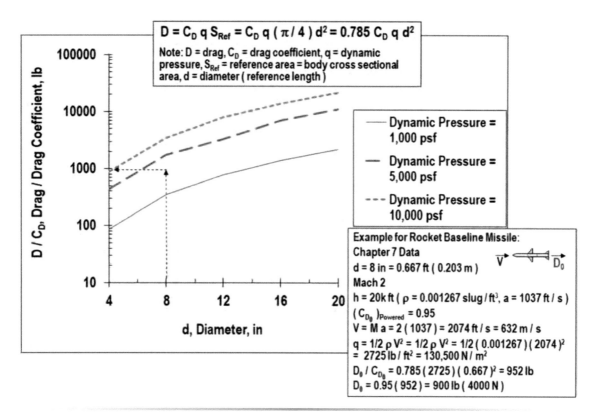

Fig. 2.7 A Small Diameter Missile Has Lower Drag.

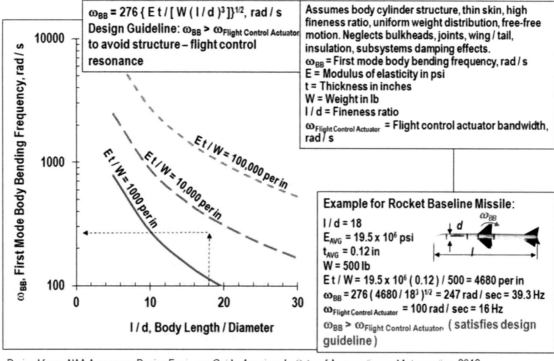

Derived from: AIAA Aerospace Design Engineers Guide, American Institute of Aeronautics and Astronautics, 2012.

Fig. 2.8 Missile Fineness Ratio May Be Limited by Resonance of Body Bending Frequency with Flight Control.

2.3 Nose Fineness and Geometry Tradeoffs

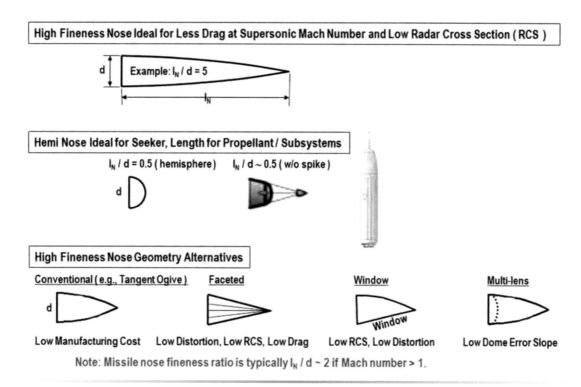

Fig. 2.9 Nose Fineness and Geometry Is a Tradeoff.
Note: See video supplement, Trident Extending Aerospike.

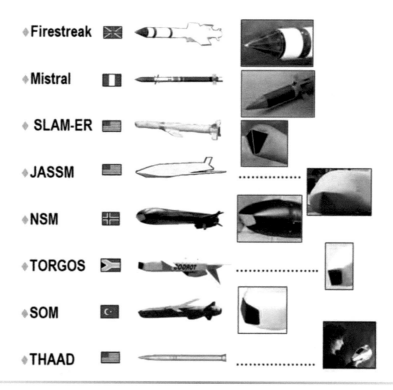

Fig. 2.10 Faceted and Flat Window Domes Can Provide Low Distortion, Low Drag, and Low Radar Cross Section.
Note: See video supplement, Faceted Dome (Mistral).

2.4 Body Drag Prediction

$(C_{D_0})_{Body} = (C_{D_0})_{Body,Friction} + (C_{D_0})_{Base} + (C_{D_0})_{Body, Wave}$

$(C_{D_0})_{Body,Friction} = 0.053\,(l/d)\,[M/(ql)]^{0.2}$. Based on Jerger reference, turbulent boundary layer, q in psf, l in ft.

$(C_{D_0})_{Base,Coast} = 0.25/M$, if $M > 1$ and $(C_{D_0})_{Base,Coast} = 0.12 + 0.13\,M^2$, if $M < 1$

$(C_{D_0})_{Base,Powered} = (1 - A_e/S_{Ref})(0.25/M)$, if $M > 1$ and $(C_{D_0})_{Base,Powered} = (1 - A_e/S_{Ref})(0.12 + 0.13\,M^2)$, if $M < 1$

$(C_{D_0})_{Body, Wave} = (1.59 + 1.83/M^2)\{\tan^{-1}[0.5/(l_N/d)]\}^{1.69}$, for $M > 1$. Based on Jerger reference, $\tan^{-1}$ in rad.

Nomenclature: $(C_{D_0})_{Body,Wave}$ = body zero-lift wave drag coefficient, $(C_{D_0})_{Base}$ = body base drag coefficient, $(C_{D_0})_{Body, Friction}$ = body skin friction drag coefficient, $(C_{D_0})_{Body}$ = body zero-lift drag coefficient, l_N = nose length, d = body diameter, l = body length, A_e = nozzle exit area, S_{Ref} = reference area = body cross-sectional area, q = dynamic pressure, $\tan^{-1}[0.5/(l_N/d)]$ in rad.

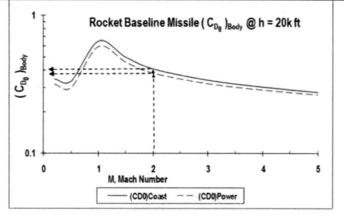

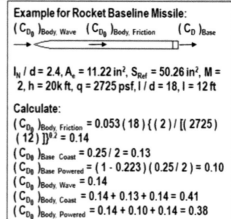

Example for Rocket Baseline Missile:
$(C_{D_0})_{Body, Wave}$ $(C_{D_0})_{Body, Friction}$ $(C_D)_{Base}$

$l_N/d = 2.4$, $A_e = 11.22\,in^2$, $S_{Ref} = 50.26\,in^2$, $M = 2$, $h = 20k\,ft$, $q = 2725\,psf$, $l/d = 18$, $l = 12\,ft$

Calculate:
$(C_{D_0})_{Body, Friction} = 0.053\,(18)\,\{(2)/[(2725)(12)]\}^{0.2} = 0.14$
$(C_{D_0})_{Base\,Coast} = 0.25/2 = 0.13$
$(C_{D_0})_{Base\,Powered} = (1 - 0.223)(0.25/2) = 0.10$
$(C_{D_0})_{Body, Wave} = 0.14$
$(C_{D_0})_{Body, Coast} = 0.14 + 0.13 + 0.14 = 0.41$
$(C_{D_0})_{Body, Powered} = 0.14 + 0.10 + 0.14 = 0.38$

Fig. 2.11 Body Maximum Zero-Lift Drag Coefficient Occurs Near Mach 1.

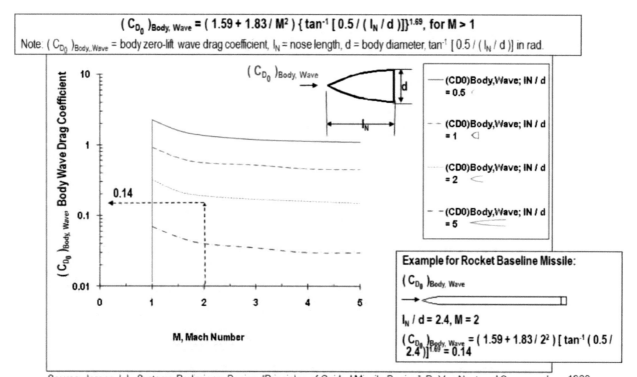

Fig. 2.12 Supersonic Body Wave Drag Is Driven by Nose Fineness.

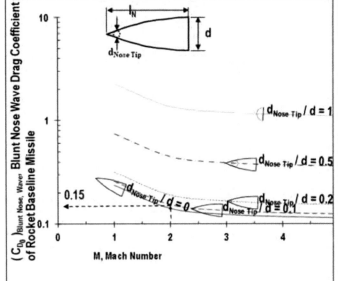

$(C_{D_0})_{Blunt\ Nose,\ Wave} = (C_{D_0})_{Sharp\ Nose,\ Wave}\,[\,1 - (d_{Nose\ Tip}/d)^2\,] + (C_{D_0})_{Hemi,\ Wave}\,(d_{Nose\ Tip}/d)^2$

$(C_{D_0})_{Body,\ Wave} = (1.59 + 1.83/M^2)\,\{\tan^{-1}[\,0.5/(l_N/d)\,]\}^{1.69}$, for $M > 1$.

Example: Rocket Baseline Missile @ $M = 2$: $l_N = 19.2$ in, $d_{Ref} = 8$ in $\Rightarrow l_N/d = 2.4$, 10% Nose Tip Bluntness

1. Compute $(C_{D_0})_{Sharp\ Nose,\ Wave}$:
$(C_{D_0})_{Wave,\ Sharp\ Nose} = (1.59 + 1.83/2^2)\,[\tan^{-1}(0.5/2.4)]^{1.69} = 0.140$ based on body reference area (body cross-sectional area)

2. Compute $(C_{D_0})_{Hemi,\ Wave}$ of hemispherical nose ($l_{NoseTip}/d_{NoseTip} = 0.5$) with $d_{NoseTip} = 0.10\,d = 0.10(8) = 0.8$ in:
$(C_{D_0})_{Wave,\ Hemi} = (1.59 + 1.83/2^2)\,\{[\tan^{-1}(0.5/(0.5))]\}^{1.69} = 2.05(0.665) = 1.36$, based on nose tip cross section area

3. Finally, compute
$(C_{D_0})_{Blunt\ Nose,\ Wave} = (C_{D_0})_{Sharp\ Nose,\ Wave}\,[\,1-(d_{Nose\ Tip}/d)^2\,] + (C_{D_0})_{Hemi,\ Wave}\,(d_{Nose\ Tip}/d)^2$
$= 0.141\,([\,1 - 0.10\,]^2 + 1.36\,(0.10)^2 = 0.140 + 0.014 = 0.153$

Note: 10% nose tip bluntness provides 9% increase in wave drag of the sharp nose

Note: Nose tip bluntness increases drag, increases nose tip strength, and decreases nose tip aero heating

Nomenclature: $d_{Nose\ Tip}$ = nose tip diameter, d = body diameter (reference length), l_N = nose length, $\tan^{-1}[0.5/(l_N/d)]$ in rad

Source: Jerger, J.J., Systems Preliminary Design, "Principles of Guided Missile Design", D. Van Nostrand Company, Inc., 1960

Fig. 2.13 Moderate Nose Tip Bluntness Causes a Relatively Small Increase in Supersonic Drag.

2.5 Boattail Tradeoffs

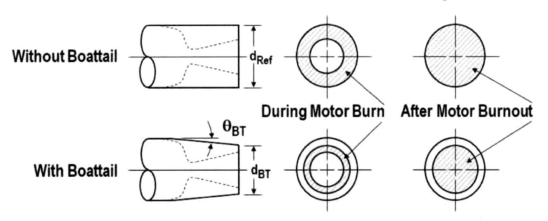

Note: Boattail angle θ_{BT} and boattail diameter d_{BT} limited by propulsion nozzle packaging, tail flight control packaging, and flow separation

Reference: Chin, S. S., *Missile Configuration Design*, McGraw-Hill Book Company, New York, 1961

Fig. 2.14 A Boattail Decreases Base Pressure Drag Area.

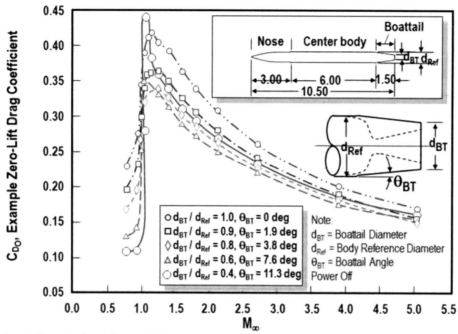

Note: Boattail angle should be < ≈ 12 deg for subsonic missile and < ≈ 7 deg for supersonic missile, to avoid flow separation.
Source: Mason, L.A., Devan, L. and Moore, F.G., "Aerodynamic Design Manual for Tactical Weapons," NSWC TR 81-156, July 1981

Fig. 2.15 A Boattail Is More Effective for a Subsonic Missile.

2.6 Body Normal Force and Lift-to-Drag Prediction

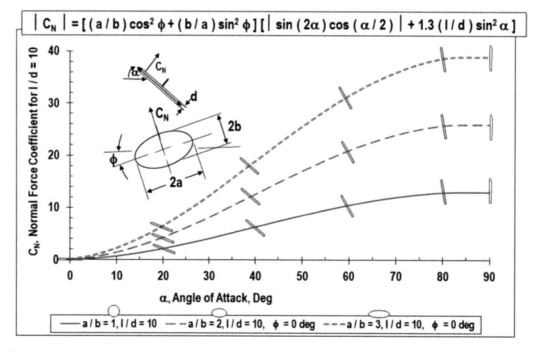

Note: Based on slender body theory (Pitts, et al) and cross flow theory (Jorgensen) references. Valid for $l/d > 5$, $d = 2(ab)^{1/2}$

Fig. 2.16 A Lifting Body Has Higher Normal Force.

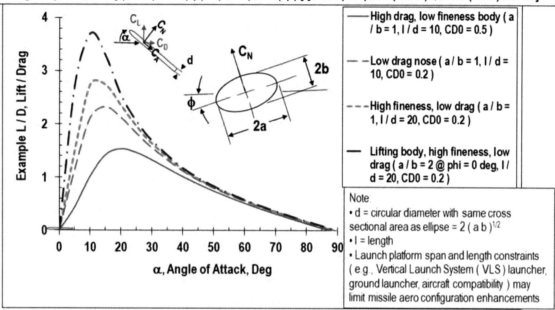

Fig. 2.17 Body Lift-to-Drag Ratio Is Impacted by Angle of Attack, Zero-Lift Drag Coefficient, Body Fineness, and Cross Section Geometry.

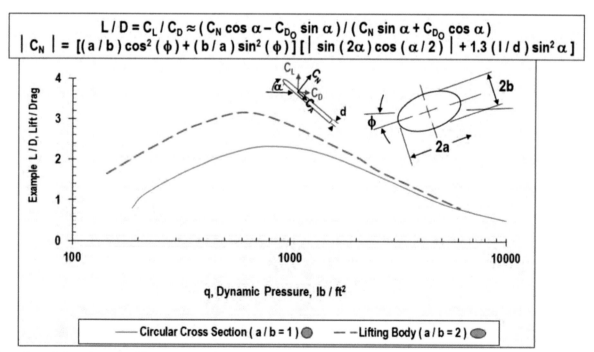

Fig. 2.18 A Lifting Body Requires Flight at Relatively Low Dynamic Pressure to Achieve High Lift-to-Drag Ratio.

26 Missile Design Guide

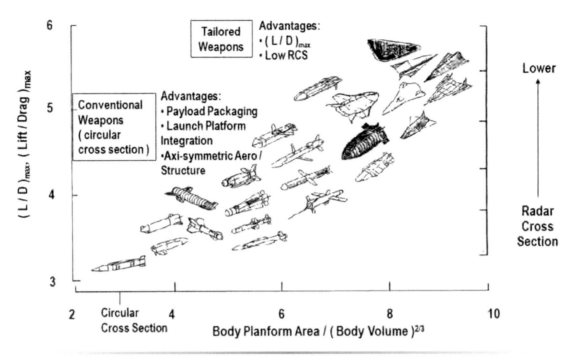

Fig. 2.19 A Lifting Body Typically Has Higher Lift-to-Drag and Lower RCS, A Circular Cross Section Body Has Higher Volumetric Efficiency.

2.7 Sign Convention of Forces, Moments, and Axes

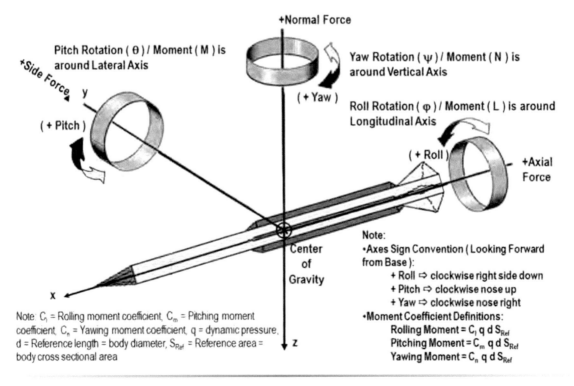

Fig. 2.20 Sign Convention of Forces, Moments, and Axes.

CHAPTER 2 Aerodynamics   27

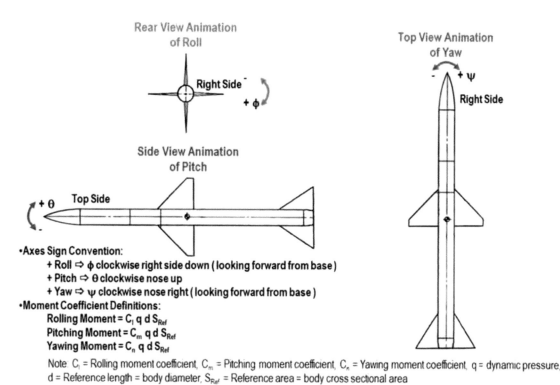

Fig. 2.21 Pitch, Yaw, and Roll Animation of Rocket Baseline Missile.
Note: Animation of Pitch, Yaw, and Roll.

2.8 Static Stability and Body Aerodynamic Center Prediction

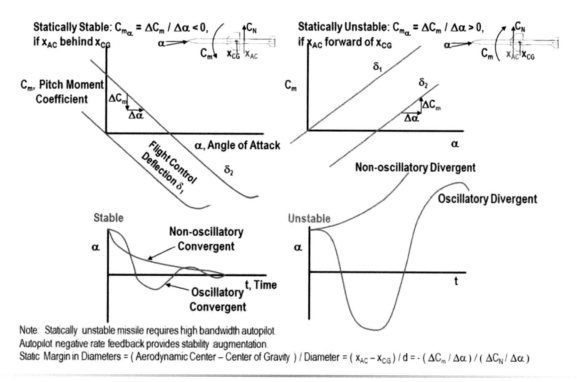

Fig. 2.22 Pitch Moment Stability $\Delta C_m/\Delta\alpha$ and Static Margin ($x_{AC} - x_{CG}$) Define Pitch Static Stability.

28 Missile Design Guide

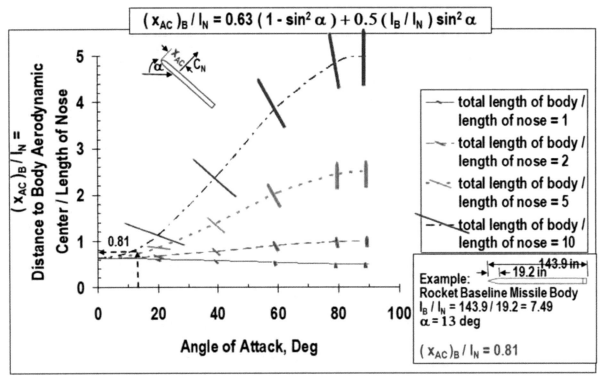

Note: Based on slender body theory (Pitts, et al) and cross flow theory (Jorgensen) for axisymmetric nose-cylinder. No flare. $(x_{AC})_B$ = location of body aerodynamic center, l_N = length of nose, α = angle of attack, l_B = total length of body.

Fig. 2.23 Body Aerodynamic Center Is Driven by Angle of Attack, Nose Length, and Body Length.

2.9 Flare Stabilizer Tradeoffs

Example of Static Margin for Body-Flare Configuration (THAAD)

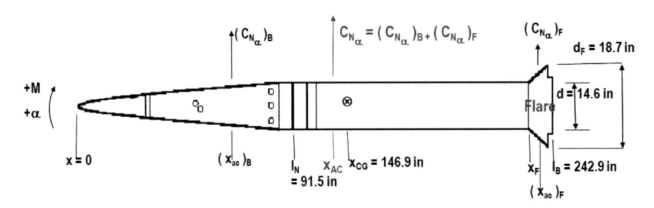

Nomenclature:
M = pitching moment, α = angle of attack, x_{AC} = location of aerodynamic center, x_{CG} = location of center of gravity, d = diameter of body, $(C_{N_\alpha})_B$ = normal force coefficient variation with angle of attack for body, $(x_{AC})_B$ = location of aerodynamic center for body, l_N = length of nose, $(C_{N_\alpha})_F$ = normal force coefficient variation with angle of attack for flare, $(x_{AC})_F$ = location of aerodynamic center for flare, l_B = length of body, d_F = diameter of flare

Fig. 2.24a An Aft Flare Increases Static Stability.

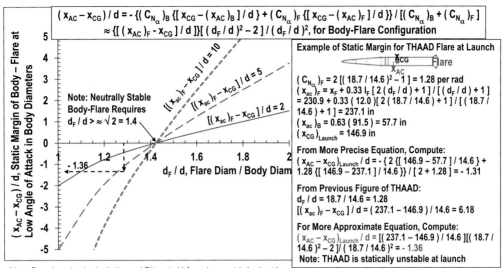

Fig. 2.24b An Aft Flare Increases Static Stability (cont).

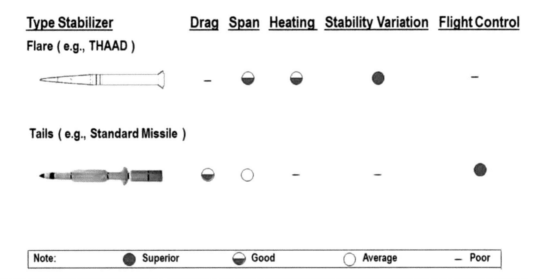

Fig. 2.25 Tail Stabilizers Have Less Drag and Provide Flight Control, Flare Stabilizer Has Less Aero Heating and Less Variation in Stability.

2.10 Wings Versus No Wings

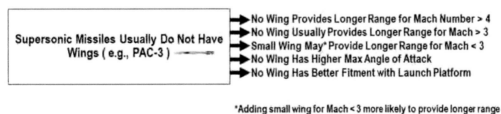

*Adding small wing for Mach < 3 more likely to provide longer range if flight is at high altitude

Fig. 2.26a Most Supersonic Missiles Do Not Have Wings.

30 Missile Design Guide

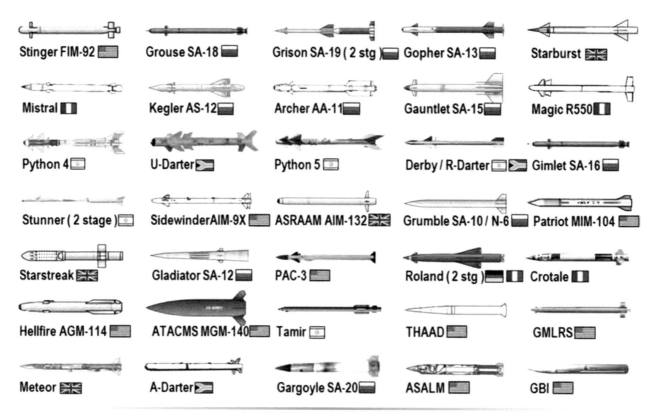

Fig. 2.26b Most Supersonic Missiles Do Not Have Wings (cont).

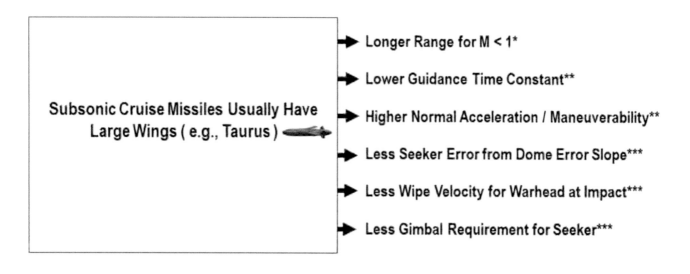

Fig. 2.27a Most Subsonic Cruise Missiles Have Relatively Large Wings.

CHAPTER 2 Aerodynamics 31

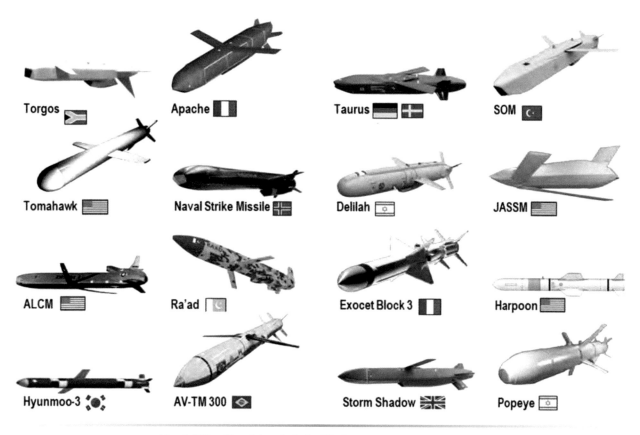

Fig. 2.27b Most Subsonic Cruise Missiles Have Relatively Large Wings (cont).

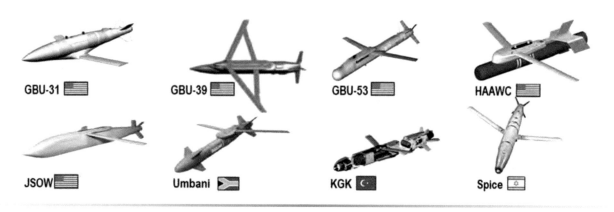

Fig. 2.28 Examples of Guided Bombs That Have Wings for Extended Range.
Note: See video supplement, GBU-39 Stormbreaker/SDB; KGK; Spice.

2.11 Normal Force Prediction for Planar Surfaces

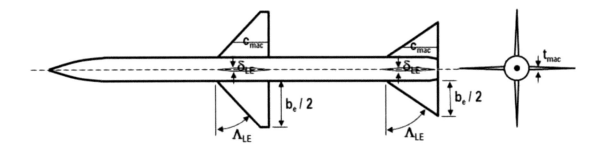

- c_{mac} = mean aerodynamic chord length
- Λ_{LE} = leading edge sweep angle
- δ_{LE} = leading edge section total angle
- t_{mac} = max thickness of mean aerodynamic chord
- b_e = span of exposed planform
- S_e = area of exposed planform
- $A_e = b_e^2 / S_e$ = aspect ratio of exposed planform

Fig. 2.29 Definition of Planar Aerodynamic Surface Geometry Parameters.

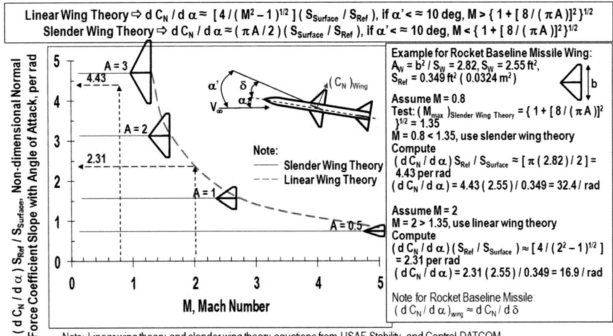

Note: Linear wing theory and slender wing theory equations from USAF Stability and Control DATCOM
Note: Slender wing theory good accuracy limited to A < ≈ 2 for small effects of compressibility, wing sweep, and taper ratio
Nomenclature: A = Aspect ratio, $S_{Surface}$ = Surface planform area, S_{Ref} = Reference area = body cross-sectional area, α = Angle of attack

Fig. 2.30 Normal Force Coefficient of a Planar Surface (Wing, Tail, Canard) Is Higher at Low Mach Number.

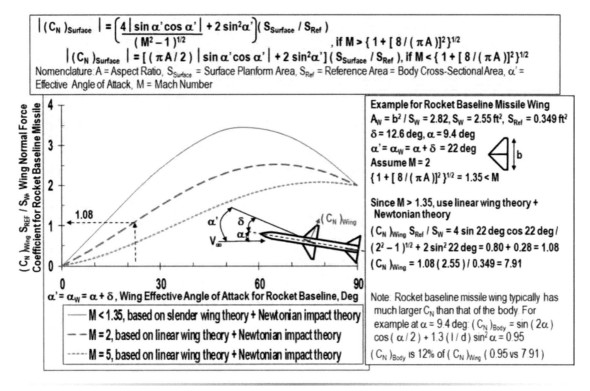

Fig. 2.31 High Normal Force for a Planar Surface Occurs at High Local Angle of Attack.

2.12 Aerodynamic Center Location and Hinge Moment Prediction for Planar Surfaces

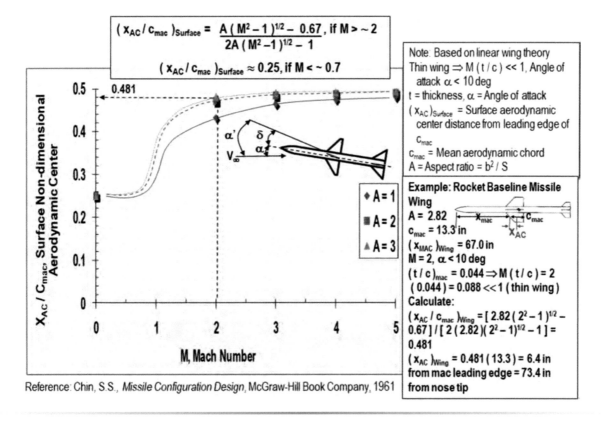

Fig. 2.32 Aerodynamic Center of a Planar Surface Moves Aft with Increasing Mach Number.

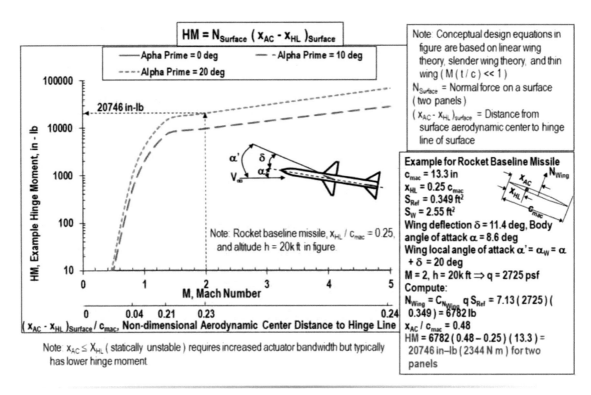

Fig. 2.33 Hinge Moment Increases with Mach Number, Aerodynamic Center Distance from Hinge Line, and Angle of Attack.

2.13 Planar Surface Drag and Lift-to-Drag Prediction

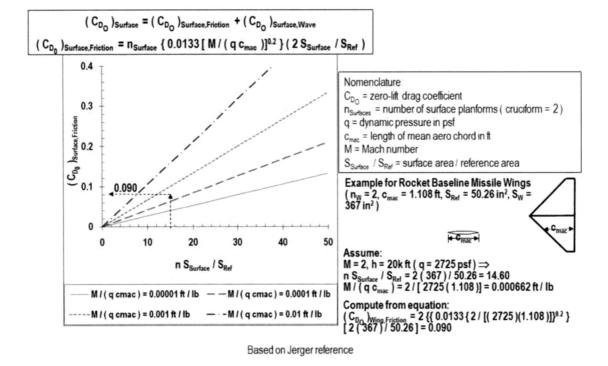

Fig. 2.34 Skin Friction Drag Is Lower for Small Surface Area.

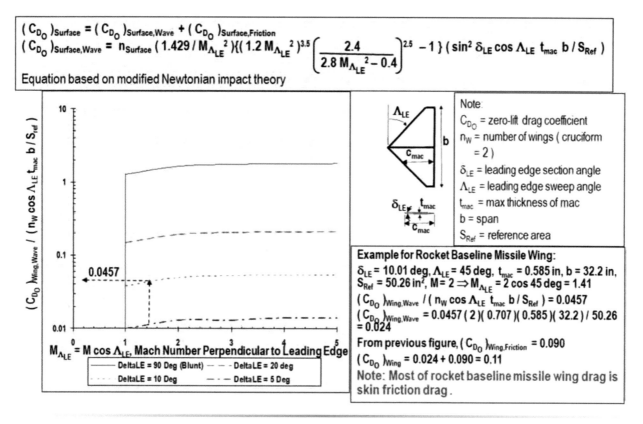

Fig. 2.35 Supersonic Drag of Planar Surface Is Smaller if Leading Edge Has Sweep and Small Section Angle.

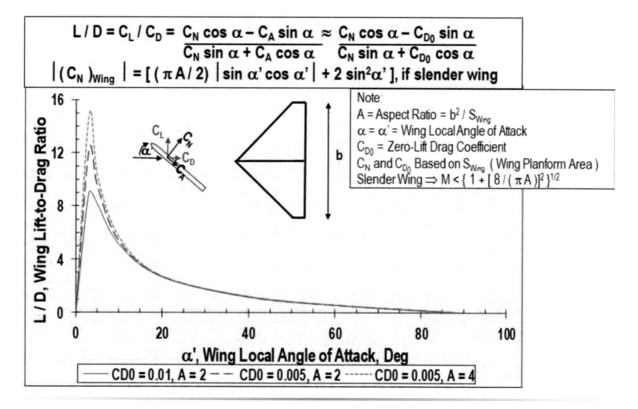

Fig. 2.36 Wing Subsonic Aero Efficiency Lift-to-Drag Is Driven by Angle of Attack, C_{D_0}, and Aspect Ratio.

2.14 Surface Planform Geometry and Integration Alternatives

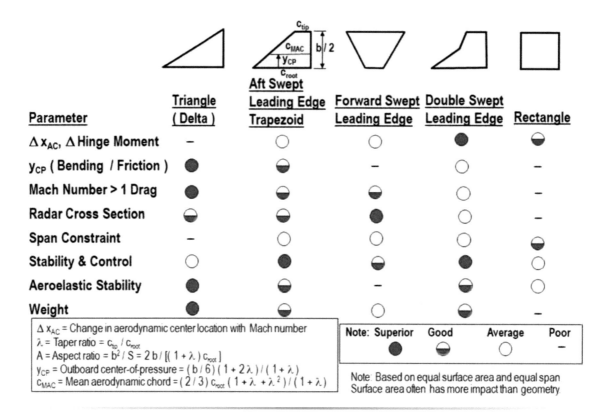

Fig. 2.37 Planar Surface (Wing, Tail, Canard) Panel Geometry Is a Tradeoff with Many Considerations.

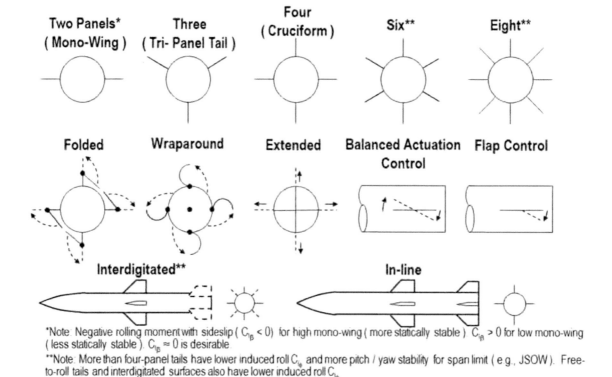

Fig. 2.38 Examples of Wing, Stabilizer, and Flight Control Surface Arrangements and Alternatives.

2.15 Flight Control Alternatives

Control Integ	Control Surfaces	Example	Control Effect	Cost	Packaging
Pitch / Yaw	2	Stinger FIM-92	–	●	●
Pitch / Roll	2	ALCM AGM-86	–	●	●
Pitch / Roll + Yaw	3	JASSM AGM-158	–	◐	●
Pitch / Yaw / Roll	3	SRAM AGM-69	–	◐	◐
Pitch / Yaw / Roll (Cruciform Most Common Type for Missiles)	4	Adder AA-12	○	○	○
Pitch + Yaw + Roll	5	Kitchen AS-4	●	–	–
Pitch / Yaw + Roll	6	Derby / R-Darter	◐	–	–
Pitch / Yaw / Roll (Blended Canard–Tail Control)	8	Stunner	●	–	–

Note: ● Superior   ◐ Good   ○ Average   – Poor

Fig. 2.39 Most Aero Flight Control Missiles Have Four Control Surfaces, with Pitch, Yaw, and Roll Control Integration.

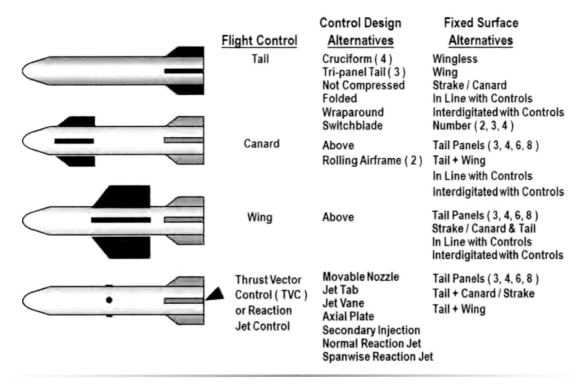

	Flight Control	Control Design Alternatives	Fixed Surface Alternatives
	Tail	Cruciform (4) Tri-panel Tail (3) Not Compressed Folded Wraparound Switchblade	Wingless Wing Strake / Canard In Line with Controls Interdigitated with Controls Number (2, 3, 4)
	Canard	Above Rolling Airframe (2)	Tail Panels (3, 4, 6, 8) Tail + Wing In Line with Controls Interdigitated with Controls
	Wing	Above	Tail Panels (3, 4, 6, 8) Strake / Canard & Tail In Line with Controls Interdigitated with Controls
	Thrust Vector Control (TVC) or Reaction Jet Control	Movable Nozzle Jet Tab Jet Vane Axial Plate Secondary Injection Normal Reaction Jet Spanwise Reaction Jet	Tail Panels (3, 4, 6, 8) Tail + Canard / Strake Tail + Wing

Fig. 2.40 There Are Many Flight Control Aerodynamic Configuration Alternatives.

Examples of Missile Flight Control

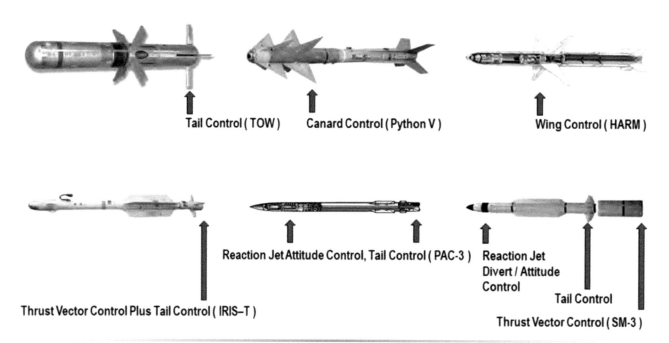

Fig. 2.41 Missile Flight Control Selection Is Typically Driven by Maneuverability, Accuracy, Weight, Volume, and Cost.

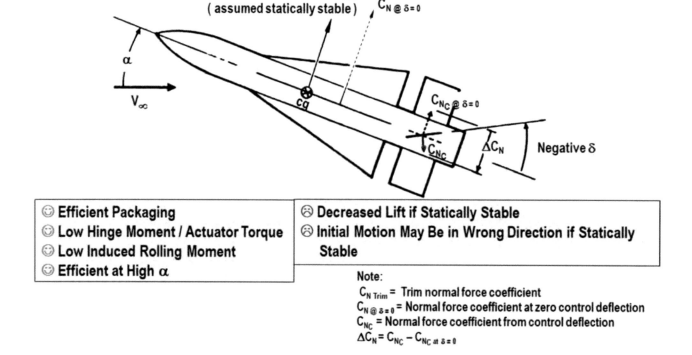

Fig. 2.42 Tail Flight Control Is Efficient at High Angle of Attack.

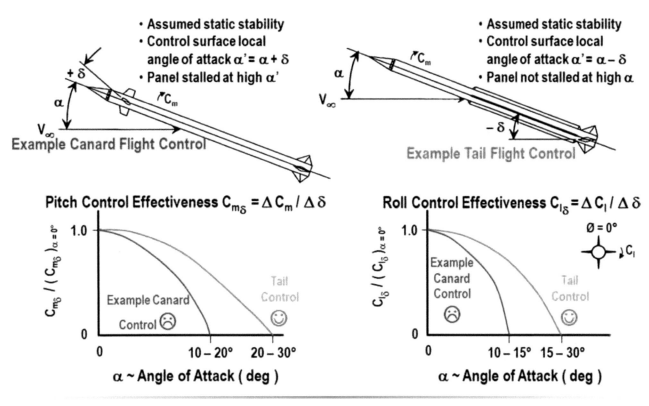

Fig. 2.43 Tail Flight Control Can Usually Operate at Higher Angle of Attack Than Canard Flight Control.

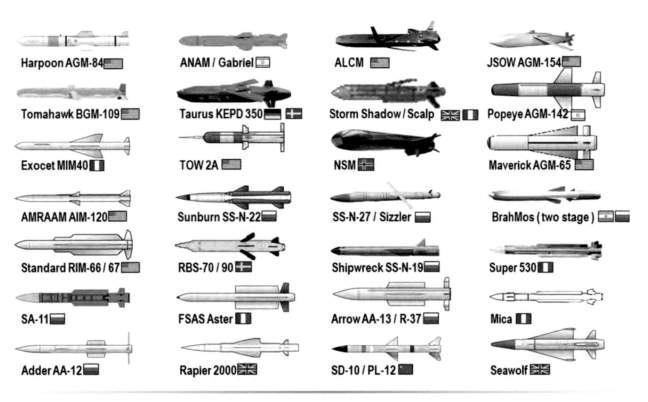

Fig. 2.44 About 70% of Tail Flight Control Missiles Have Wings.

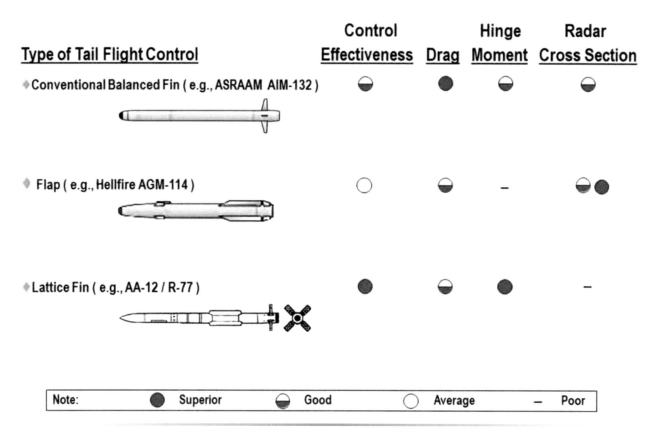

Fig. 2.45 Tail Flight Control Alternatives—Conventional Balanced Actuation Fin, Flap, and Lattice Fin.

◆ Advantages ☺

- High control effectiveness at low subsonic and high supersonic Mach number
- Low hinge moment
- Short chord length

◆ Disadvantages ☹

- High radar cross section (cavities, normal leading edges)
- High drag at transonic Mach number (choked flow)
- High drag at low supersonic Mach number (shock wave – boundary layer interaction)
- Higher leading edge heat transfer at supersonic Mach number

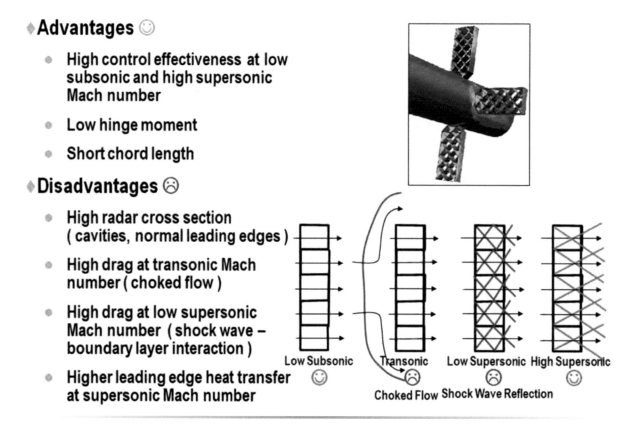

Fig. 2.46 Lattice Fin Flight Control Has Advantages for Low Subsonic and High Supersonic Missiles.
Note: See video supplement, Lattice Fin Flight Control.

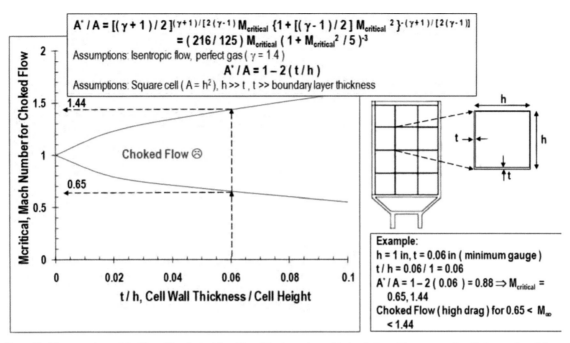

Fig. 2.47 Lattice Fin Choked Flow Is Driven by Lattice Section Thickness and Transonic Mach Number.

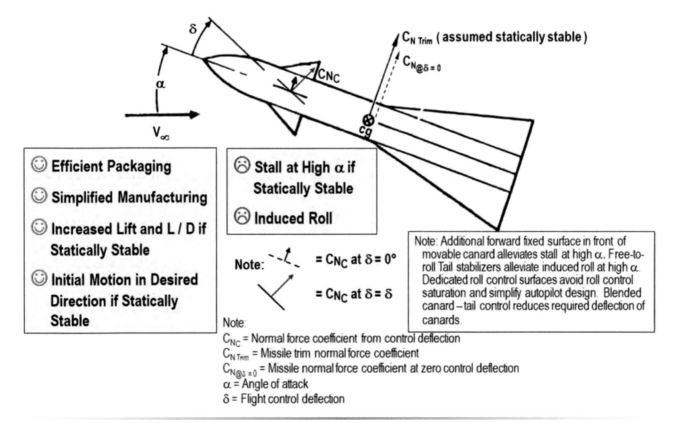

Fig. 2.48 Conventional Canard Flight Control Is Efficient at Low Angle of Attack, but Stalls at High Angle of Attack with Induced Roll.

42 Missile Design Guide

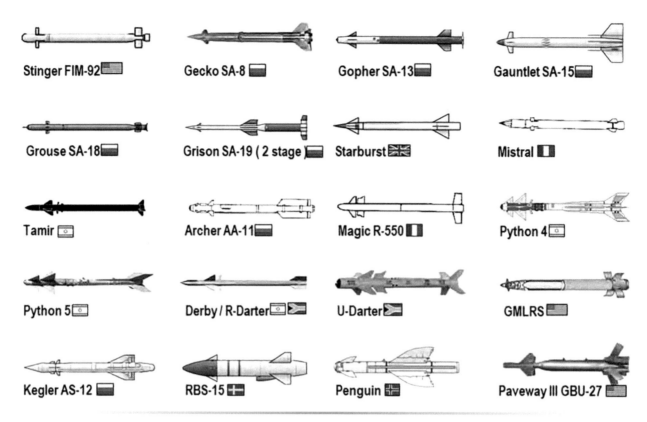

Fig. 2.49 Most Canard Flight Control Missiles Are Wingless and Most Are Supersonic.

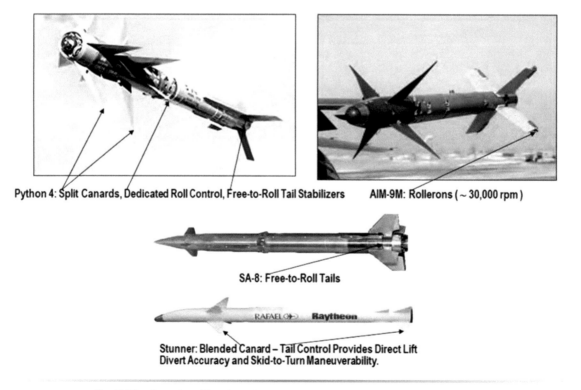

Fig. 2.50 Examples of Aerodynamic Approaches that Enhance Maneuverability and Accuracy of Canard Flight Control.
Note: See video supplement, DAGR Free-to-Roll Tail Stabilizers.

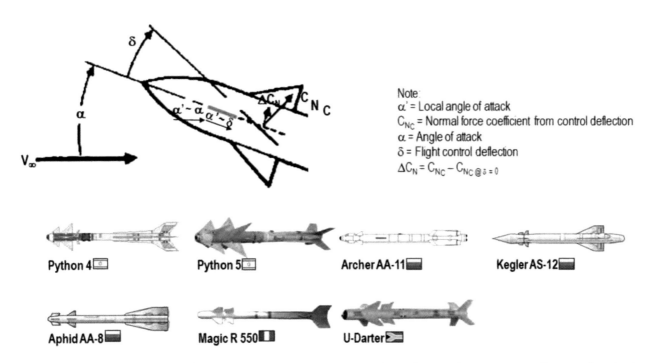

Note:
α' = Local angle of attack
C_{NC} = Normal force coefficient from control deflection
α = Angle of attack
δ = Flight control deflection
$\Delta C_N = C_{NC} - C_{NC @ \delta = 0}$

Note: Forward fixed surface reduces local angle-of-attack for movable canard, providing lower hinge moment and higher stall angle of attack. Forward surface also provides a fixed, symmetrical location for vortex shedding from the body.

Python 4 also has free-to-roll tail stabilizers and dedicated roll control ailerons.

Fig. 2.51 Split Canard Flight Control Provides Maneuverability at High Angle of Attack with Lower Hinge Moment.

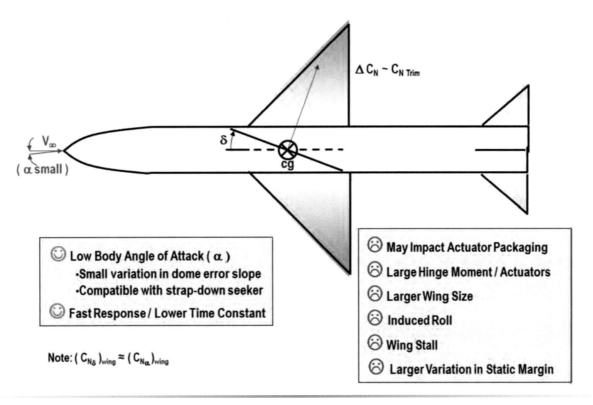

☺ **Low Body Angle of Attack (α)**
 · Small variation in dome error slope
 · Compatible with strap-down seeker
☺ **Fast Response / Lower Time Constant**

Note: $(C_{N\delta})_{wing} \approx (C_{N\alpha})_{wing}$

☹ May Impact Actuator Packaging
☹ Large Hinge Moment / Actuators
☹ Larger Wing Size
☹ Induced Roll
☹ Wing Stall
☹ Larger Variation in Static Margin

Fig. 2.52 A Wing Flight Control Advantage Is Low Body Rotation. Disadvantages Are Larger Hinge Moment, Induced Roll, and Tendency to Stall.

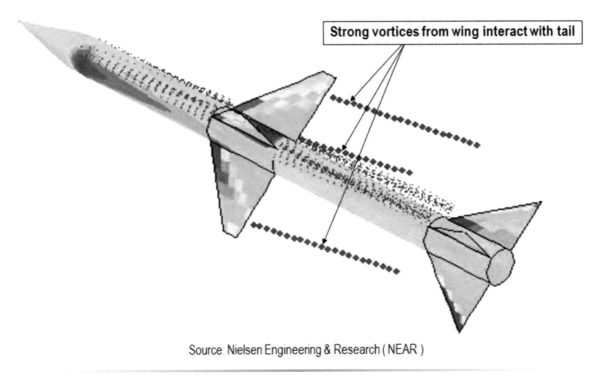

Fig. 2.53 Wings Are Susceptible to Strong Vortex Shedding.
Note: See video supplement, Vortices from Delta Wing at High Angle of Attack.

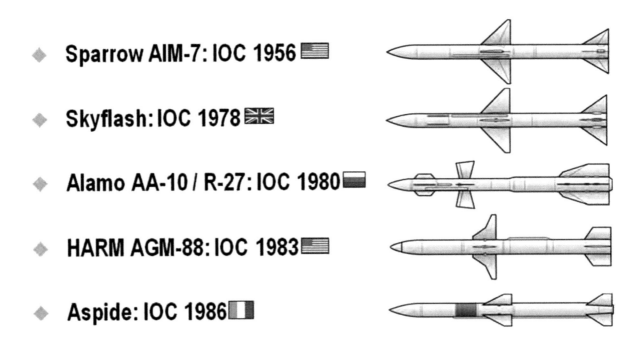

Note: IOC = Initial operational capability

Fig. 2.54 Current Wing Flight Control Missiles Are Supersonic and Are Old Technology.

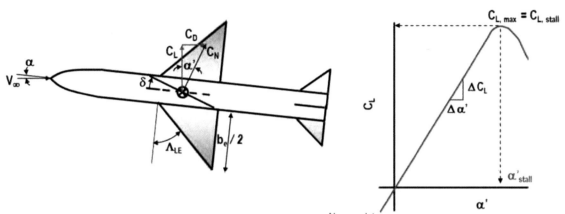

Fig. 2.55 Missile Aerodynamic Flight Control Typically Stalls at a Surface Local Angle of Attack of Approximately 22 Deg.

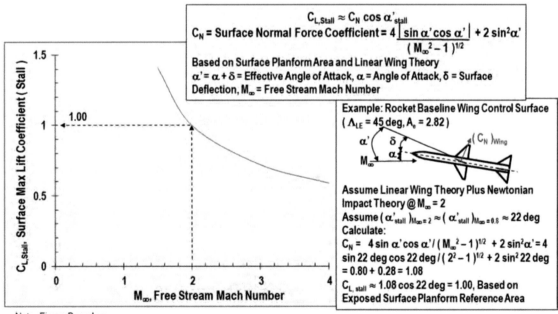

Fig. 2.56 Surface Maximum Lift (Stall) Deceases with Supersonic Mach Number.

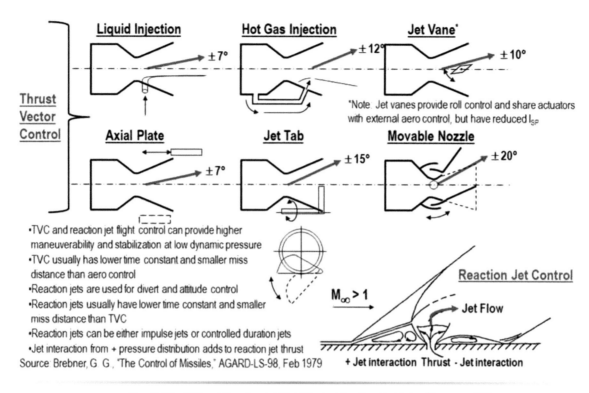

Fig. 2.57 TVC and Reaction Jet Flight Control Can Provide Higher Maneuverability and Stabilization at Low Dynamic Pressure.

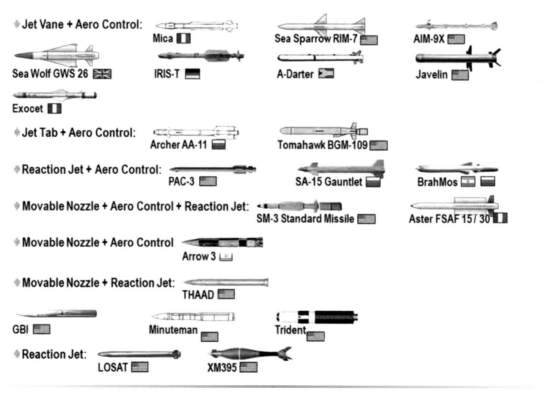

Fig. 2.58 Most Missiles with TVC or Reaction Jet Flight Control Also Use Aerodynamic Flight Control.
Note: See video supplement, TVC (Aster, Javelin, Exocet) and Reaction Jet Flight Control (Gauntlet, BrahMos, PAC-3).

2.16 Maneuver Law Alternatives

Skid-To-Turn (STT)
- Advantage ☺: Fast response
- Disadvantage ☹: Usually limited to axisymmetric cruciform missiles with low aspect ratio
- Feature:
  - Usually small roll attitude / rate commands from autopilot

Bank-To-Turn (BTT)
- Advantage ☺: Higher maneuverability for mono-wing, noncircular / lifting bodies, and airbreathers
- Disadvantages ☹:
  - Time to roll
  - Roll rate limited by gain for radome error slope stability
- Features
  - Large roll attitude commands from autopilot
  - Small sideslip

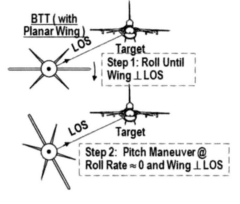

Note: LOS is line-of-sight

Fig. 2.59a Skid-to-Turn Is the Most Common Maneuver Law for Missiles.

Rolling Airframe (RA)
- Advantages ☺:
  - Requires fewer gyros / accelerometers / actuators
  - Compatible with rosette scan / pseudo Image seeker
- Disadvantages ☹:
  - Reduced maneuverability for aero control
  - Requires higher rate gyros / actuators / seeker tracking
  - Higher drag with coning flight trajectory
  - Requires precision geometry and thrust alignment
  - Induces radial stress
  - Thrust varies with roll rate
- Features
  - Bias roll rate (~ 10 Hz) from bias roll moment
  - Can use "bang-bang" / impulse steering
  - Compensates for thrust offset

Divert
- Advantages ☺:
  - Lower time constant
  - Less effect of radome error slope
  - Often has smaller miss distance
- Disadvantages ☹:
  - Usually higher cost
  - May not provide sufficient maneuverability
- Features
  - Direct lift / side force w/o rotation
  - Either wing, blended canard – tail, or divert reaction jet control

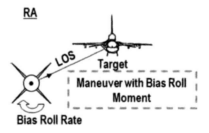

Note: LOS is line-of-sight

Fig. 2.59b Skid-to-Turn Is the Most Common Maneuver Law for Missiles (cont).

48 Missile Design Guide

Skid-To-Turn (STT):   Bank-to-Turn (BTT):   Rolling Airframe (RA):   Divert:
Evolved Sea Sparrow   JASSM   SeaRAM   MKV

Fig. 2.60 Examples of Skid-to-Turn, Bank-to-Turn, Rolling Airframe, and Divert Maneuvering.
Note: See video supplement, Evolved Sea Sparrow, JASSM, SeaRAM, and MKV Flight Trajectories.

Type Inlet	Location	Propulsion	Example Missile
Twin	Side	Ramjet	ASMP
"	"	"	C-101 C-301
"	"	Turbojet	Taurus KEPD-350
"	Cheek	Ducted Rocket	HSAD Meteor
Single	Bottom Scoop	Scramjet	X-51 SED
"	"	Ramjet	ASALM
"	"	Turbojet	Tomahawk RBS-15 SOM

Note: Bank-to-turn maneuvering maintains low sideslip ($\beta \approx 0$ deg) with better inlet efficiency

Fig. 2.61a Non-Cruciform Inlets Require Bank-to-Turn Maneuvering.

Type Inlet	Location	Propulsion	Example Missile
Single	Bottom Scoop (cont)	Turbojet	NSM TORGOS Ra'ad
"	"	"	Storm Shadow Sizzler
			Delilah Kh-35
"	"	"	Hyunmoo III
"	"	"	Babur Sea Eagle
"	Bottom Flush	"	JASSM Harpoon
"	"	"	Gabriel
"	Top	Turbofan	ALCM / CALCM

Note: Bank-to-turn maneuvering maintains low sideslip ($\beta \approx 0$ deg) with better inlet efficiency

Fig. 2.61b Non-Cruciform Inlets Require Bank-to-Turn Maneuvering (cont).

2.17 Roll Angle and Control Surface Sign Convention

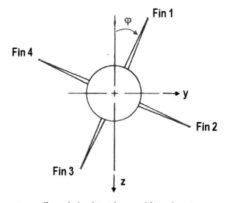
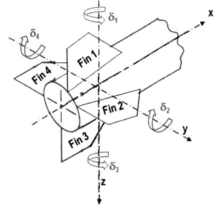

+ φ roll angle looking forward from base

Fins 1 and 3 have trailing edge right for + deflection.
Fins 2 and 4 have trailing edge up for + deflection.
The above figure shows fins + deflection.

δ_e = Equivalent elevator deflection (+ δ_e produces + (up) pitching moment) = $[(\delta_2 + \delta_4)/2]\cos\varphi - [(\delta_1 + \delta_3)/2]\sin\varphi$
δ_r = Equivalent rudder deflection (+ δ_r produces + (right) yawing moment) = $[(\delta_2 + \delta_4)/2]\sin\varphi + [(\delta_1 + \delta_3)/2]\cos\varphi$
δ_a = Equivalent aileron deflection (+ δ_a produces + (clockwise) rolling moment) = $(\delta_2 + \delta_3 - \delta_1 - \delta_4)/4$

Note: For minimum total fin deflection (⇨ lowest total hinge moment, lowest drag, highest control effectiveness)
$\delta_1 = \delta_r - \delta_a, \delta_2 = \delta_e + \delta_a, \delta_3 = \delta_r + \delta_a, \delta_4 = \delta_e - \delta_a$

Note: Other sign conventions may be used for + deflection of flight control surfaces. For each missile program, it is important that the selected sign convention be consistent, well understood, and well documented.

Fig. 2.62 Typical Sign Convention for Cruciform Missile Roll Angle and Flight Control Surface Deflection.

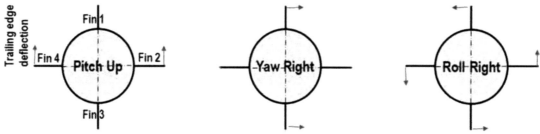

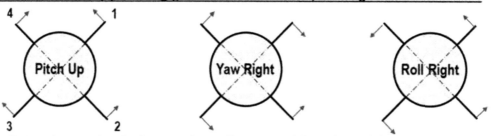

Note: + roll orientation sometimes has lower trim drag and less static stability and control effectiveness in pitch and yaw.
+ roll often has statically unstable roll moment derivative ($C_{l\phi} > 0$) in supersonic flight.
X roll orientation usually has better launch platform compatibility, higher L/D, higher static stability and control effectiveness in pitch and yaw. X roll often has statically unstable roll moment derivative ($C_{l\phi} > 0$) in subsonic flight.

Fig. 2.63 X Roll Orientation Flight Is Usually Better Than + Roll Orientation Flight.

2.18 Trim and Static Stability Considerations

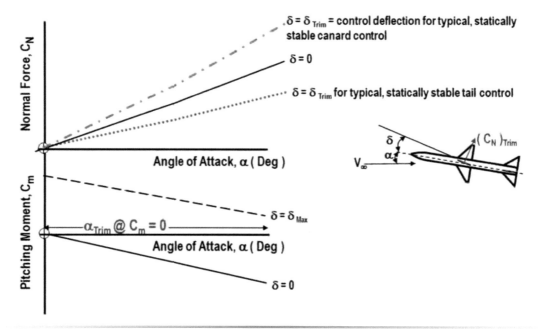

Fig. 2.64 Trimmed Normal Force Is Defined at Zero Pitching Moment.

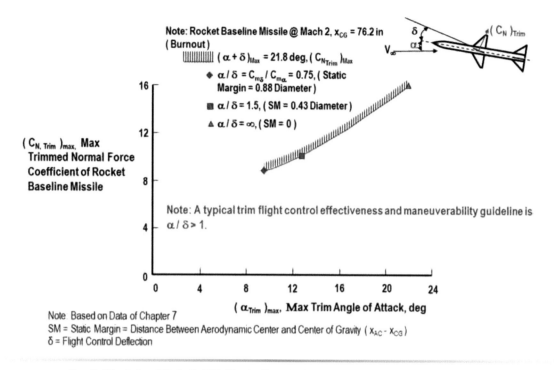

Fig. 2.65 Relaxed Static Stability Margin Allows Higher Trim Angle of Attack and Higher Normal Force.

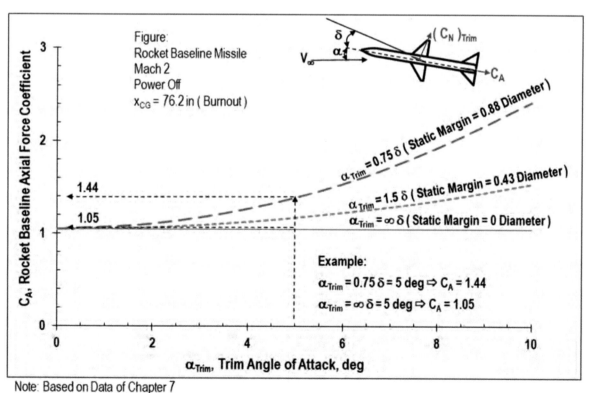

Fig. 2.66 Relaxed Static Stability Margin Reduces Drag.

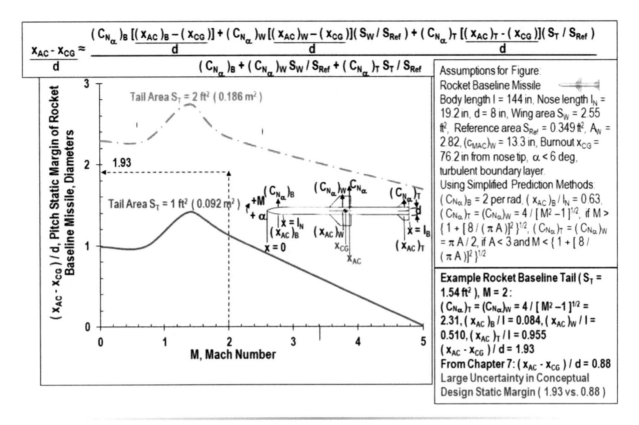

Fig. 2.67 Missile Static Margin Is Driven by Tail Area and Static Margin Prediction Has Large Uncertainty.

2.19 Stability and Control Conceptual Design Criteria

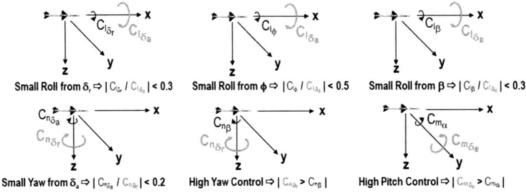

Note:
δ_r = Rudder Deflection, $C_{l_{\delta_r}}$ = Rolling Moment Coefficient Derivative from Yaw Control Deflection, $C_{l_{\delta_a}}$ = Rolling Moment Coefficient Derivative from Roll Control Deflection
ϕ = Roll Angle, C_{l_ϕ} = Rolling Moment Coefficient Derivative with Roll Angle
C_{l_β} = Rolling Moment Coefficient Derivative from Sideslip, β = Sideslip Angle
$C_{n_{\delta_a}}$ = Yawing Moment Coefficient Derivative from Roll Control Deflection, $C_{n_{\delta_r}}$ = Yawing Moment Coefficient Derivative from Yaw Control Deflection, δ_a = Aileron Deflection
C_{n_β} = Yawing Moment Coefficient Derivative from Sideslip
C_{m_α} = Pitching Moment Coefficient Derivative from Angle of Attack, $C_{m_{\delta_e}}$ = Pitching Moment Coefficient Derivative from Pitch Control Deflection

Fig. 2.68 Stability & Control Cross Coupling Requires Higher Flight Control Effectiveness.

M2-F2 Lifting Body

X-24B Lifting Body

Fig. 2.69 Stability & Control Cross Coupling of a Lifting Body Requires Higher Flight Control Effectiveness. *Note*: See video supplement, M2-F2 Lifting Body Flight Test.

Chapter 3 Propulsion

3.1 Introduction

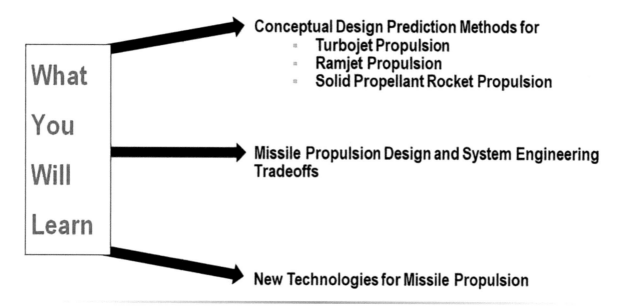

Fig. 3.1   Chapter 3 Propulsion—What You Will Learn.

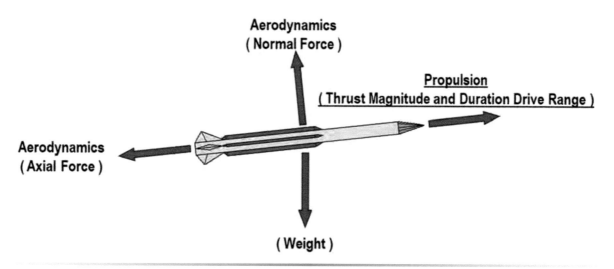

Fig. 3.2   Propulsion Thrust Magnitude and Duration Drive Missile Flight Range.

3.2 Propulsion Alternatives Assessment

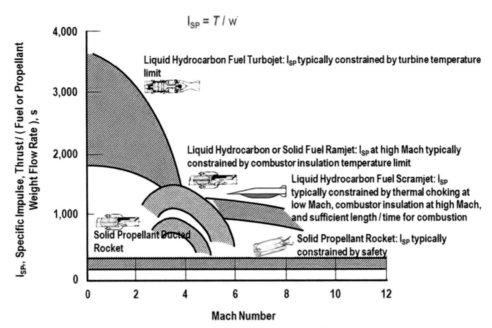

Note: Above values of I_{SP} are in English units (s). To convert to metric units (m / s) multiply by metric gravitational constant (9.81 m / s^2).

Fig. 3.3 Each Type of Air Breathing Propulsion Has an Optimum Mach Number for Propulsion Efficiency.

$$R = (L / D) I_{sp} V \ln [W_L / (W_L - W_P)], \text{Breguet Range Equation}$$

Parameter	Typical Value for 2000 lb Precision Strike Missile			
	Subsonic Turbojet w Hydrocarbon Fuel	Ramjet w Liquid Hydrocarbon Fuel	Scramjet w Hydrocarbon Fuel	Solid Propellant Rocket
L / D, Lift / Drag	10	5	3	5
I_{sp}, Specific Impulse	3000 s	1300 s	1000 s	250 s
V_{AVG}, Average Velocity	1000 ft / s	3500 ft / s	6000 ft / s	3000 ft / s
W_P / W_L, Cruise Propellant or Fuel Weight / Launch Weight	0.3	0.2	0.1	0.4
R, Max Cruise Range	1800 nm (3330 km)	830 nm (1537 km)	310 nm (574 km)	250 nm (463 km)
$t_{250 nm}$, Time to 250 nm Range	25 m	7 m	4 m	8 m

Note: Ramjet and Scramjet missiles booster propellant for Mach 2 to 4+ take-over speed not included in W_P for cruise. Solid Propellant Rockets require thrust magnitude control (e.g., pintle, pulse, or gel motor) for effective cruise. Max range for a rocket is usually a lofted boost-glide, semi-ballistic, or ballistic flight profile, instead of cruise flight. Multiple stages may be required for rocket range greater than 200 nm.

Fig. 3.4 Cruise Range Is Driven by Lift-to-Drag Ratio, Specific Impulse, Velocity, and Propellant or Fuel Weight Fraction.

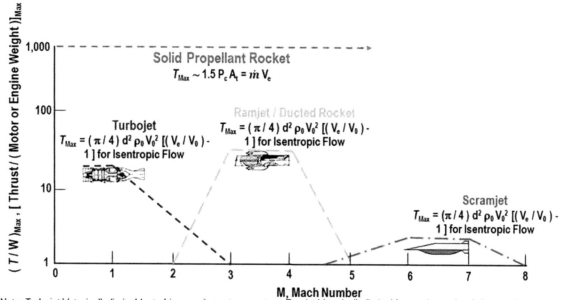

Fig. 3.5 Solid Propellant Rocket Propulsion Has Higher Thrust-to-Weight Capability Than Air-Breathing Propulsion.
Note: See video supplement, Spartan and Sprint Rockets During Launch/Turn.

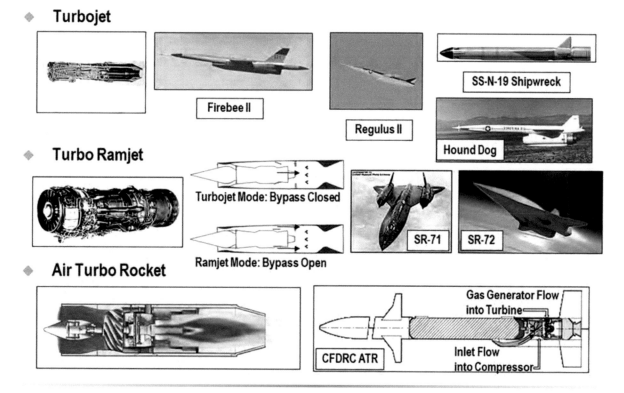

Fig. 3.6 Turbine-Based Missile Propulsion Is Capable of Subsonic to Supersonic Cruise.

3.3 Turbojet Flow Path, Components, and Nomenclature

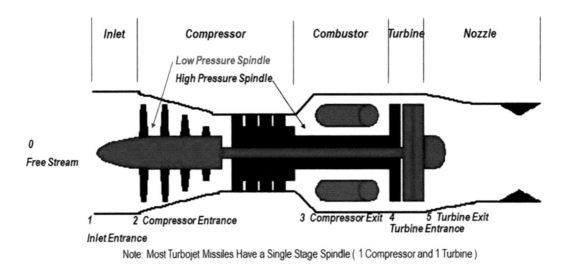

Fig. 3.7 Schematic of Turbojet Flow Path, Components, and Nomenclature.
Note: Video Animation Courtesy of NASA.

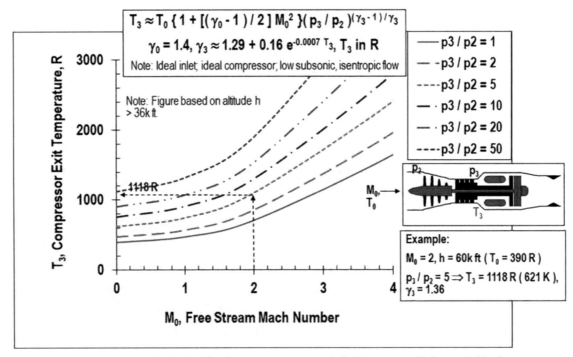

T_3 = Compressor exit temperature in Rankine, T_0 = free stream temperature in Rankine, γ = specific heat ratio, M_0 = free stream Mach number, p_3 = compressor exit pressure, p_2 = compressor entrance pressure

Fig. 3.8 High Temperature Compressors Are Required to Achieve High Pressure Ratio at High Mach Number.

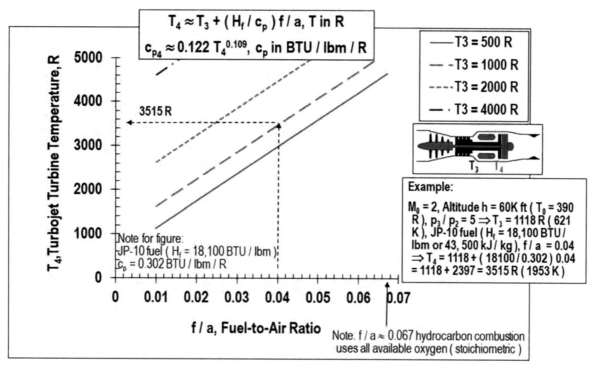

Fig. 3.9 A High Temperature Turbine Is Required for a High-Speed Turbojet Missile.

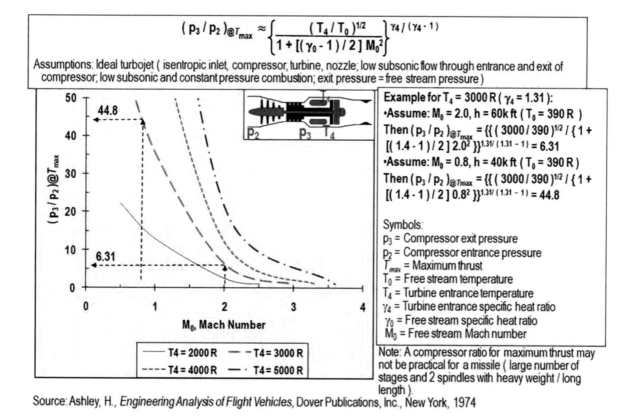

Fig. 3.10 Compressor Pressure Ratio and Turbine Temperature Limit Turbojet Maximum Thrust.

Advanced High Temperature Materials Allow Mach 3+ Short Duration (< 30 Minutes) Turbojet

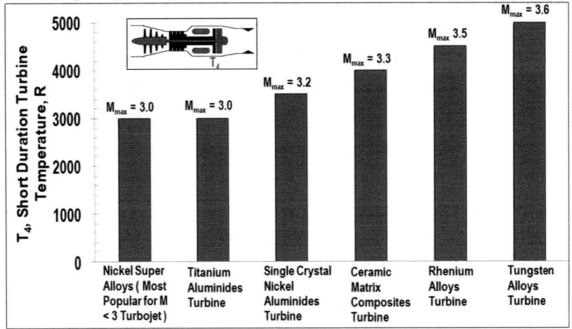

Note:
Above figure based on altitude h = 36k ft (ft (T_0 = 390 R). Turbine maximum temperature limits the max available turbojet Mach number and thrust. M_{max} at (p_3 / p_2) = 1, T_4 = turbine temperature = combustor temperature. $M_{max} = \{ 5 [(T_4 / T_0)^{1/2} - 1]\}^{1/2}$

Fig. 3.11a Approaches for Mach 3+ Missile Turbojet Propulsion.

Other Approaches for Mach 3+ Turbojet

Approach	Thrust	I_{SP}	Length	Weight	Cost	Risk
	Measures of Merit					
Actively Cooled Turbine[1]	○	○	○	○	○	–
Afterburner[2]	○	–○	–	–	○	○
Insulated Turbine Blades[3]	–	–	◒	◒	◒	–

● *Superior* ◒ *Good* ○ *Average* – *Poor*

Note:

1 Max achievable coolant rate is a risk for highly cooled hollow turbine blades. Candidate approaches include compressor air, water, or fuel through the hollow turbine blades.

2 Afterburner adds length and weight. Achievable specific impulse (I_{SP}) is design dependent.

3 Insulation failure, char, and wall leakage are a risk for insulated turbine blades.

Fig. 3.11b Approaches for Mach 3+ Missile Turbojet Propulsion (cont).

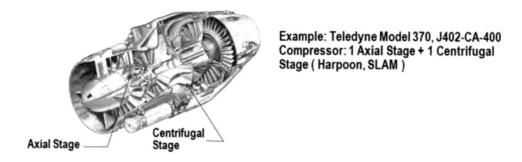

Example: Teledyne Model 370, J402-CA-400
Compressor: 1 Axial Stage + 1 Centrifugal Stage (Harpoon, SLAM)

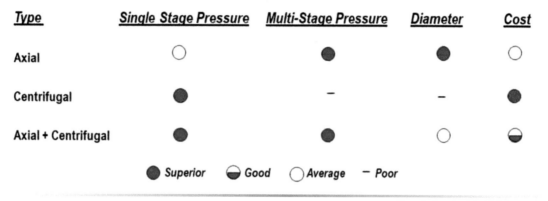

Fig. 3.12 Compressor Alternatives Are Multi-Stage Axial, Single Centrifugal, Multi-Stage Axial+Centrifugal.

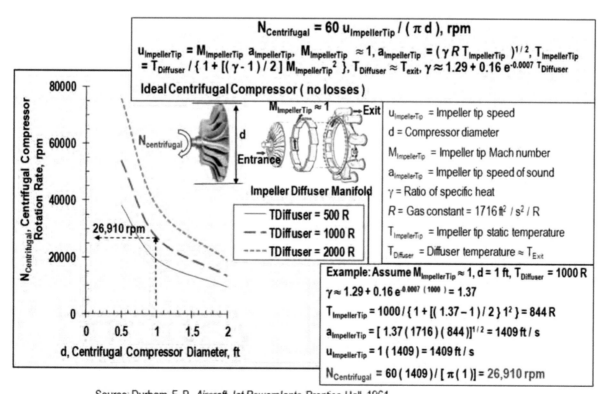

Source: Durham, F. P., *Aircraft Jet Powerplants*, Prentice-Hall, 1961

Fig. 3.13 A Small Diameter Turbojet Requires a High-Speed Compressor.

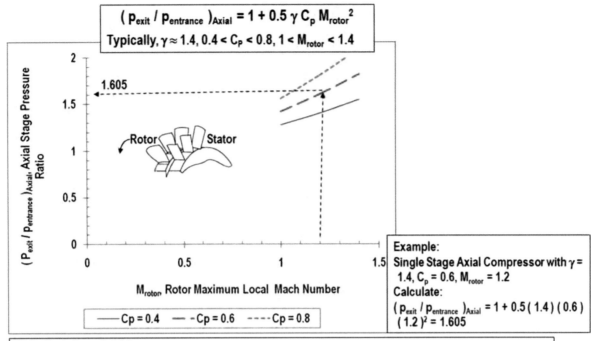

Fig. 3.14 Rotors of a High-Pressure Axial Compressor Have High Local Mach Number.
Note: See video supplement, http://www.lerc/nasa.gov

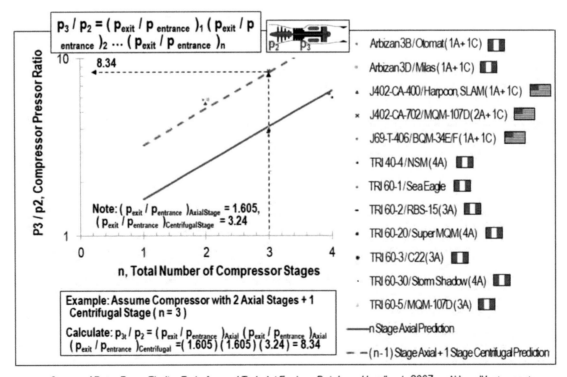

Fig. 3.15 Multi-Stage Compressors Provide Higher Compressor Pressure Ratio.

3.4 Turbojet Thrust Prediction

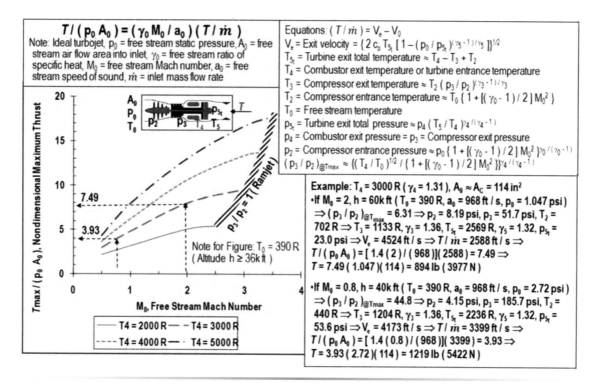

Fig. 3.16 Turbojet Thrust Is Limited by Inlet Flow and Turbine Maximum Allowable Temperature.

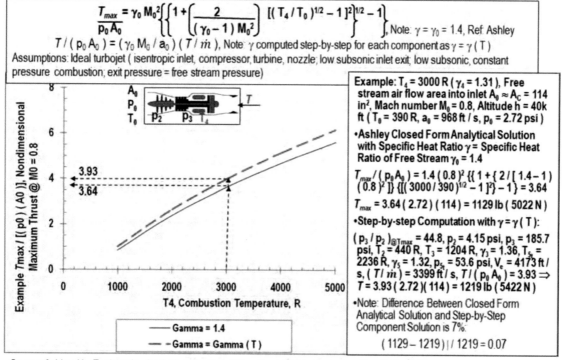

Source: Ashley, H., *Engineering Analysis of Flight Vehicles*, Dover Publications, Inc., New York, 1974

Fig. 3.17 Assumption Specific Heat Ratio=1.4 Has Sufficient Accuracy for Turbojet Conceptual Design Thrust.

3.5 Turbojet Specific Impulse Prediction

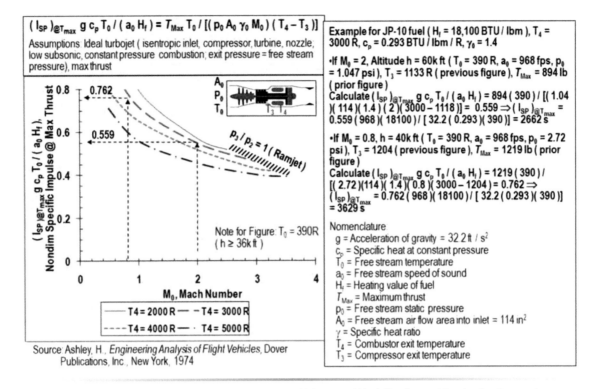

Fig. 3.18 Turbojet Specific Impulse Decreases with Increased Mach Number and Combustion Temperature.

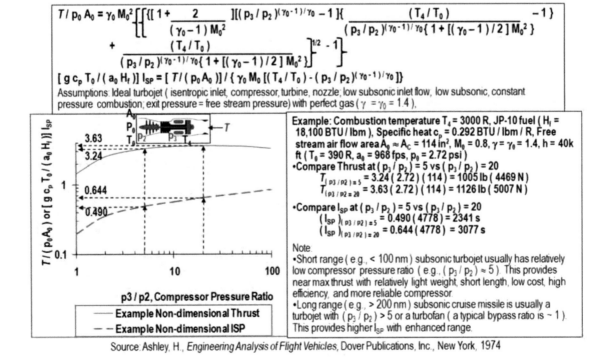

Fig. 3.19 Turbojet Has Relatively High Thrust @ Relatively Low Compressor Pressure Ratio but High Specific Impulse Requires High Pressure Ratio.

3.6 Subsonic Turbojet Propulsion Efficiency

- **Conceptual Design Prediction for Thrust, Specific Impulse, and Efficiency**
  - $T_{predicted} = \eta \, (A_0/A_C) \, T_{ideal}$
  - $(I_{SP})_{predicted} = \eta \, (I_{SP})_{ideal}$
  - $\eta = \eta_{Inlet} \, \eta_{Compressor} \, \eta_{Combustor} \, \eta_{Turbine} \, \eta_{Nozzle}$ (Typically, $0.6 < \eta < 0.8$)
- **Inlet Efficiency (Subsonic)**
  - Scoop Inlet: $\eta_{Inlet} \approx 0.94$
  - Flush Inlet: $\eta_{Inlet} \approx 0.91$
  - Capture Efficiency: $(A_0/A_C)_{cruise} \approx 1$, $(A_0/A_C)_{accelerating} > 1$, $(A_0/A_C)_{decelerating} < 1$
- **Compressor Efficiency**
  - $\eta_{Compressor} = 0.85$ to 0.95, depending upon number of compressor stages
- **Combustor Efficiency**
  - $\eta_{Combustor} = 0.95$ to 0.98
- **Turbine Efficiency**
  - $\eta_{Turbine} = 0.85$ to 0.98 for single stage, function of number of compressor stages
- **Nozzle Efficiency**
  - $\eta_{Nozzle} \approx 0.99$

Note: A_0 = Free stream air flow area into inlet, A_C = Inlet capture area

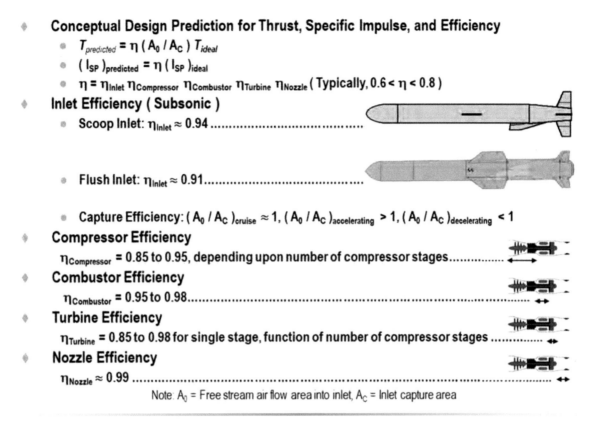

Fig. 3.20 Typical Values of Subsonic Turbojet Propulsion Efficiency.

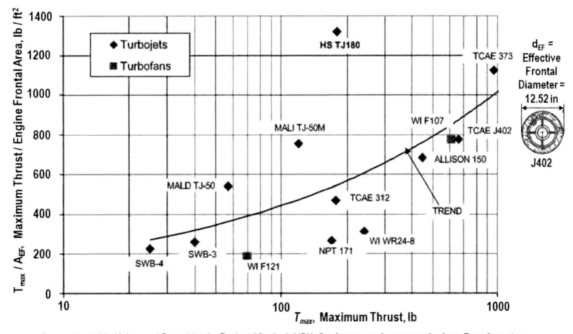

Source: Lyon, M., "Advanced Propulsion for Tactical Missiles", NDIA Conference on Armaments for Army Transformation, June 2001.

Fig. 3.21 Higher Thrust Turbojet and Turbofan Expendable Engines Usually Have More Efficient Performance.

66 Missile Design Guide

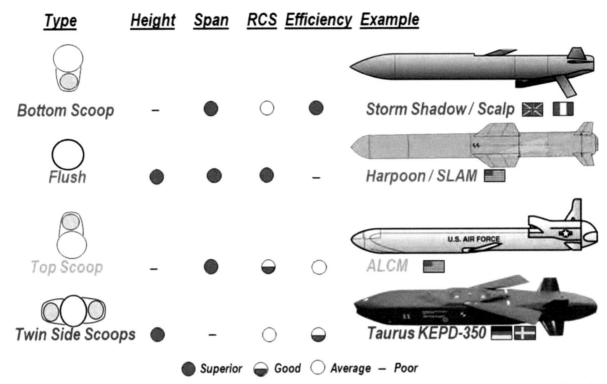

Note: Radar cross section (RCS) based on ground radar view into inlet. Inlet efficiency from inlet pressure ratio and max angle of attack.

Fig. 3.22   Turbojet Missile Inlet Best Location Is Driven by Launch Platform Fitment, Radar Cross Section, and Pressure Recovery.

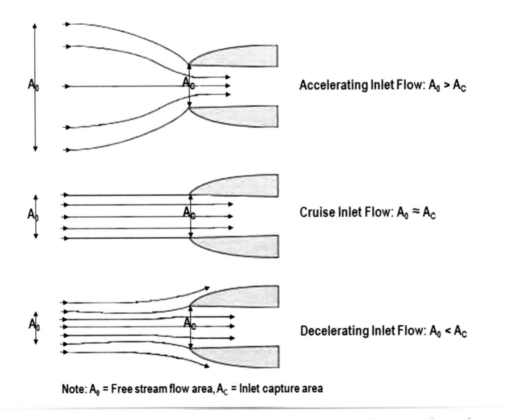

Note: A_0 = Free stream flow area, A_C = Inlet capture area

Fig. 3.23   During Subsonic Cruise, Free Stream Flow into Inlet is Approximately Equal to Inlet Capture Geometry.

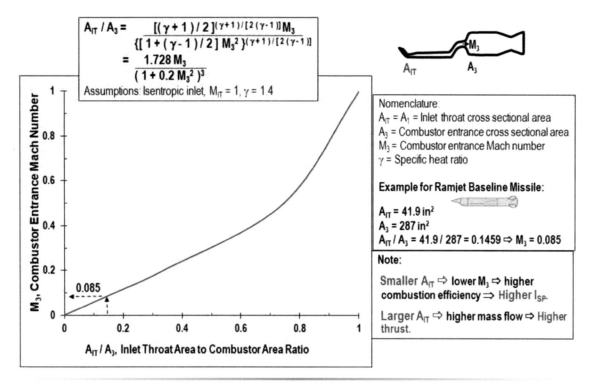

Fig. 3.32 Ramjet Inlet Throat Area Is a Driver for Combustion Mach Number, Mass Flow, Specific Impulse, and Thrust.

3.11 Ramjet Combustor Design Considerations

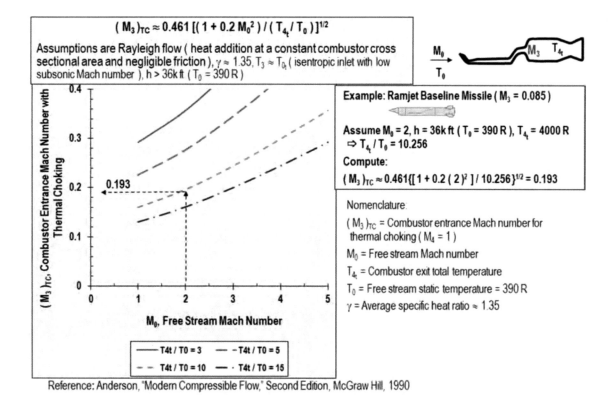

Fig. 3.33 A Ramjet in Low Supersonic Flight with High Temperature Combustion May Be Susceptible to Thermal Choking.

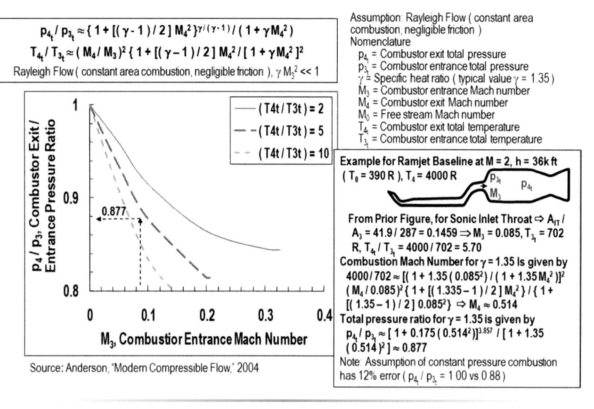

Fig. 3.34 A Relatively Low Mach Number into Combustor Is Desirable for High Ramjet Combustion Efficiency.

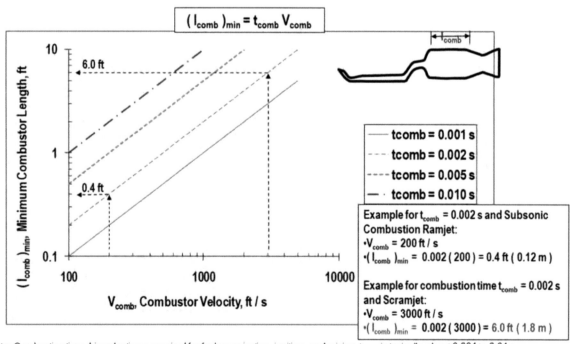

Fig. 3.35 Required Length and Weight of the Ramjet Combustor Is a Function of Combustor Velocity.

3.12 Ramjet Booster Integration

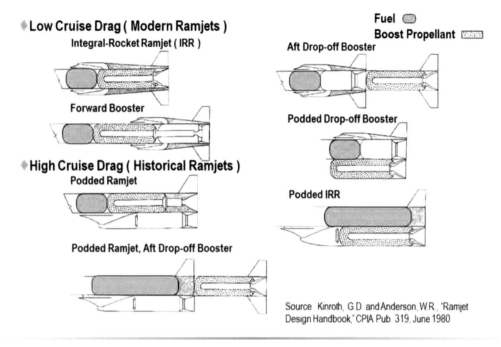

Fig. 3.36 Ramjet Engine and Booster Integration Options.

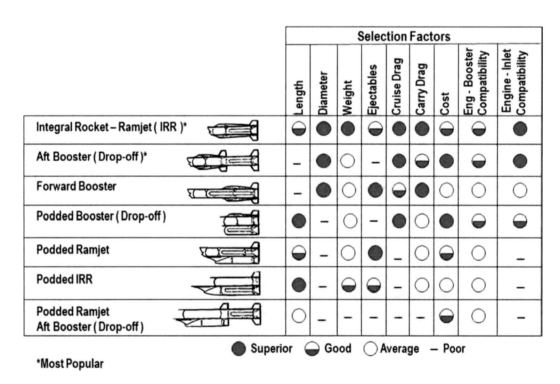

*Most Popular

Fig. 3.37 Ramjet Engine, Booster, Inlet Integration Trades.

74 Missile Design Guide

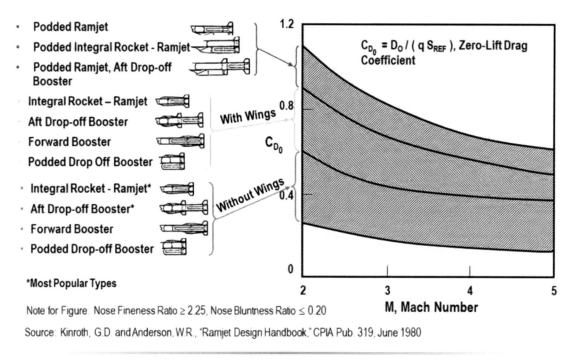

Fig. 3.38 Ramjets with Internal Boosters and No Wings Have Lower Drag.

3.13 Ramjet Inlet Options

Type Inlet	Sketch	Placement
Nose*		Annular nose inlet
Chin*		Forward underside in nose compression field – partial annular
Forward Cruciform Annular		Forward in nose compression field - cruciform (four) annular inlets
Aft Cruciform Annular*		Aft-cruciform (four) annular inlets
Under Wing Annular		In planar wing compression field - twin annular inlets
Twin Two-dimensional*		Aft - twin cheek-mounted two dimensional inlets
Underslung Annular		Aft underside - full annular
Underslung Two-dimensional		Aft underside - belly mounted two dimensional inlet
Aft Cruciform Two-dimensional*		Aft - cruciform (four) two dimensional inlets

*Most Popular

Source: Kinroth, G.D. and Anderson, W.R., "Ramjet Design Handbook," CPIA Pub. 319, June 1980.

Fig. 3.39 Ramjet Inlet Options.

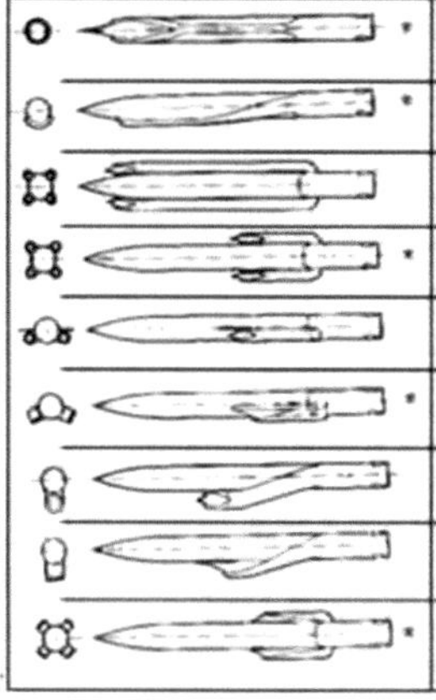

Type Inlet	Selection Factors							Preferred Steering	Preferred Flight Cont.	Primary Mission Suitability
	Pressure Recovery	Carriage Envelope	Max Angle of Attack	Weight	Drag	Warhead Shrouding	Inlet Cost			
O ⟶*	◐	●	○	◐	●	—	◐	STT	T, C	ATS, STA
◌ ⟶*	●	●	●	○	●	○	○	BTT	T	ATS, ATA, STA
✕ ⟶	○	○	◐	—	○	—	—	STT	T	ATS, ATA, STA
✕ ⟶*	○	◐	○	◐	○	●	○	STT	T	ATS
◌ ⟶	◐	—	●	●	○	●	◐	BTT	T	ATS, ATA, STA
◌ ⟶*	◐	○	◐	◐	◐	◐	○	BTT	T	ATS, ATA, STA
◯ ⟶	○	—	○	●	○	◐	●	BTT	T	ATS
◯ ⟶	—	—	◐	◐	◐	◐	◐	BTT	T	ATS, ATA, STA
✕ ⟶*	○	◐	○	◐	◐	●	○	STT	T	ATS

*Most Popular Types   ● Superior   ◐ Good   ○ Average   — Poor

Note:
BTT = Bank to Turn, STT = Skid to Turn
C = Canard Flight Control, T = Tail Flight Control
ATS = Air-to-Surface, ATA = Air-to-Air, STA = Surface-to-Air

Source: Kinroth, G.D. and Anderson, W.R., "Ramjet Design Handbook," CPIA Pub. 319, June 1980

Fig. 3.40   Ramjet Inlet Concept Trades.

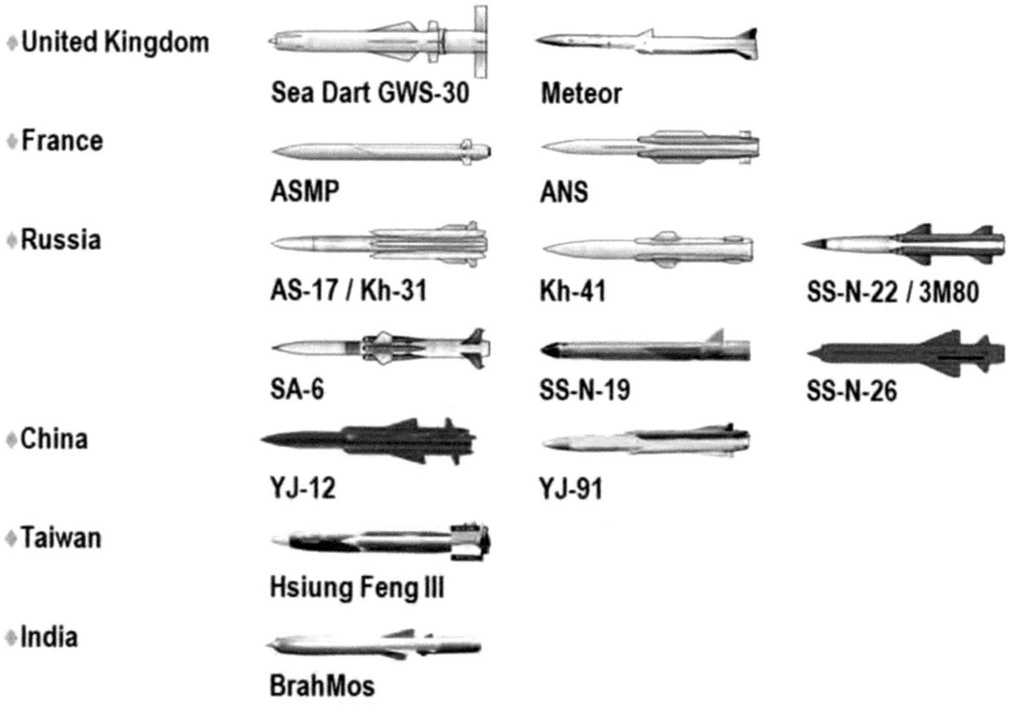

♦ **United Kingdom**
    Sea Dart GWS-30   Meteor

♦ **France**
    ASMP   ANS

♦ **Russia**
    AS-17 / Kh-31   Kh-41   SS-N-22 / 3M80
    SA-6   SS-N-19   SS-N-26

♦ **China**
    YJ-12   YJ-91

♦ **Taiwan**
    Hsiung Feng III

♦ **India**
    BrahMos

- Aft inlets have lower inlet volume and do not degrade lethality of forward located warhead.
- Nose Inlet may have higher flow capture, higher pressure recovery, smaller carriage envelope, and lower drag.

Fig. 3.41   Examples of Inlets for Supersonic Air-Breathing Missiles.

3.14 Supersonic Inlet/Airframe Integration

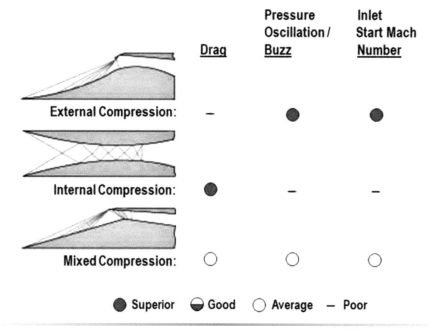

Fig. 3.42 Supersonic Inlet-Airframe Integration Tradeoffs Include Drag, Pressure Oscillation, and Inlet Start. *Note*: See video supplement, Mach 1.7 Inlet Buzz. NASA Glenn 8ft × 6ft Wind Tunnel. Frame Rate 2000 Hz.

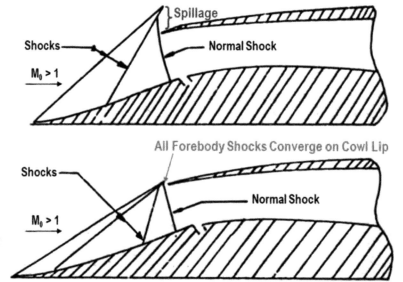

Fig. 3.43 A Supersonic External Compression Inlet with Shock Wave(s) on the Cowl Lip Prevents Spillage.

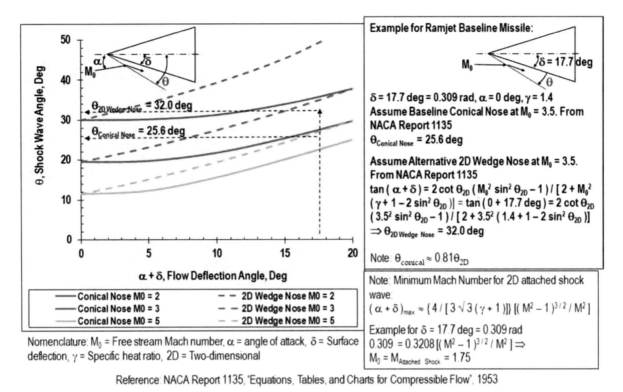

Fig. 3.44 Nose Shock Wave Angle Is Driven by Type of Shock Wave, Angle of Attack, Surface Deflection, and Mach Number.

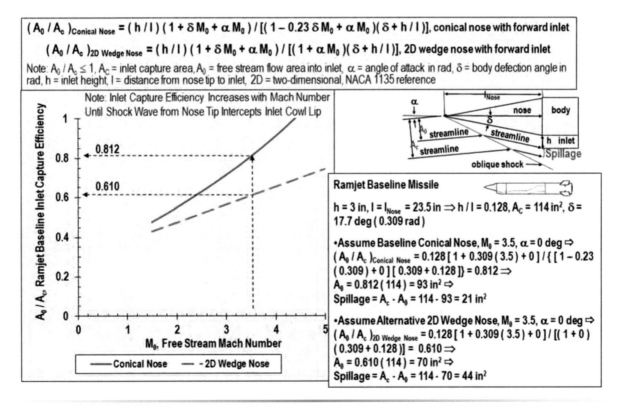

Fig. 3.45 A Conical Nose Inlet Typically Has Higher Capture Efficiency Than a Two-Dimensional Wedge Nose Inlet.

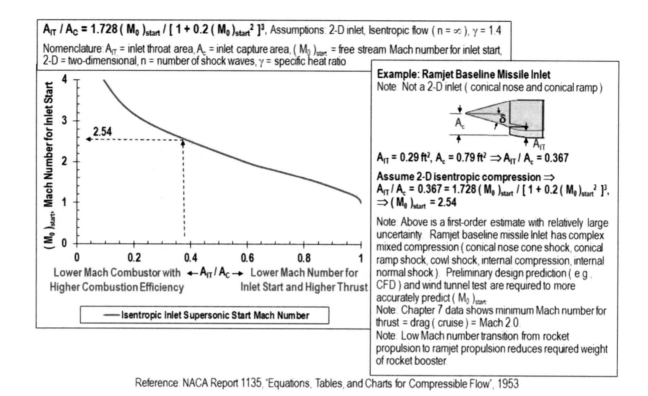

Fig. 3.46 Inlet Throat Area Is a Driver for Supersonic Inlet Start Mach Number, Combustion Efficiency, and Thrust.

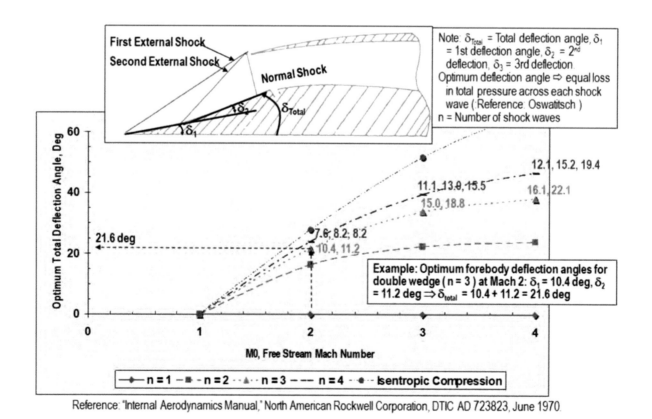

Fig. 3.47 Optimum Forebody Deflection Angle(s) for Best Pressure Recovery Increases with Mach Number.

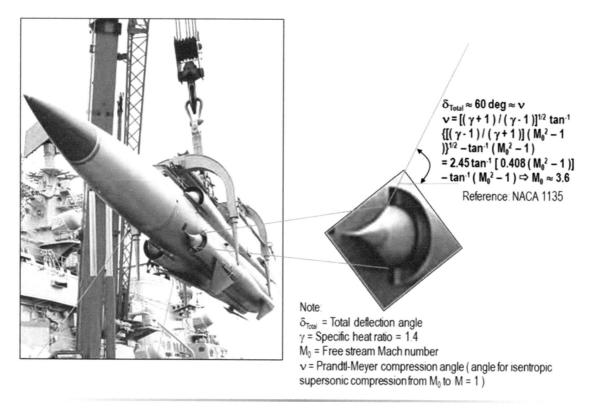

Fig. 3.48 Example of Near-Isentropic Supersonic Compression Inlet—3M80 (SS-N-22 Sunburn).

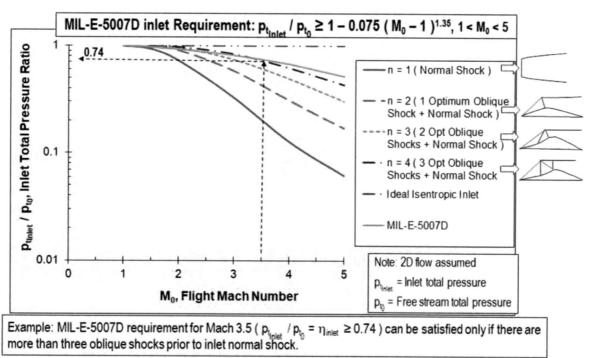

Fig. 3.49 Oblique Shocks Prior to the Inlet Normal Shock Are Required to Satisfy MIL-E-5007D.

3.15 Fuel Alternatives

Type	Fuel	Density, lbm / in³	Volumetric Performance, BTU / in³	Low Observables
Hydrocarbon	Turbine (JP-4, JP-5, JP-7, JP-8, JP-10)	~ 0.031	~ 559	●
↓	Liquid Ramjet (RJ-4, RJ-5, RJ-6, RJ-7)	~ 0.037	≈ 648	●
	HTPB	~ 0.034	606	○
	Slurry (40% JP-10 / 60% carbon)	~ 0.049	801	○
	Solid Carbon (graphite)	~ 0.075	1132	–
Metal	Slurry (40% JP-10 / 60% aluminum)	~ 0.072	866	–
	Slurry (40% JP-10 / 60% boron carbide)	~ 0.050	1191	–
	Solid Mg	~ 0.068	1200	–
↓	Solid Al	~ 0.101	1300	–
	Solid Boron	~ 0.082	2040	–

● Superior   ◐ Good   ○ Average   – Poor

Fig. 3.50   High Density Fuel Has Higher Volumetric Performance, but Also Has Higher Observables.

3.16 Solid Propellant Rocket Motor Flow Path, Components, and Nomenclature

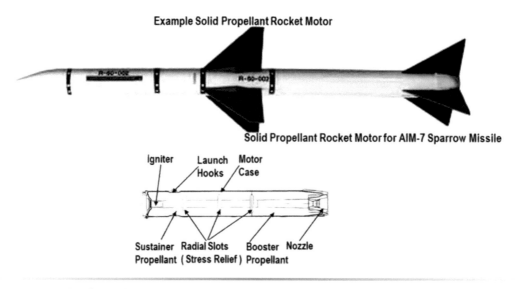

Fig. 3.51   Solid Propellant Rocket Motor Design Components, Considerations, and Tradeoffs.
Note: See video supplement, The Rocket, Circa 1967.

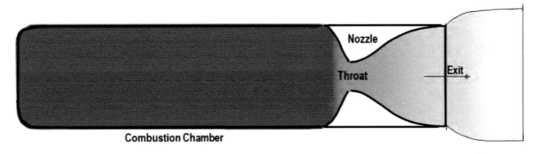

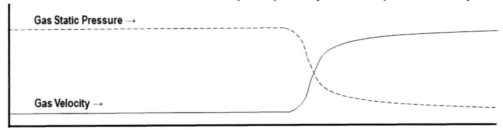

Fig. 3.52 A Rocket Generates Thrust by Converting High Pressure Combustion into High Velocity Exhaust. *Note*: Overexpanded plume may interfere with command guidance.

3.17 Rocket Motor Performance Prediction

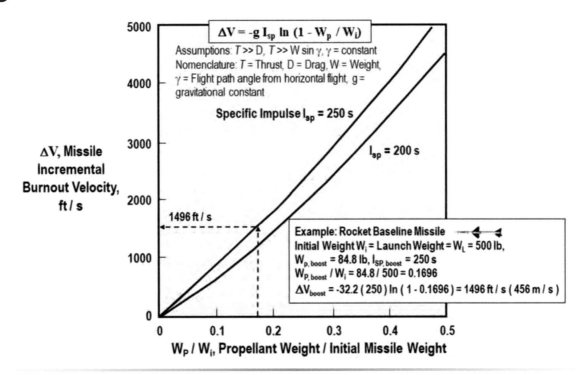

Fig. 3.53 High Propellant Weight Fraction and High Specific Impulse Increase Rocket Burnout Velocity.

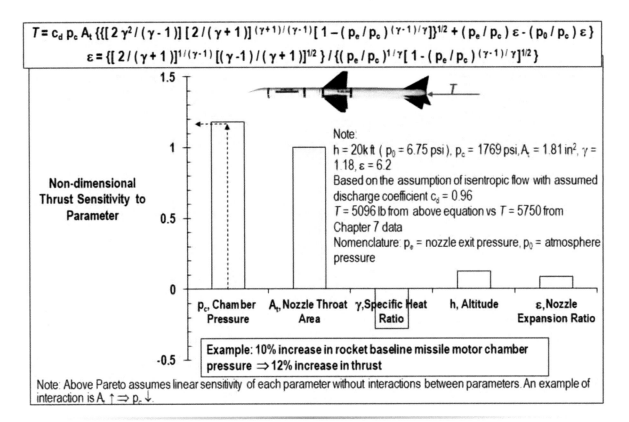

Fig. 3.54 Rocket Baseline Pareto Shows Thrust Drivers Are Chamber Pressure and Nozzle Throat Area.

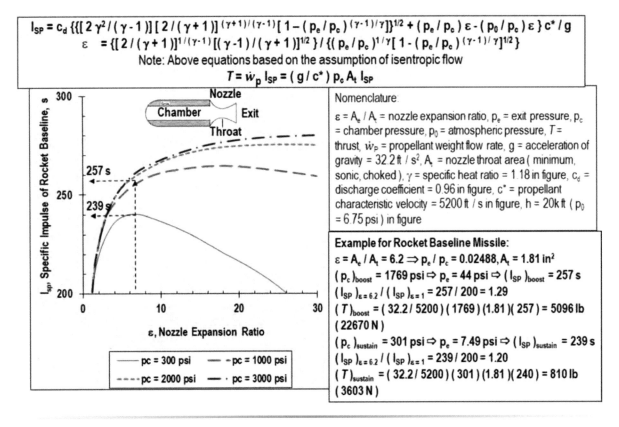

Fig. 3.55 A High Specific Impulse Rocket Requires High Chamber Pressure and Optimum Nozzle Expansion.

3.18 Rocket Motor Sizing Process

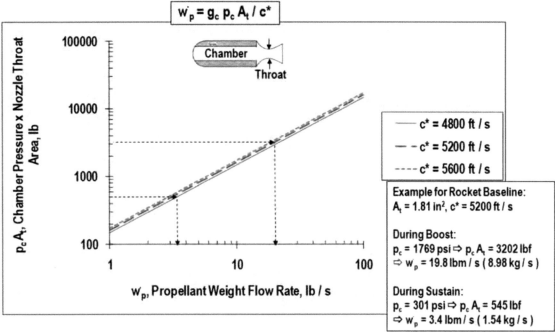

Fig. 3.56 High Propellant Weight Flow Rate Is Driven by High Chamber Pressure and Nozzle Throat Area.

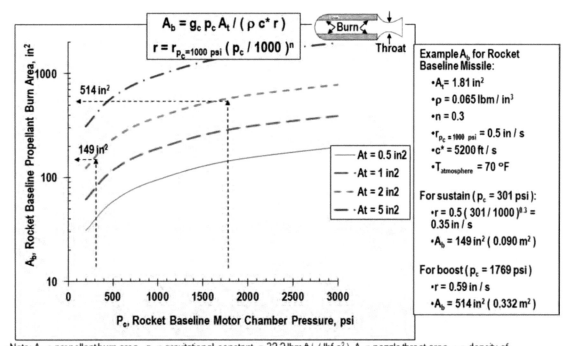

Fig. 3.57 Rocket Motor Chamber Pressure Is Driven by Propellant Burn Area and Nozzle Throat Area.

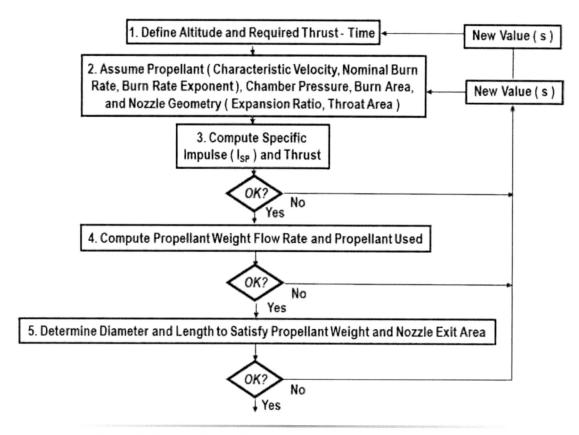

Fig. 3.58 Conceptual Design Sizing of a Solid Propellant Rocket Motor Is an Iterative Process.

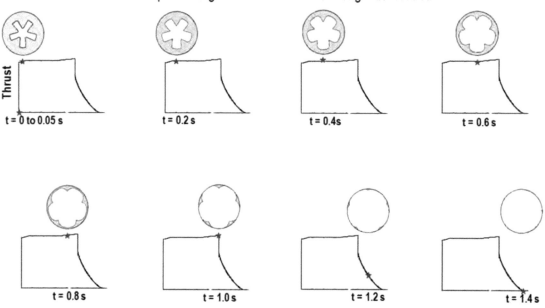

Fig. 3.59 Solid Propellant Rocket Motor Pressure, Weight Flow Rate, and Thrust Are Proportional to Propellant Burn Area.

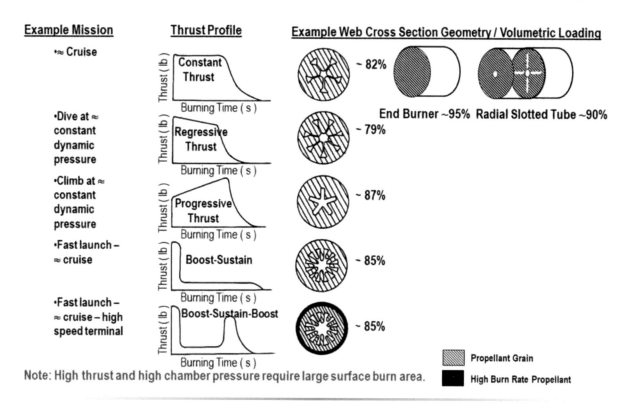

Fig. 3.60   Missile Thrust—Time Requirements Drive Solid Propellant Rocket Grain Cross Section Geometry.

3.19 Solid Propellant Rocket Motor Production Alternatives

- **Case Bonded Propellant Motor***
  - Production Process
    - Mix Propellant Slurry
    - Pour Propellant Slurry into Motor Case
    - Place Motor into Oven for Curing / Bonding
- **Extruded Propellant Motor**
  - Production Process
    - Extrude Propellant
    - Load Propellant into Cartridge
    - Load Propellant Cartridge into Motor Case

*Most Missiles Use Case Bonded Motors (higher volumetric loading)

Solid propellant Case Bonded Rocket Motor Production. Photo Courtesy of BAE.

Extrusion Production of Star Web Propellant. Photo Courtesy of BAE.

Fig. 3.61a   Solid Propellant Motor Production Alternatives—Case Bonded Propellant vs Extruded Propellant.

Measure of Merit	Volume Loading / Total Impulse	Production Rate	Thermal Stress
Case Bonded Propellant Motor*	✓		
Extruded Propellant Motor		✓	✓

*Most Missiles Use Case Bonded Motors.

Fig. 3.61b Solid Propellant Motor Production Alternatives—Case Bonded Propellant vs Extruded Propellant (cont).
Note: See video supplement, Solid Propellant Case Bonded Rocket Motor-Circa 1962.

3.20 Solid Propellant Rocket Thrust Magnitude Control

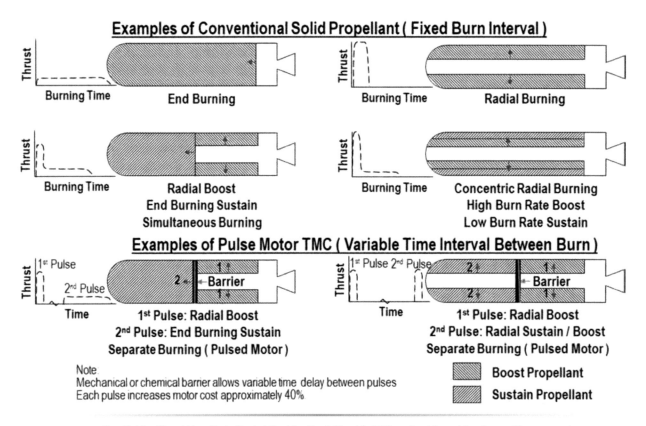

Fig. 3.62 Thrust Magnitude Control Provides Flexibility of Solid Propellant Propulsion Energy Management.

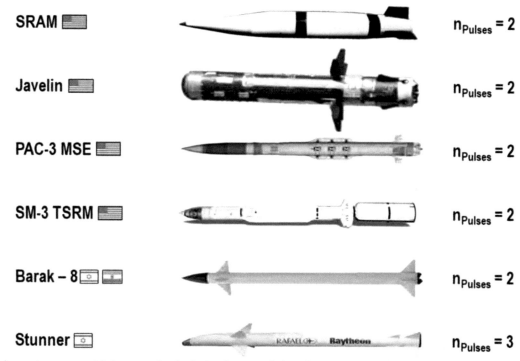

Note: Pulse motor can provide lower peak velocity (⇒ lower peak drag, longer range, and lower peak aero heating). Pulse motor may provide higher terminal velocity / maneuverability. Pulse motor may eliminate need for a two-stage missile. Pulse motor can also provide lower launch noise / blast for a tube launched missile (e.g., Javelin).

Fig. 3.63 Examples of Solid Propellant Rocket Missiles with Pulse Motor Thrust Magnitude Control.

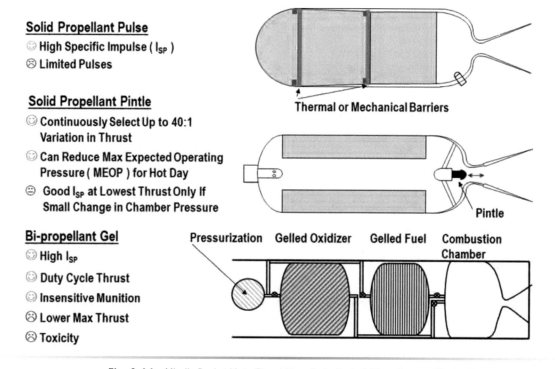

Solid Propellant Pulse
☺ High Specific Impulse (I_{SP})
☹ Limited Pulses

Solid Propellant Pintle
☺ Continuously Select Up to 40:1 Variation in Thrust
☺ Can Reduce Max Expected Operating Pressure (MEOP) for Hot Day
☺ Good I_{SP} at Lowest Thrust Only If Small Change in Chamber Pressure

Bi-propellant Gel
☺ High I_{SP}
☺ Duty Cycle Thrust
☺ Insensitive Munition
☹ Lower Max Thrust
☹ Toxicity

Fig. 3.64 Missile Rocket Motor Thrust Magnitude Control Alternatives and Tradeoffs.

3.21 Solid Propellant Alternatives

Type	I_{SP}, Specific Impulse, s	ρ, Density, lbm / in³	Burn Rate @ 1000 psi, in / s	Explosive Safety	Toxicity	Observables
Min Smoke. No Al fuel or AP oxidizer. Either composite with nitramine oxidizer (CL-20, ADN, AN, HMX, RDX) or double base (nitrocellulose with nitroglycerin). Very low contrail (H_2O).	– ○ 220 – 255	– ○ 0.055 - 0.062	○ ◐ 0.25 - 2.0	– ○	◐	◐
Reduced Smoke. No Al (binder is fuel). AP oxidizer. Low contrail (HCl).	○ 250 - 260	○ 0.062	◐ 0.1 - 1.5	○	○	○
High Smoke Example: 18% Al fuel, 70% AP oxidizer, 12% HTPB binder ⇒ High smoke (Al_2O_3).	◐ 260 – 265	◐ 0.065	● 0.1 - 3.0	◐	○	–

● Superior   ◐ Good   ○ Average   – Poor

Note: Propellant specific impulse is driven by flame temperature and molecular weight. $I_{SP} \approx 25 \, (T_{Flame} / M)^{1/2}$, T_{Flame} in Kelvin.

Fig. 3.65 Solid Propellant Primary Tradeoffs Are Performance, Explosive Safety, Toxicity, and Observables.

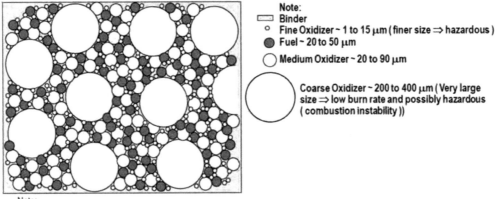

Example of Typical Composite Solid Propellant Mixture
- Source: Merrill Beckstead, BYU, Lecture on "Fundamentals of Combustion and Detonation", April 2000
- ~ 12% Hydroxyl-terminated polybutadiene binder (HTPB) Binder
- ~ 18% Aluminum Fuel
- ~ 70% Ammonium Perchlorate Oxidizer

Note:
- Binder
- Fine Oxidizer ~ 1 to 15 μm (finer size ⇒ hazardous)
- Fuel ~ 20 to 50 μm
- Medium Oxidizer ~ 20 to 90 μm
- Coarse Oxidizer ~ 200 to 400 μm (Very large size ⇒ low burn rate and possibly hazardous (combustion instability))

Note:
- Different sizes of particles enhance loading efficiency and alleviate crack propagation.
- Above particles represented as spheres. Actual particles are more irregular, because of crystal cracking.
- Insufficient wetting of particles by binder causes voids, which promote crack propagation through the binder.
- Differences in coefficients of thermal expansion α lead to thermal stress ($\sigma_{TS} = \alpha \, E \, \Delta T$) / voids / cracks

Fig. 3.66 Efficient Loading of Solid Propellant Particles Prevents Crack Propagation through the Binder.

- **Binder Characteristics**
  - Polymer (e.g., HTPB, GAP)
  - Liquid During High Temperature Mixing and Casting
  - Bonds Oxidizer / Fuel of Composite Propellant
  - Provides Propellant Strength

HTPB vs GAP Binder Tradeoffs	HTPB	GAP
Strength, Stress, and Shock Resistance	✓	
Aging Resistance	✓	
Energy Added for Combustion		✓
Burn Rate		✓

Note: Binder Requires ≈ 10 to 15% of Propellant Weight for Efficient Mixing / Casting
HTPB: Hydroxyl-terminated polybutadiene binder
GAP: Glycidyl azide polymer

Fig. 3.67 Missile Solid Propellant Binder Tradeoffs Include Elasticity, Aging, Energy Added, and Burn Rate.

3.22 Solid Propellant Aging

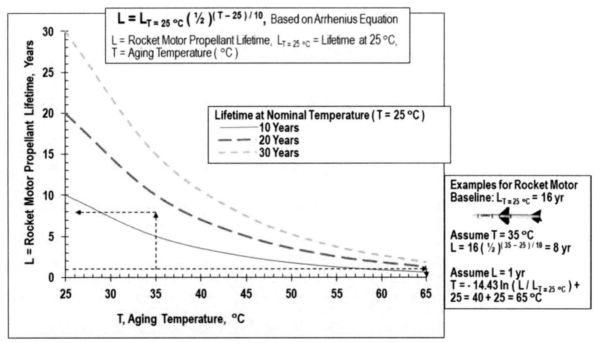

Note: Elevated temperature tests, with accelerated aging, are used to verify rocket motor propellant lifetime. Test measurements following accelerated aging tests include measurement of loss in weight (from moisture loss), chemistry composition, oxidation, strain, and brittleness.
This approach is also applicable to evaluate aging for other energetic subsystems (e.g., battery, squib, warhead).

Fig. 3.68 Solid Propellant Rocket Motor Lifetime May Be Driven by a High Temperature Environment.

Type of Environment	Typical Min Temp	Typical Max Temp	Typical Max Shock / Vibration @ Max / Min Temp
Depot (e.g., Bunker) 	-9 °F	120 °F	100 g @ 10 ms (~ 0.5 m drop onto concrete)
Transportation 	-20 °F	130 °F	100 g @ 10 ms (~ 0.5 m drop) 7 g @ 300 Hz, 10 g @1000 Hz)
Field 	-60 °F	160 °F	100 g @ 10 ms (~ 0.5 m drop) 7 g @ 300 Hz, 10 g @1000 Hz)

- **Propellant Failure from Shock / Vibration / Strain @ T_{min}**
  - Grain Cracking (From Δ Thermal Expansion: Grain / Case, Fuel / Oxidizer / Binder; Brittle Propellant)
  - Insulation Liner De-bond (From Δ Thermal Expansion of Grain / Case)
  - Voids (From Δ Thermal Expansion of Fuel / Oxidizer / Binder)
- **Propellant Failure from Shock / Vibration / Strain @ T_{max}**
  - Grain Cracking (From Low Modulus of Elasticity, Propellant Decomposition Gas Bubbles)
  - Does not satisfy Insensitive Munition (IM) requirements

Note:
- Propellant failure more likely from repeated cycling at min / max temperature
- Propellant failure more likely at low humidity (loss of moisture) and high humidity (propellant specie migration)

Fig. 3.69 Solid Propellant Rocket Motor Lifetime May Be Driven by Shock and Strain in a Low Temperature Environment.

3.23 Solid Propellant Rocket Combustion Stability

Motor Geometry Drives 1st Mode Longitudinal Acoustic Frequency and Nozzle Damping*

- $f_{1st} = a / (2 \, l_{case}) \approx 19200 / l_{case}$
- Long length motor case ⇒ lower acoustic frequency ⇒ flight control feedback coupling and motor case structural coupling / resonance
- Small throat diameter ⇒ more reflection of acoustic waves off aft dome close-out ⇒ higher amplitude pressure oscillation

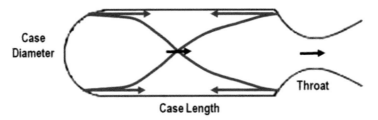

*Note. Combustion stability may be a driver for maximum expected operating pressure (MEOP) and required factor of safety (FOS). Units of frequency f_{1st} are Hz and units of case length l_{case} are inches. The equation $f_{1st} \approx 19200 / l_{case}$ is based on typical values of solid propellant combustion speed of sound (a), temperature, gas constant, and specific heat ratio.

Example Calculation: From the above equation, the rocket baseline missile rocket motor 1st mode longitudinal acoustic frequency is $f_{1st} \approx 19200 / 55 = 349$ Hz

Fig. 3.70a A Driver of Solid Propellant Rocket Motor Combustion Stability Is Motor Geometry.

Motor Geometry Drives Required Burn Rate for Stable Ignition, w/o Chuffing

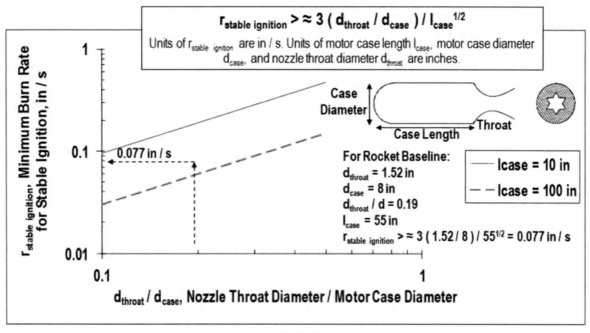

Note: $r_{stable\ ignition}$ based on correlation of data from AIAA 2007-5803. Data for typical minimum smoke and composite solid propellant, with 10 in < l_{case} < 100 in.

Reference: Blomshield, Fred S., "Lessons Learned in Solid Rocket Combustion Instability", AIAA 2007-5803, July 2007

Fig. 3.70b A Driver of Solid Propellant Rocket Motor Combustion Stability Is Motor Geometry (cont).

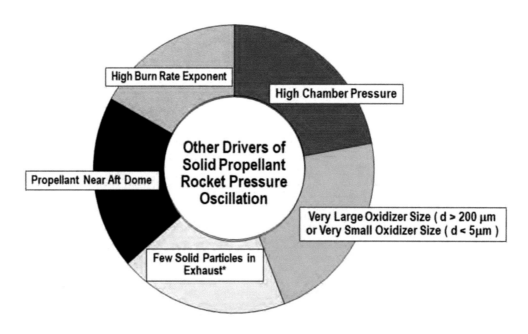

*Optimum size of exhaust particles for combustion stability: $d_{optimum} = 1.84\ l_{case}^{1/2}$. Units of $d_{optimum}$ in μm. Units of l_{case} in inches. Example calculation of size of exhaust particles that provide minimum pressure fluctuation for rocket motor baseline:
$d_{optimum} = 1.84\ l_{case}^{1/2} = 1.84\ (55)^{1/2} = 14$ μm

Fig. 3.71 Other Drivers of Rocket Combustion Stability Include Chamber Pressure, Oxidizer Size, and Exhaust Particles.

3.24 Rocket Motor Case and Nozzle Material Alternatives

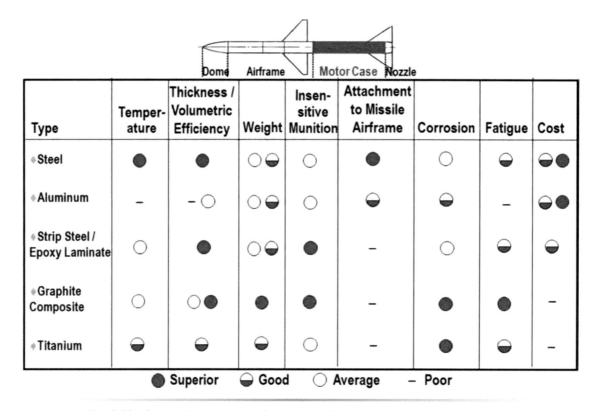

Fig. 3.72 Steel and Aluminum Motor Cases Are Low Cost but a Composite Motor Case Is Light Weight.

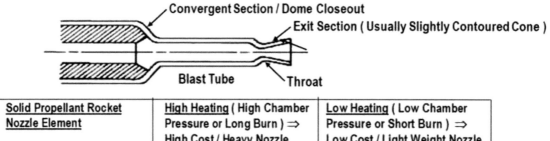

Solid Propellant Rocket Nozzle Element	High Heating (High Chamber Pressure or Long Burn) ⇒ High Cost / Heavy Nozzle	Low Heating (Low Chamber Pressure or Short Burn) ⇒ Low Cost / Light Weight Nozzle
♦ Typical Structure Housing Material Alternatives	♦ Steel ♦ Phenolic	♦ Cellulose / Phenolic ♦ Aluminum
♦ Typical Throat Insert Material Alternatives	♦ Tungsten Insert ♦ Rhenium Insert ♦ Molybdenum Insert	♦ Cellulose / Phenolic Insert ♦ Silica / Phenolic Insert ♦ Graphite Insert ♦ Carbon – Carbon Insert
♦ Typical Exit Section, Convergent Section, and Blast Tube Insert Material Alternatives	♦ Silica / Phenolic Insert ♦ Graphite / Phenolic Insert ♦ Silicone Elastomer Insert	♦ No Insert ♦ Glass / Phenolic Insert

Note: Above does not include insulation material.

Fig. 3.73 Heat Transfer Drives Solid Propellant Rocket Nozzle Materials, Weight, and Cost.

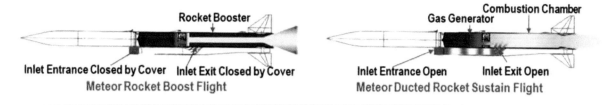

Fig. 3.74 Ducted Rocket Propulsion Often Provides Longer Standoff Range Than Rocket Propulsion.
Note: See video supplement, Meteor Ducted Rocket Missile.

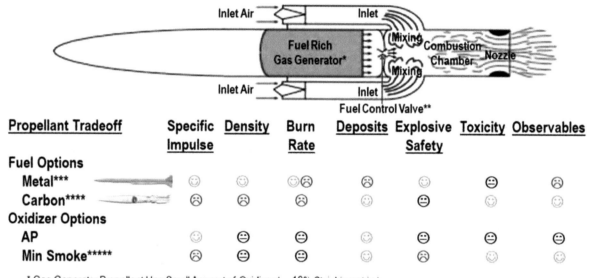

* Gas Generator Propellant Has Small Amount of Oxidizer (~ 10% Stoichiometric)
**Typical Fuel Control / Thrust Magnitude Control Is a Pintle or Valve in Gas Generator Throat. If Excess Fuel from Gas Generator < ~ 30% Propellant Weight ⇒ Behaves More Like Rocket (Higher Burn Rate, Higher Burn Temp, Lower I_{SP}). If Excess Fuel from Gas Generator > ~ 70% Propellant Weight ⇒ Behaves More Like Ramjet (Higher I_{SP}, Lower Burn Rate, Lower Burn Temp)
***Typical Metal Fuels: B, Al, Mg. Typical Metal Fuel Propellant Mix: 50% B, 10% HTPB, 40% AP
****Typical Carbon Based Fuels: C, HTPB. Typical Carbon Based Fuel Mix: 65% HTPB, 35% AP
*****Typical Min Smoke Oxidizers: CL-20, ADN, HMX, RDX, AN, Double Base (Nitrocellulose / Nitroglycerine)
Reference: Stowe

Fig. 3.75 Ducted Rocket Design Tradeoffs Include Type of Propellant Fuel and Oxidizer.

3.25 Ducted Rocket Design Considerations

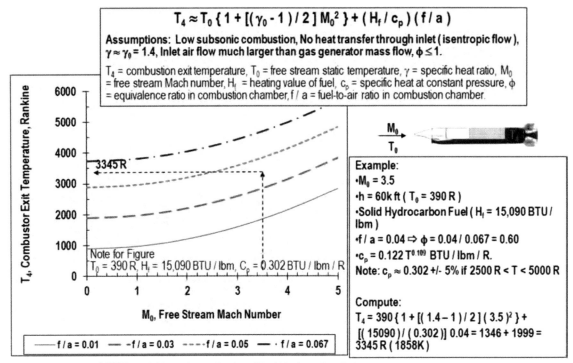

Fig. 3.76 Ducted Rocket Combustion Temperature Increases with Mach Number and Fuel Flow.

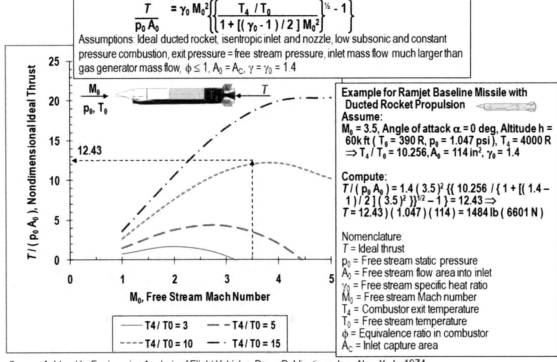

Fig. 3.77 Ducted Rocket High Thrust Occurs for High Mach Number, High Combustion Temperature, and High Inlet Flow.

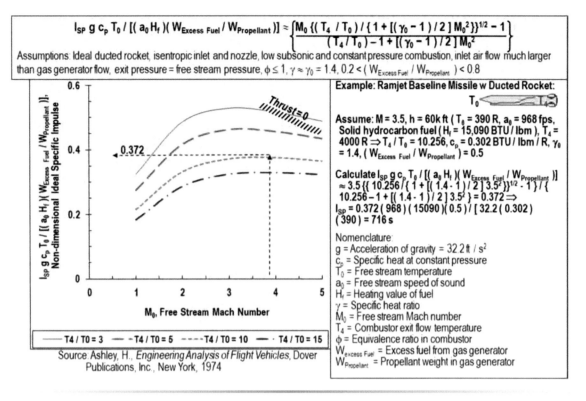

Fig. 3.78 Ducted Rocket High Specific Impulse Occurs with a High Heating Value Fuel @ Mach 3 to 4 Flight.

Chapter 4 Weight

4.1 Introduction

Fig. 4.1 Chapter 4 Weight—What You Will Learn.

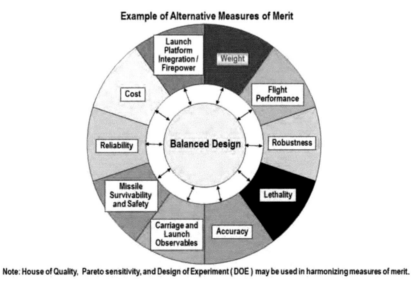

Fig. 4.2 A Balanced Missile Design Requires Harmonized Mission Requirements and Measures of Merit.

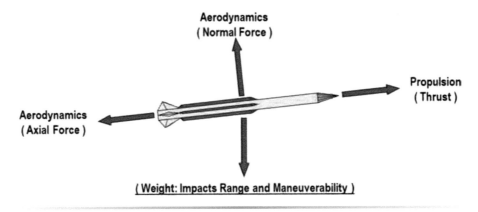

Fig. 4.3 Weight Impacts Missile Flight Range and Maneuverability.

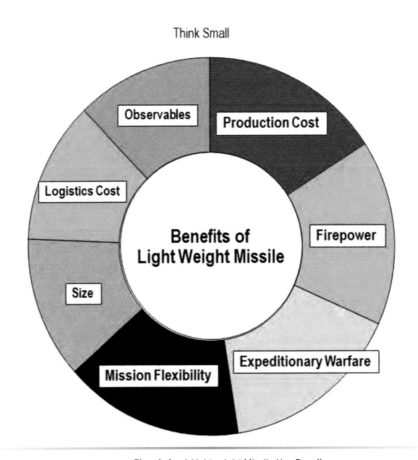

Fig. 4.4 A Lightweight Missile Has Payoff.

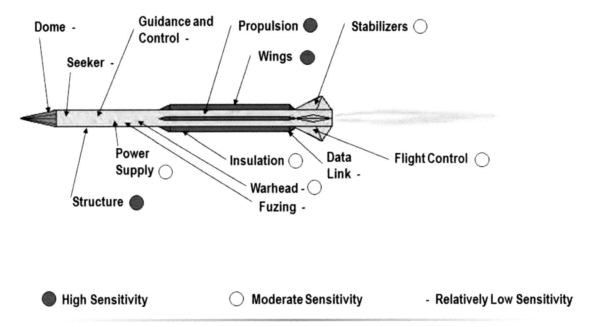

Fig. 4.5 Weights of Missile Subsystems Often Drive Flight Performance.

4.2 Missile Weight Prediction

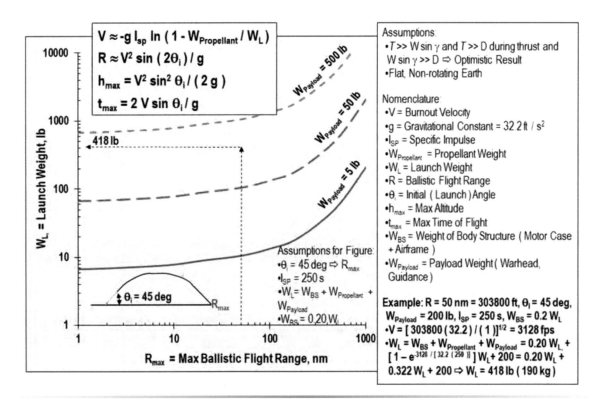

Fig. 4.6 Ballistic Missile Launch Weight Is Driven by Range, Payload Weight, Propellant Weight, and Specific Impulse.

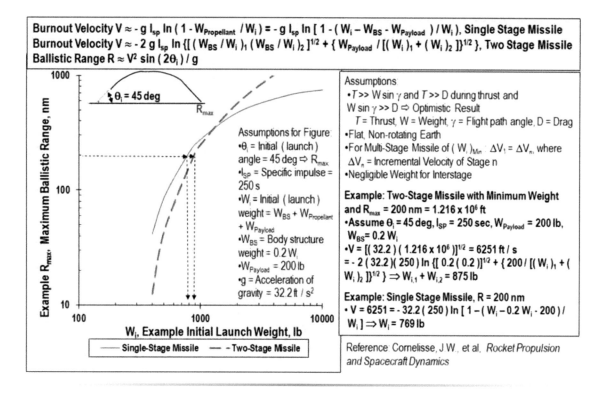

Fig. 4.7 Staging Provides Range and Weight Payoff for Long Range Ballistic Missiles.

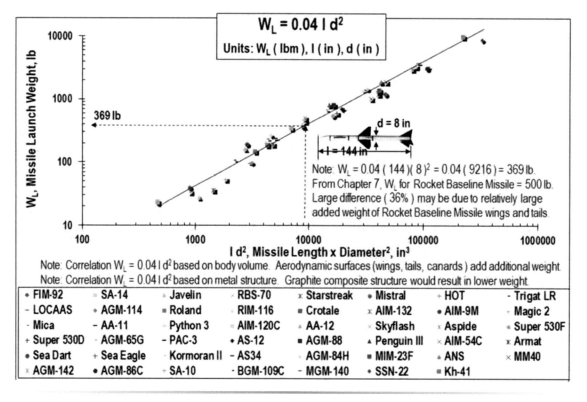

Fig. 4.8 A First-Order Estimate of Missile Weight Can Be Derived from Body Geometry Dimensions.

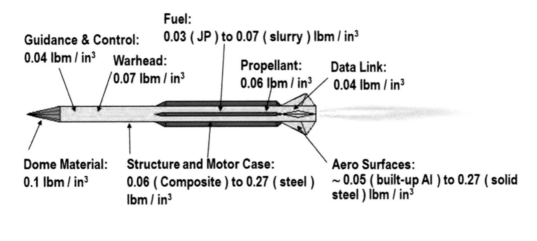

Fig. 4.9 Most Subsystems for Missiles Have a Weight Density of about 0.05 lbm per in³.

4.3 Center-of-Gravity and Moment-of-Inertia Prediction

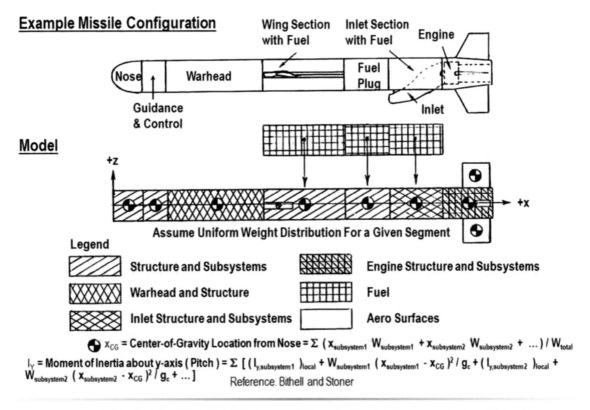

Fig. 4.10 Modeling Missile Weight, Balance, and Moment-of-Inertia Is Based on a Build-up of Subsystems.

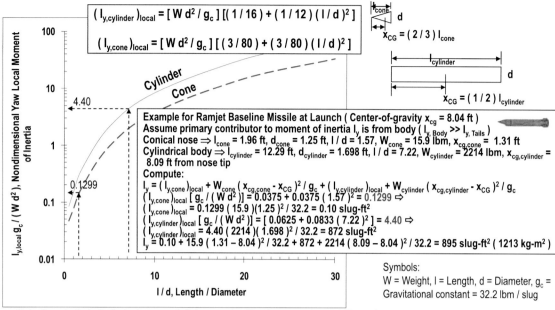

Fig. 4.11 A Missile with a High Fineness Body Has a Larger Moment-of-Inertia.

4.4 Missile Airframe Structure Manufacturing Processes

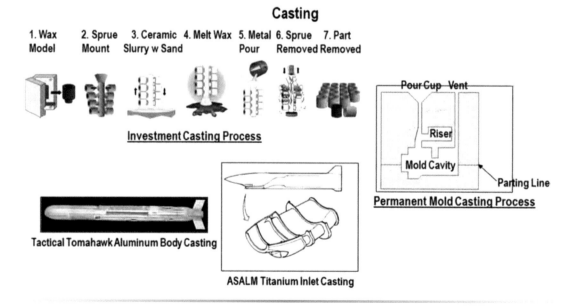

Fig. 4.12a Examples of Missile Structure Manufacturing Processes.
Note: See video supplement, Investment Casting.

CHAPTER 4 Weight 103

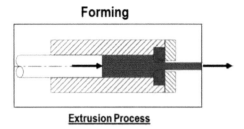

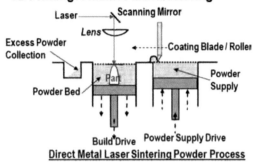

Fig. 4.12b Examples of Missile Structure Manufacturing Processes (cont).
Note: See video supplement, Forging, Ring/Strip Rolling; 3D Printing Using DMLS.

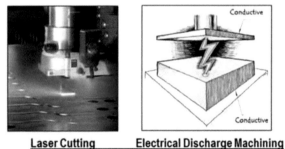

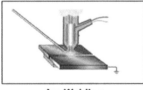 

Fig. 4.12c Examples of Missile Structure Manufacturing Processes (cont).
Note: See video supplement, Cutting, Milling, Drilling, EDM; Resistance/Arc, Laser, Friction Welding.

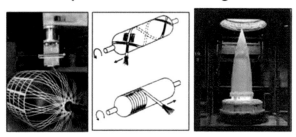
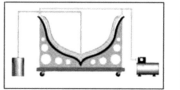

Fig. 4.12d Examples of Missile Structure Manufacturing Processes (cont).
Note: See video supplement, Carbon Fiber Manufacturing; Composite Filament Winding, RTM, Vacuum Bagging.

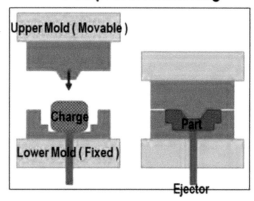
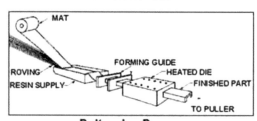
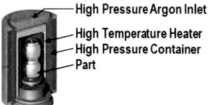

Fig. 4.12e Examples of Missile Structure Manufacturing Processes (cont).
Note: See video supplement, Compression Molding.

Type of Joint Attachment	Max Load	Thermal Stress			Fatigue	Inspection / Disassembly
		Metal – Metal	Graphite – Graphite	Metal - Graphite		
Mechanical Fastener	●	○	○	◡	○	●
Adhesive Bonding	◡	◐	●	○	●	◡

● Superior ◐ Good ○ Average ◡ Poor

Fig. 4.13 Mechanical Fastener Versus Adhesive Bonding Is a Tradeoff for Structural Joint Attachment.

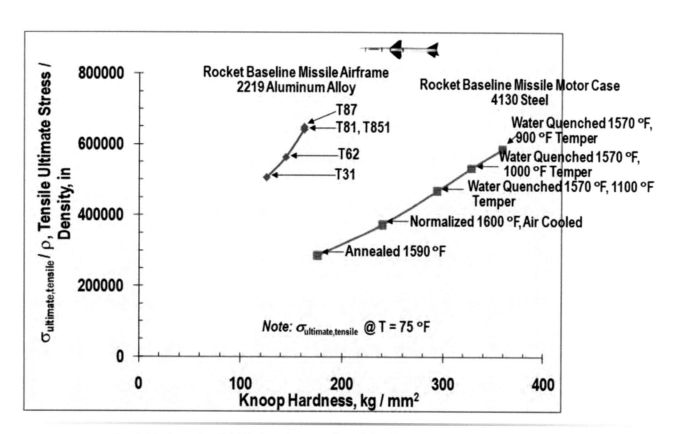

Fig. 4.14 Increasing Metal Hardness Increases Strength, but Machining Is More Difficult and More Expensive.

Missile Airframe Geometry Alternatives	Missile Structure Concept Alternatives	Examples of Structure Manufacturing Process Alternatives										
		Graphite Composites						Metals				
		Vacuum Assist RTM	Compression Mold	Filament Wind	Pultrusion	Thermal Form	Vacuum Bag / Autoclave	Cast	3D Print / Additive	HIP	High Speed Machine	Forming	Strip Laminate
Non-Axisymmetric Body Airframe	Monocoque	●	◐	●	●	◐	●	●	●	●	—	◐	
	Integrally Hoop Stiffened	●	◐				◐	●	●		—		
	Integrally Longitudinal Stiffened	●	◐		●		●	●	●		—		
Axisymmetric Body Airframe	Monocoque	●		●			●	●	●	●		●	●
	Integrally Hoop Stiffened	●					◐	●	●		◐		
	Integrally Longitudinal Stiffened	●		●			●	●	●		◐		
Aerodynamic Surface	Solid	●	●			●	●	●	●	●		●	
	Sandwich	◐	◐				◐		●			○	

Note: Manufacturing process cost is a function of recurring cost (unit material, unit labor) and non-recurring cost (tooling)
Note: RTM: Resin Transfer Molding, HIP: Hot Isostatic Press
Note: ● Very Low Parts Count ◐ Low Parts Count ○ Moderate Parts Count — High Parts Count

Fig. 4.15 Missile Structure Parts Count Is Driven by the Manufacturing Process.

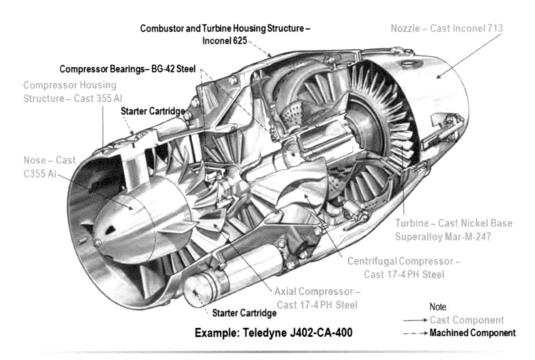

Fig. 4.16 Missile Turbojet Engine Parts Are Often Castings (Lower Parts Count and Lower Cost).

4.5 Missile Airframe Material Alternatives

Type	Material	Tensile Stress (σ_{TU}/ρ)	Buckling Stability ($\sigma_{Buckling}/\rho$)	Max Short–Life Temp	Thermal Stress	Joining	Fatigue	Cost	Weight
Metallic	Aluminum 2219	○	◐	– ○	–	◐	–	●○	○
↓	Steel PH 15-7 Mo	◐	–	●	○	●	◐	●○	–
↓	Titanium Ti-6Al-4V	◐	○	●	◐	○	◐	–	○
Composite	S994 Glass / Epoxy and S994 Glass / Polyimide	◐	– ○	○	◐	○	●	●	◐
↓	Glass or Graphite Reinforce Molding	–	– ○	○	◐	○	●	○	◐
↓	Graphite / Epoxy and Graphite Polyimide	●	○	○ ◐	●	–	●	–	●

Note: ● Superior ◐ Good ○ Average – Poor

Fig. 4.17 Missile Airframe Material Alternatives Include Aluminum, Steel, Titanium, and Composite.

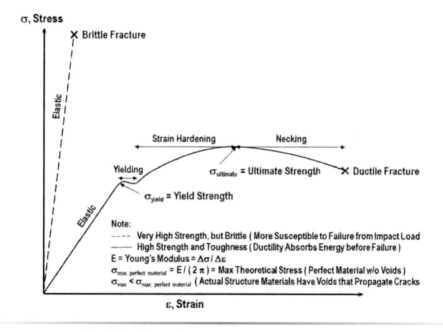

Fig. 4.18 Missile Structure Drivers Include Strength and Toughness.
Note: See video supplement, Stress-Strain.

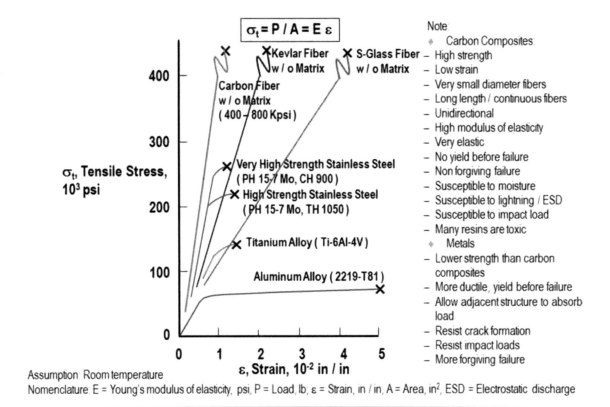

Fig. 4.19 Example of Strength—Elasticity Comparison of Missile Structure Material Alternatives.

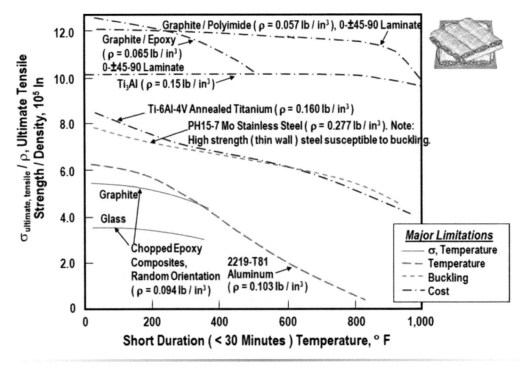

Fig. 4.20 Laminate Graphite Composite Provides a High Strength-to-Weight Structure.

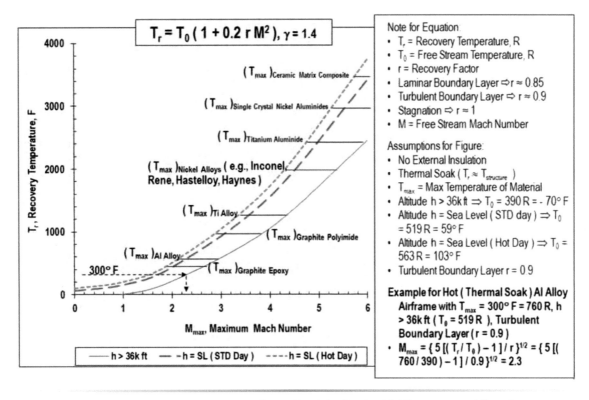

Fig. 4.21 A Hypersonic Missile without External Insulation Requires High Temperature and Heavy Structure.

4.6 Missile Structure/Insulation Trades

Example Structure / Insulation Concepts	Concept	T_{max}	k	c	ρ	α
Hot Metal Structure (e.g., Al Heat Sink) without Insulation	1	300 - 600	0.027	0.22	0.103	0.000722
Hot Metal Structure (e.g., Al Heat Sink)	2	300 - 600	0.027	0.22	0.103	0.000722
Internal Insulation (e.g., Min-K)		2000	0.0000051	0.24	0.012	0.00000106
Self-insulating Composite Structure (e.g., Graphite Polyimide)	3	1100	0.000109	0.27	0.057	0.00000410
Ext Insulation (e.g., Micro-Balloon Quartz)		1200	0.0000131	0.28	0.012	0.00000226
Cold Metal Structure (e.g., Al Heat Sink)	4	300 – 600	0.027	0.22	0.103	0.000722
Internal Insulation (e.g., Min-K)		2000	0.0000051	0.24	0.012	0.00000106

Note:
- Missiles use passive thermal protection (no active cooling) based on structure heat sink, insulation, and possibly phase change material (e.g., paraffin to cool electronics).
- Small thickness for insulation allows more propellant / fuel for diameter constrained missiles (e.g., VLS launcher).
- Weight and cost are application specific.
- T_{max} = max temp capability, °F; k = thermal conductivity, BTU / s / ft / °F; c = specific heat or thermal capacity, BTU / lbm / °F; ρ = density, lbm / in³; α = thermal diffusivity = k / (ρ c), ft² / s
- Insulation thickness driven by thermal diffusivity = k / (ρ c). Insulation weight driven by k ρ / c.

Fig. 4.22 Missile Structure Insulation Concepts and Trades for Short Duration Flight.

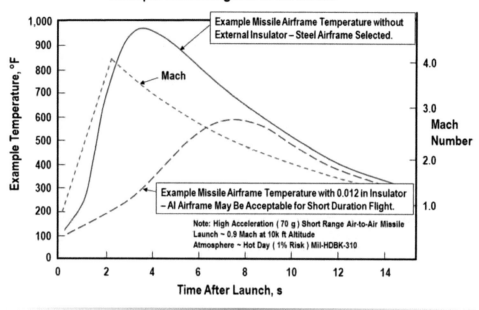

Fig. 4.23 External Structure Insulation Has High Payoff for Short Duration Flight at High Mach Number.

4.7 High Temperature Insulation Materials

Type	Min Thickness	Max Temp	Max Mach	Min Weight	Strain/ Shock	Strength	Out-Gassing	Cost
Phenolic Composites	◐	◐	●	○	○	○●	◐	○
Low Density Composites	●	○	○◐	◐	●	○	○	○●
Plastics	○	-	-	○	●	-	-	●
Porous Ceramics	-	○	◐	-	-	●	●	○
Bulk Ceramics	-	◐	●	-	-	●	●	○●
Graphites	-	●	●	○	-	●	●	-

Note: ● Superior   ◐ Good   ○ Average   - Poor

Fig. 4.24 There Are Many Considerations for High Temperature Missile Insulation.

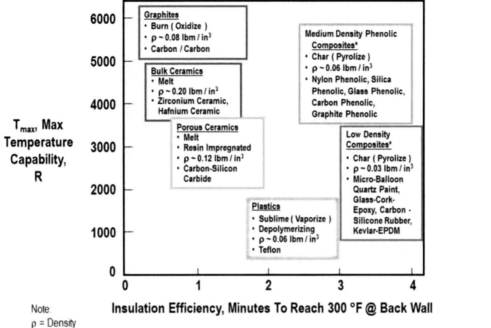

Fig. 4.25 Maximum Temperature and Insulation Efficiency Are Drivers for High Temperature Insulation.

- **Paint Insulation Usually Best if Insulation Thickness < ≈ 0.5 in**
  - Spray or Trowel Multiple Coats

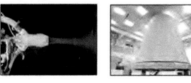

Spray-on Insulation for NASA SLS

- **Tile Insulation Usually Best if Insulation Thickness > ≈ 0.5 in**
  - Bond Insulation to Airframe Using High Temperature Adhesive (e.g., Silicone)
  - Insulation Expansion Joints May Be Required to Alleviate Thermal Stress from Airframe

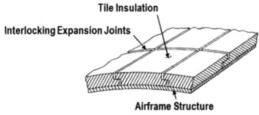

Fig. 4.26 Required Insulation Thickness Is a Consideration for Integrating External Insulation with Airframe.

4.8 Missile Aerodynamic Heating/Thermal Response Prediction

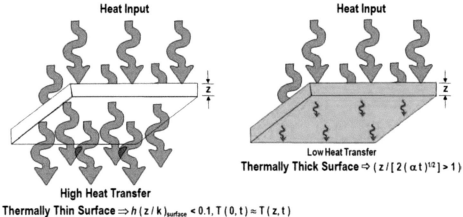

Fig. 4.27 A Thermally Thin Surface Has High Heat Transfer, A Thermally Thick Surface Has Low Heat Transfer.

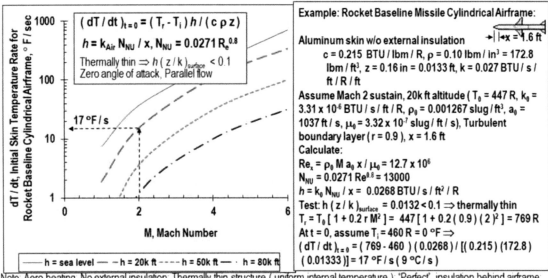

Fig. 4.28a A "Thermally Thin" Surface in Aero Heating Has Rapid Temperature Rise.

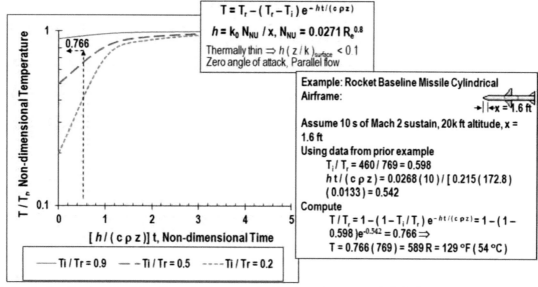

Fig. 4.28b A "Thermally Thin" Surface in Aero Heating Has Rapid Temperature Rise (cont).

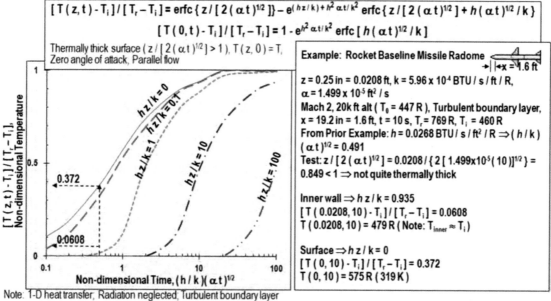

Fig. 4.29a A "Thermally Thick" Surface in Aero Heating Has Large Internal Temperature Gradient.

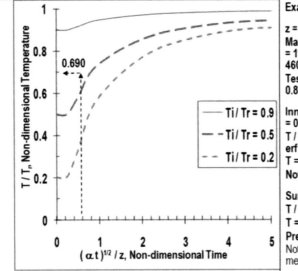

Fig. 4.29b A "Thermally Thick" Surface in Aero Heating Has Large Internal Temperature Gradient (cont).

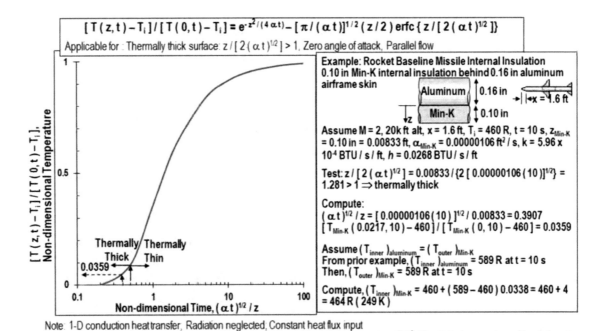

Fig. 4.30 Airframe Internal Insulation Temperature and Required Thickness Can Be Predicted Assuming Constant Flux Conduction.

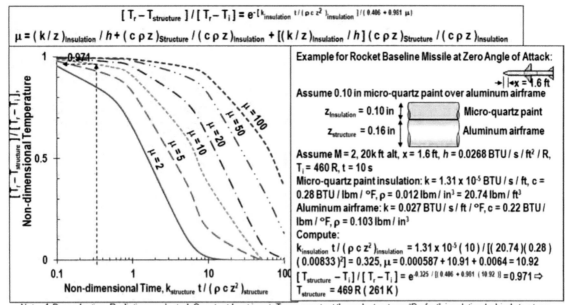

Fig. 4.31 External Insulation Greatly Reduces Airframe Structure Temperature in Short Duration Flight.

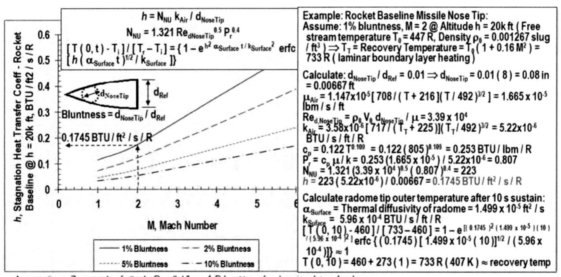

Fig. 4.32 A Sharp Nose Tip or Sharp Leading Edge Has High Aerodynamic Heating in Hypersonic Flight.

4.9 Localized Aerodynamic Heating and Thermal Stress

IRdome / Radome
- Dome material selection and thickness driven by transmission quality
- Large internal temperature gradients due to low thermal conduction
- Thermal stress at attachment (different coefficients of thermal expansion)
- Low tensile strength (dome material selection driven by transmission quality)
- Dome fails in tension

Sharp Leading Edge / Nose Tip
- Hot stagnation temperature on leading edge
- Small radius prevents use of external insulation
- Colder heat sink material as chord thickness increases leads to leading edge warp
- Shock wave interaction with adjacent body structure

Body Joint
- Hot missile shell
- Colder frames or bulkheads
- Causes premature buckling

Note: σ_{TS} = Thermal stress from restraint in compression or tension = $\alpha E \Delta T$
Nomenclature: α = coefficient of thermal expansion, E = modulus of elasticity, $\Delta T = T_2 - T_1$ = temperature difference

Example: Thermal Stress σ_{TS} for Rocket Baseline Missile Pyroceram Dome ($\alpha = 3 \times 10^{-6}$ / R, $E = 13.3 \times 10^6$ psi, $\sigma_{max} = 25{,}000$ psi)
Assume Mach Number M = 2, Altitude h = 20k ft, Time t = 10 s. From prior figures: $\Delta T = T_{OuterWall} - T_{InnerWall}$ = 575 − 479 = 96 R
(Jerger Reference), ΔT = 769 − 531 = 238 R (Carslaw and Jaeger Reference)
Then $\sigma_{TS} = 3 \times 10^{-6}$ (13.3×10^6) (96) = 3830 psi (Jerger), $\sigma_{TS} = 3 \times 10^{-6}$ (13.3×10^6) (238) = 9500 psi (Carslaw and Jaeger)

Note: Carslaw and Jaeger reference less accurate because of approximation surface temperature ≈ recovery temperature [$T(0,t) \approx T_r$]

Fig. 4.33a A Missile Design Concern Is Localized Aerodynamic Heating and Thermal Stress.

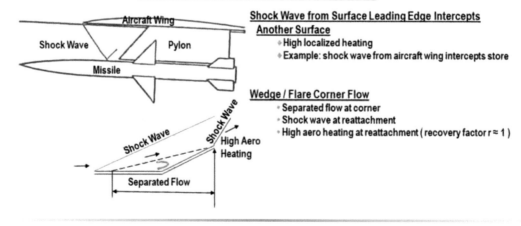

Fig. 4.33b A Missile Design Concern Is Localized Aerodynamic Heating and Thermal Stress (cont).
Note: See video supplement, 24 Deg Flare at Mach 2.3.

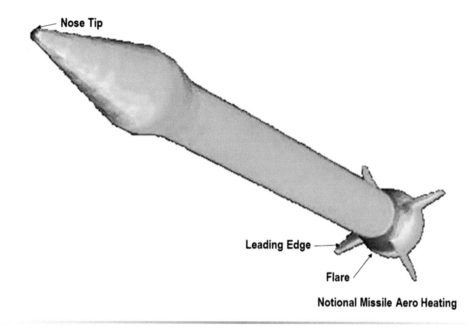

Fig. 4.34 Examples of Aerodynamic Hot Spots.
Note: See video supplement, Radiometric Imagery-SM 3 Flight.

4.10 Missile Structure Design

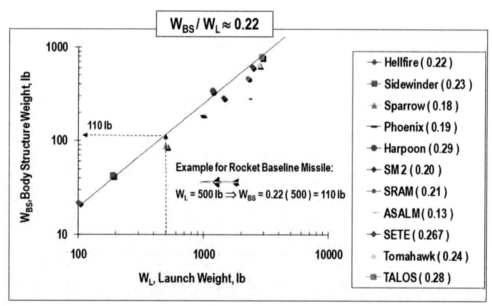

Note: W_{BS} includes all load carrying body structure. If motor case, engine, or warhead case carry external loads then they are included in W_{BS}. W_{BS} does not include tail, wing, or other surface weight.
Note: Above based on metal structure. Graphite composite structure would result in lower body structure weight fraction.

Fig. 4.35 Missile Metal Body Structure Weight Is about 22% of the Missile Launch Weight.

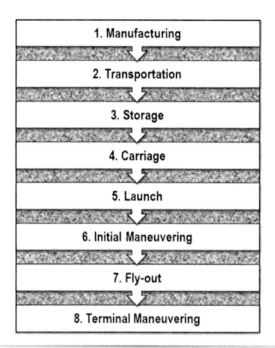

Fig. 4.36 Missile Structure Is Based on Considering the Cradle-to-Grave Environment.

Contributors to Body Structure Thickness	Cylindrical Body Structure Thickness Equation
Min Gauge for Manufacturing (e.g., drilling, machining, bending)	$t = 0.7\, d\, [(\, p_{ext}\, /\, E\,)\, l\, /\, d\,]^{0.4}$, $t \approx 0.06$ in, if $p_{ext} \approx 10$ psi
Localized Buckling in Bending*	$t = 2.9\, r\, \sigma\, /\, E$
Localized Buckling in Axial Compression	$t = 4.0\, r\, \sigma\, /\, E$
Thrust Force	$t = T\, /\, (\, 2\, \pi\, \sigma\, r\,)$
Maneuver Bending Moment*	$t = M\, /\, (\, \pi\, \sigma\, r^2\,)$
Internal Pressure*	$t = p\, r\, /\, \sigma$

Reference: Atkinson, J.R. and Staton, R.N., "Missile Body Weight Prediction", SAWE 1497, May 1982

*Note: Often a major contributor to required thickness
Note: Does not include factor of safety (FOS)

t = thickness, d = diameter, P_{ext} = external pressure, l = length, r = radius, σ = strength, E = modulus of elasticity

Fig. 4.37 Missile Body Structure Required Thickness Is Based on Considering Many Design Conditions.

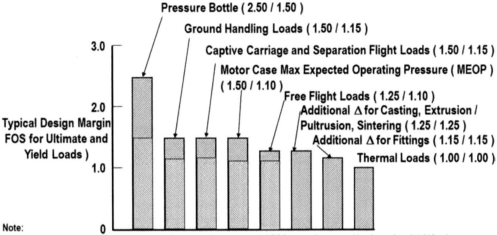

Fig. 4.38 Higher Structure Design Factor of Safety Is Required for Hazardous Subsystems and Hazardous Flight Conditions.

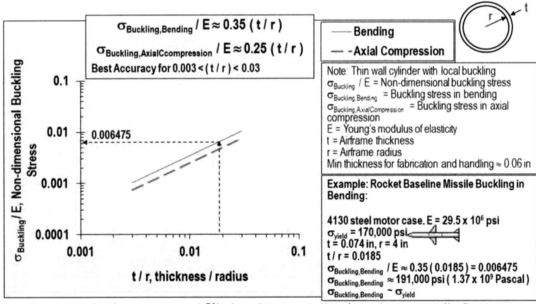

Fig. 4.39 Localized Buckling May Be a Concern for a Thin Wall Structure.

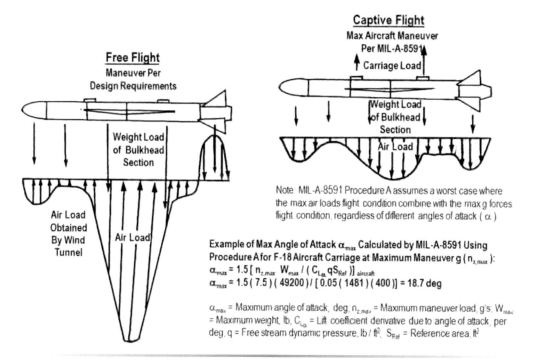

Fig. 4.40 MIL-A-8591 Provides a Conceptual Design Procedure to Estimate Captive Carriage Maximum Flight Load.

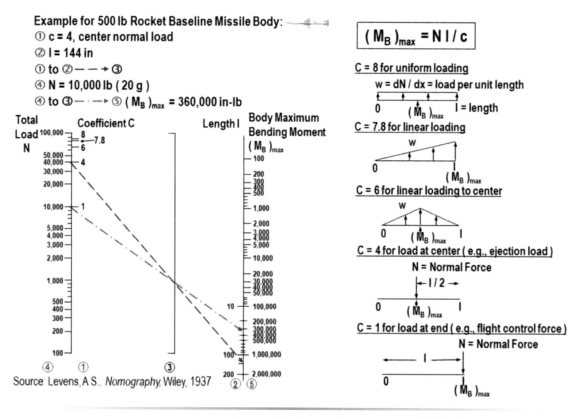

Fig. 4.41 Maximum Body Bending Moment Depends Upon Load Distribution.

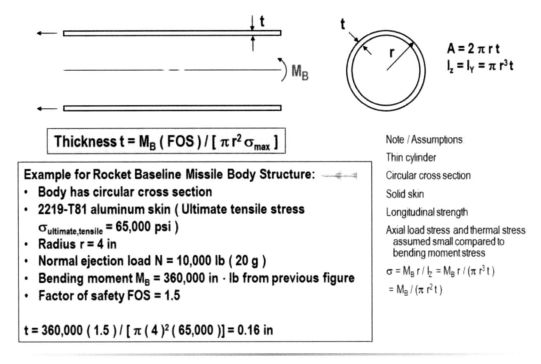

Fig. 4.42 Body Bending Moment May Drive Body Structure Thickness and Weight.

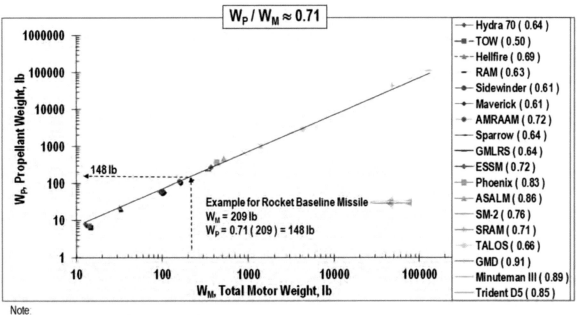

Fig. 4.43 For a Typical Solid Propellant Rocket Motor, about 71% of the Motor Weight Is Propellant.

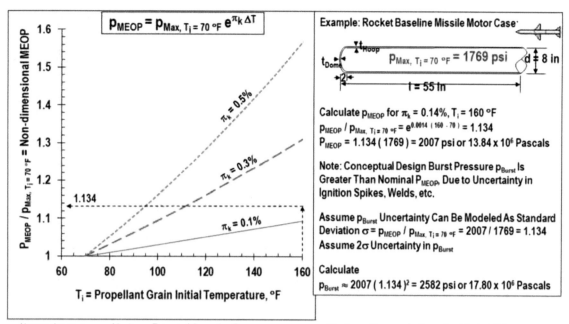

Fig. 4.44 Maximum Expected Operating Pressure (MEOP) Increases with Propellant Grain Initial Temperature.

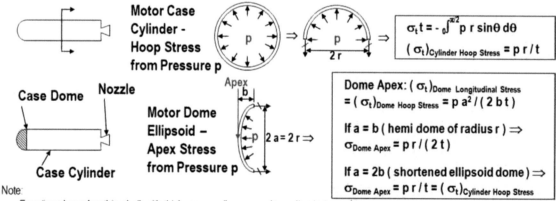

Fig. 4.45a Solid Propellant Rocket Motor Case Required Thickness and Weight Is Usually Driven by Internal Pressure.

CHAPTER 4 Weight   123

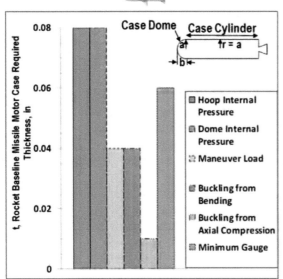

Required Thickness for Rocket Baseline Missile Motor Case
- Ellipsoid Forward Dome (a / b = 2, a = r = 4 in)
- Aft Cylinder (r = Radius = 4 in)
- 4130 Steel Motor Case (ρ = Density = 0.283 lb / in^3, (σ_t)$_{ultimate}$ = Ultimate Stress = 190,000 psi

Calculate Required Thickness from Internal Pressure with Factor of Safety FOS = 1.5

$(t_{Hoop})_{InternalPressure}$ = (FOS) p_{burst} r / σ_t = 1.5 (2582)(4.0)/ 190000 = 0.08 in

$(t_{Dome\ Apex})_{InternalPressure}$ = (FOS) p_{burst} a^2 / (2 b σ_t) = 1.5 (2582) (4^2) / [2 (2)(190000)] = 0.08 in

Calculate Required Thickness from Maneuver Bending
$t_{ManeuverBending}$ = (FOS) M$_B$ / (π σ r^2)
Assume 30 g maneuver n at launch (W$_L$ = 500 lb) with uniform loading (c = 8) over length l of missile (144 in) and Factor of Safety FOS = 1.5 Then
Normal Force N = n W = 30 (500) = 15,000 lb
Bending Moment M$_B$ = N l / c = 15000 (144) / 8 = 270,000 in-lb
$\Rightarrow$ t$_{ManeuverBending}$= 1.5 (270000)/ [π (190000) 4^2] = 0.04 in

Total Required Thickness Under Combined Loads of Internal Pressure + Maneuver Load Is

t_{Hoop} = $t_{ManeuverBending}$ + $t_{InternalPressure}$ = 0.04 + 0.08 = 0.12 in
t_{Dome} = $t_{ManeuverBending}$ + $t_{InternalPressure}$ = 0.04 + 0.08 = 0.12 in

Note: Above case thickness drivers are internal pressure and maneuver loads, more important than buckling and min gauge.

Fig. 4.45b Solid Propellant Rocket Motor Case Required Thickness and Weight Is Usually Driven by Internal Pressure (cont).

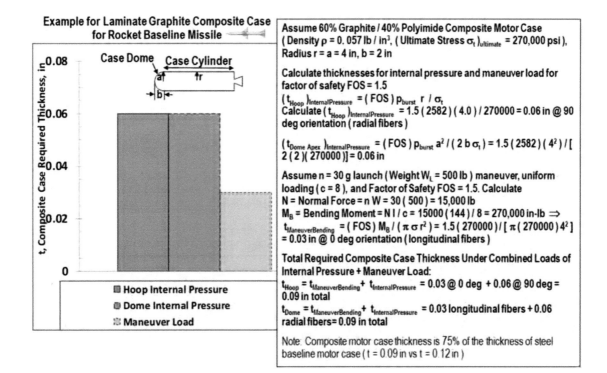

Fig. 4.45c Solid Propellant Rocket Motor Case Required Thickness and Weight Is Usually Driven by Internal Pressure (cont).

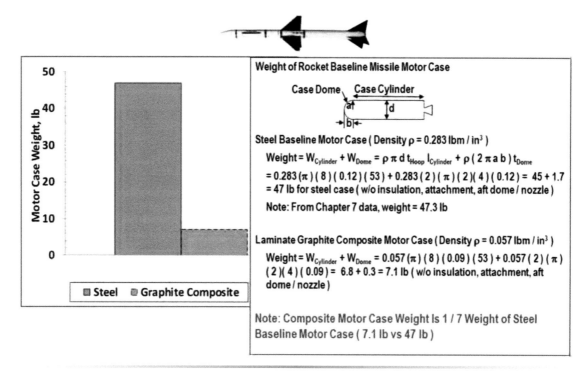

Fig. 4.46 Laminate Graphite Composite Rocket Motor Case for Rocket Baseline Missile Is Lighter Weight.

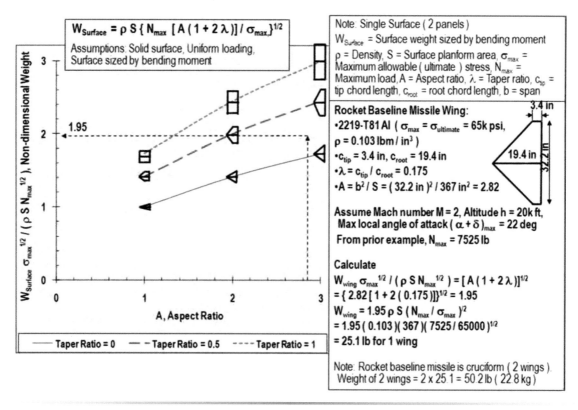

Fig. 4.47 A Low Aspect Ratio Delta Surface Planform Has Lighter Weight Structure.

4.11 Seeker Dome Alternatives

Multi-Mode Seeker Dome Material	Density (g/cm^3)	Dielectric Constant	SWIR/MWIR and LWIR Bandpass		Transverse Strength (10^3 psi)	Thermal Expansion ($10^{-6}/°F$)	Erosion, Knoop (kg/mm^2)	Max Temp (°F) w/o EO/EM Degradation
Zinc Sulfide (Z_nS)	4.05	8.4 ○	●	●	18 ○	4 ○	350 ○	700 ○
Sapphire (Al_2O_3) / Spinel ($MgAl_2O_4$)	3.68	8.5 ○	●	⌣	28 ○	3 ○	1650 ●	1800 ●
Quartz / Fused Silica (SiO_2)	2.20	3.7 ●	○	-	8 ⌣	0.3 ●	600 ◐	2000 ●
Diamond (C)	3.52	5.6 ◐	○	●	400 ●	1 ●	8800 ●	3500 ●
Mag. Fluoride (M_gF_2)	3.18	5.5 ◐	●	⌣	7 ⌣	6 ⌣	420 ◐	1000 ◐

● Superior ◐ Good ○ Average ⌣ Poor - Very Poor

Note:
RF = Radar Frequency, IR = Infrared, SWIR = Short Wave Infrared (< 3 μm Wavelength), MWIR = Mid Wave Infrared (3 to 5 μm Wavelength), LWIR = Long Wave Infrared (8 to 14 μm Wavelength), EO = Electro Optical, EM = Electro Magnetic

Fig. 4.48 Multi-mode Seeker Dome Material Is Driven by Transmission and Flight Environment.

Infrared Seeker Dome Material	Density (g/cm^3)	SWIR/MWIR and LWIR Bandpass		Transverse Strength (10^3 psi)	Thermal Expansion ($10^{-6}/°F$)	Erosion, Knoop (kg/mm^2)	Max Temp (°F) w/o EO Degradation
Zinc Sulfide (Z_nS)	4.05	●	●	18 ○	4 ○	350 ○	700 ○
Zinc Selenide (Z_nS_e)	5.16	●	●	8 ⌣	4 ○	150 ○	600 ○
Mag. Fluoride (M_gF_2)	3.18	●	⌣	7 ⌣	6 ⌣	420 ◐	1000 ◐
Germanium (Ge)	5.33	○	○	15 ○	4 ○	780 ◐	200 ⌣
Sapphire (Al_2O_2) / Spinel ($MgAl_2O_4$)	3.68	●	⌣	28 ○	3 ○	1650 ●	1800 ●
Diamond (C)	3.52	○	●	400 ●	1 ●	8800 ●	3500 ●
Alon ($Al_{23}O_{27}N_5$)	3.67	●	-	44 ◐	3 ○	1900 ●	1800 ●
Quartz / Fused Silica (SiO_2)	2.20	○	-	8 ⌣	0.3 ●	600 ◐	2000 ●
Yttria (Y_2O_3)	5.01	●	⌣	23 ○	4 ○	700 ◐	1800 ●

● Superior ◐ Good ○ Average ⌣ Poor - Very Poor

Note:
IR = Infrared, SWIR = Short Wave Infrared (< 3 μm Wavelength), MWIR = Mid Wave Infrared (3 to 5 μm Wavelength), LWIR = Long Wave Infrared (8 to 14 μm Wavelength), EO = Electro Optical

Fig. 4.49 Infrared Seeker Dome Material Is Driven by IR Transmission and Flight Environment.

Radar Seeker Dome Material	Density (g / cm³)	Dielectric Constant	Transverse Strength (10³ psi)	Thermal Expansion (10⁻⁶ / °F)	Erosion, Knoop (kg / mm²)	Max Temp (°F) w/o EM Degradation
Quartz / Fused Silica (SiO_2)	2.20	3.7 ●	8 ︶	0.3 ●	600 ◐	2000 ●
Silicon Nitride (SiN_4)	3.18	6.1 ◐	90 ●	2 ◐	2200 ●	2700 ●
Diamond (C)	3.52	5.6 ◐	400 ●	1 ●	8800 ●	3500 ●
Pyroceram	2.55	5.8 ◐	25 ○	3 ○	700 ◐	2200 ●
Polyimide	1.54	3.2 ●	17 ○	40 ︶	70 ︶	700 ○
Mag. Fluoride (MgF_2)	3.18	5.5 ◐	7 ︶	6 ︶	420 ◐	1000 ◐

● Superior ◐ Good ○ Average ︶ Poor - Very Poor

Note:
RF = Radar Frequency, EM = Electro Magnetic

Fig. 4.50 Radar Seeker Radome Material Is Driven by RF Transmission and Flight Environment.

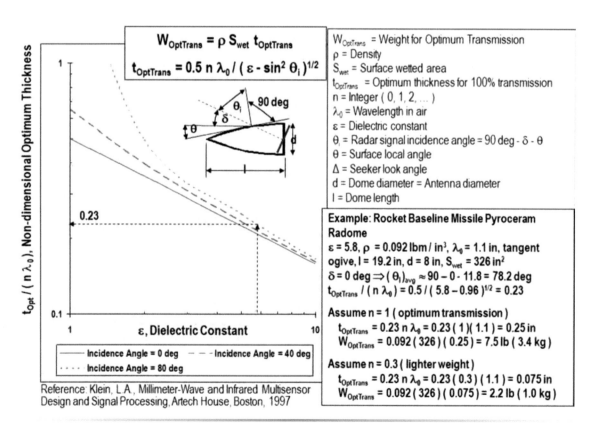

Fig. 4.51 A Driver for Radome Weight Is the Optimum Thickness Required for Efficient Transmission.

4.12 Missile Power Supply and Flight Control Actuators

Measure of Merit	Generator	Lithium Battery (Non-Rechargeable)	Lithium Battery (Rechargeable)	Thermal Battery
Weight for Long Time of Flight	●	○	○	○
Weight for Short Time of Flight	○	○	○	●
Max Acceleration	∽	◐	◐	●
Storage Life	●	○ ◐	∽ ◐	●
Voltage Stability	○	●	●	○
Max / Min Temp	○	∽	∽	○
Safety	●	○	○	◐

● Superior   ◐ Good   ○ Average   ∽ Poor

Note: Generator provides highest energy with light weight for long time of flight (e.g., cruise missile).
Lithium battery provides nearly constant voltage suitable for electronics. Relatively high energy with light weight. A disadvantage is military environment maximum temperature specification (e.g., 160 deg F) typically exceeds safe maximum allowable temperature.
Thermal battery provides highest power with light weight for short time of flight and safe operation in military environment (most popular type of battery for missiles).

Fig. 4.52 Missile Electrical Power Supply Drivers Include Weight, Environment, and Safety.

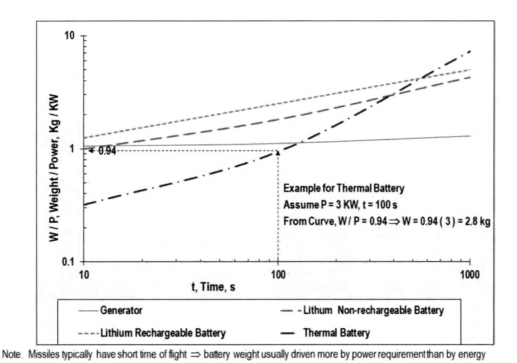

Example for Thermal Battery
Assume P = 3 KW, t = 100 s
From Curve, W/P = 0.94 ⇒ W = 0.94(3) = 2.8 kg

Note: Missiles typically have short time of flight ⇒ battery weight usually driven more by power requirement than by energy

Fig. 4.53 A Thermal Battery Has Lighter Weight for Short Time of Flight—Generator Has Lighter Weight for Long Flight Time.

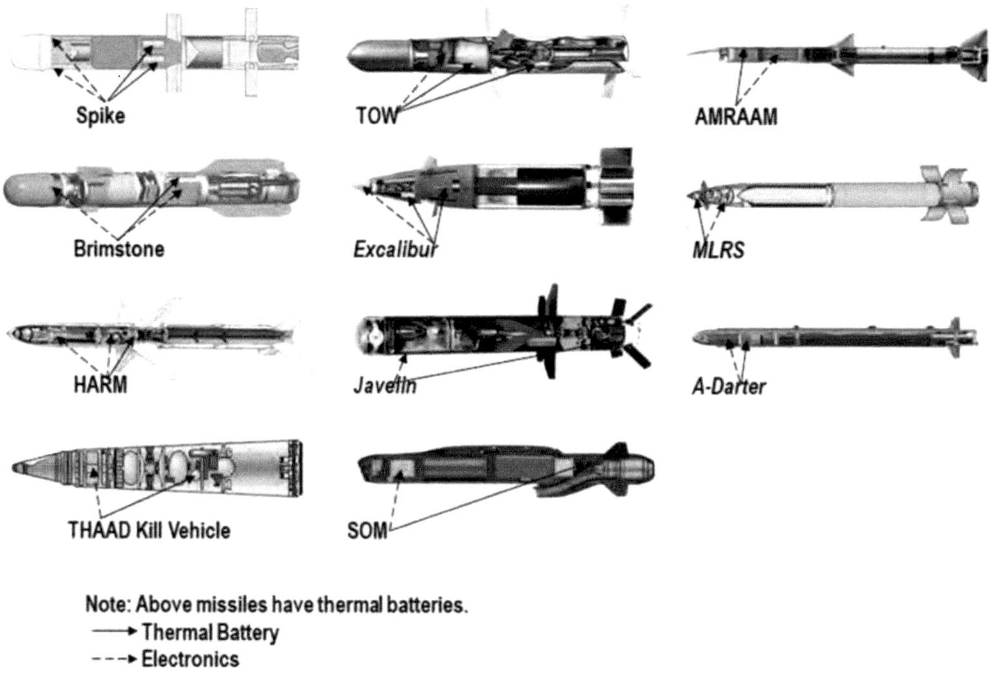

128 Missile Design Guide

Fig. 4.54 Missile Power Supply Is Usually a Thermal Battery and Most Batteries Are Located Near Electronics. *Note*: See video supplement, TUBITAK-SAGE Video: Thermal Battery for SOM.

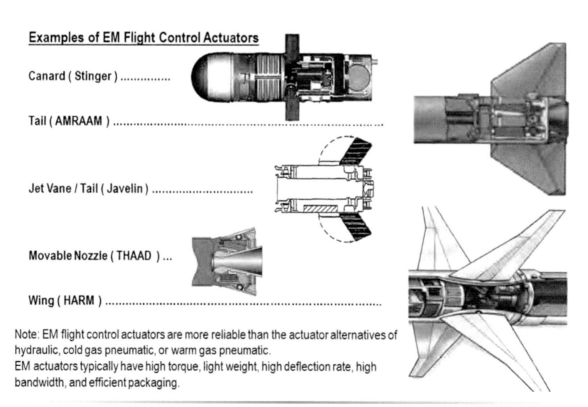

Fig. 4.55 Most Missiles Use Electromechanical Flight Control Actuators.

Chapter 5 Flight Performance

5.1 Introduction

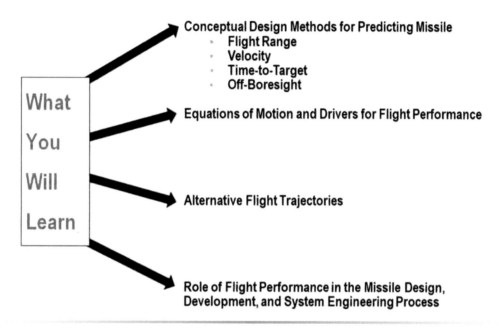

Fig. 5.1 Chapter 5 Flight Performance—What You Will Learn.

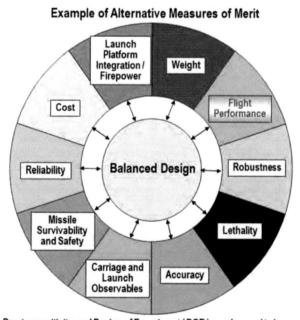

Fig. 5.2 A Balanced Missile Design Requires Harmonized Mission Requirements and Measures of Merit.

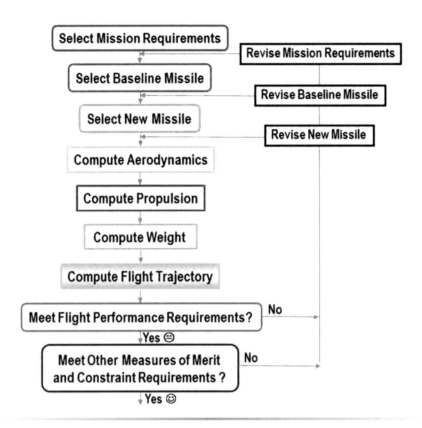

Fig. 5.3 Missile Conceptual Design and System Engineering Requires Broad, Creative, Rapid, and Iterative Evaluations.

5.2 Missile Flight Performance Envelope

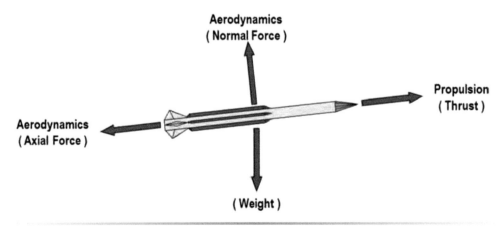

Fig. 5.4 Flight Performance and Trajectory Are Driven by Forces (Aerodynamics, Propulsion, Weight) on the Missile.

Sir Isaac Newton (Philosophiae Naturalis Principia Mathematica, July 5, 1687):
$$F = d(mV)/dt \approx \dot{m} V + m a \approx m a$$

Fig. 5.5 Flight Trajectory Is Driven by Aerodynamic, Propulsion, and Gravity Forces. *Note*: See video supplement, Aerodynamic Lift Magnitude/Orientation for Cruise, Climb, Dive, and Turn Trajectories.

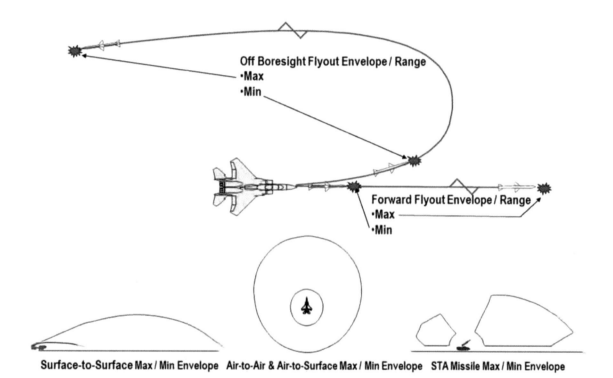

Fig. 5.6 Missile Flight Envelope May Be Characterized by Maximum Range, Minimum Range, and Off Boresight.

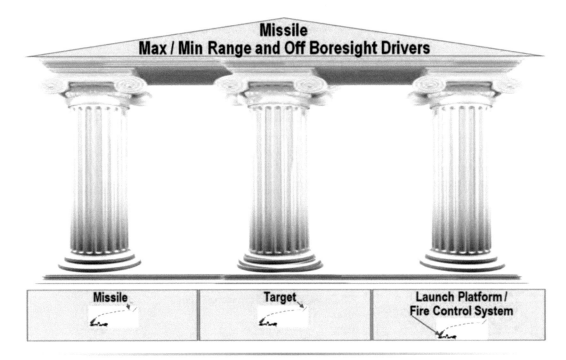

Fig. 5.7a Missile Flight Envelope Is Defined by the Missile, Target, Launch Platform, and Fire Control System.

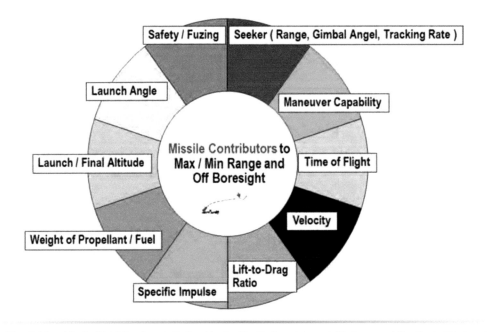

Fig. 5.7b Missile Flight Envelope Is Defined by the Missile, Target, Launch Platform, and Fire Control System (cont).

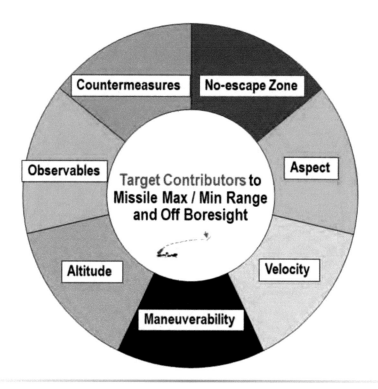

Fig. 5.7c Missile Flight Envelope Is Defined by the Missile, Target, Launch Platform, and Fire Control System (cont).

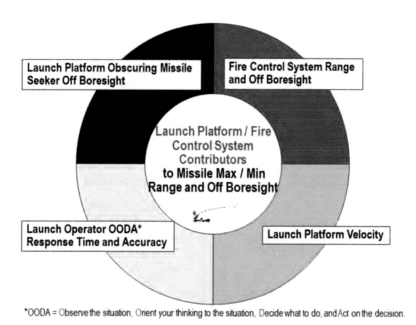

*OODA = Observe the situation, Orient your thinking to the situation, Decide what to do, and Act on the decision.

Fig. 5.7d Missile Flight Envelope Is Defined by the Missile, Target, Launch Platform, and Fire Control System (cont).

Example of Impact of Human Launch Operator's OODA (Observe–Orient–Decide–Act) Response Time on Missile Launch Range

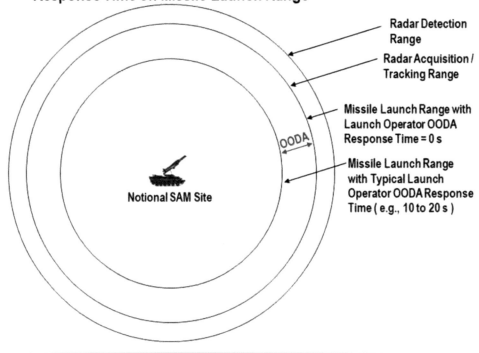

Fig. 5.7e Missile Flight Envelope Is Defined by the Missile, Target, Launch Platform, and Fire Control System (cont).

A Shoot-Look-Shoot Firing Doctrine Usually Provides Higher Probability of Kill (P_K) Than a Single Shot Probability of Kill (P_K)$_{SS}$ But Usually Requires Either
- Multiple Launch Sites
- Launch Site with a Larger Max Range, Smaller Min Range, and Larger Off Boresight

A Salvo Firing Doctrine (e.g., Shoot – Shoot) Usually Provides Higher P_K Than a Single Shot (P_K)$_{SS}$ But Requires More Missiles

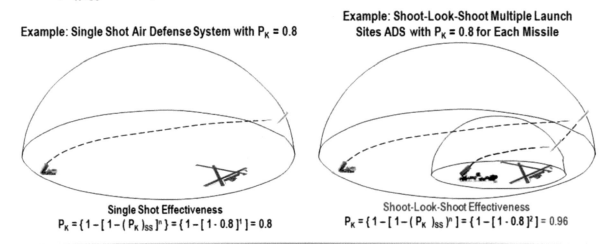

Example: Single Shot Air Defense System with $P_K = 0.8$

Single Shot Effectiveness
$P_K = \{1 - [1 - (P_K)_{SS}]^n\} = \{1 - [1 - 0.8]^1\} = 0.8$

Example: Shoot-Look-Shoot Multiple Launch Sites ADS with $P_K = 0.8$ for Each Missile

Shoot-Look-Shoot Effectiveness
$P_K = \{1 - [1 - (P_K)_{SS}]^n\} = \{1 - [1 - 0.8]^2\} = 0.96$

Fig. 5.7f Missile Flight Envelope Is Defined by the Missile, Target, Launch Platform, and Fire Control System (cont).

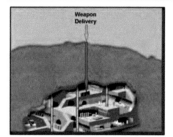
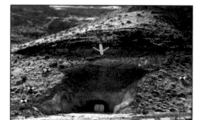

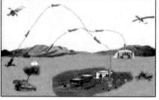

Fig. 5.8 Example Targets That Require Extreme Terminal Flight Trajectory Shaping.

5.3 Equations of Motion Modeling

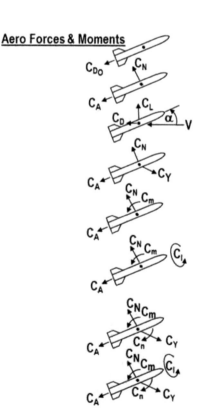

- **Conceptual Design Force and Moment Modeling**
  - 1-DOF [axial force (C_{D_O}), thrust, weight]
    - Simple, analytical equations
  - 2-DOF [normal force (C_N), axial force, thrust, weight]
    - Includes trim forces in body axes
  - 2-DOF [lift force (C_L), drag (C_D), thrust, weight]
    - Includes trim forces in stability axes
  - 3-DOF point mass [3 aero forces (normal, axial, side), thrust, weight]
    - Allows first order evaluation of off boresight envelope
  - 3-DOF pitching [2 aero forces (normal, axial), 1 aero moment (pitching), thrust, weight]
    - Allows first order sizing of tail stabilizers
  - 4-DOF [2 aero forces (normal, axial), 2 aero moments (pitching, rolling), thrust, weight]
    - Allows first order evaluation of a rolling airframe
- **Preliminary Design Force and Moment Modeling**
  - 5-DOF [3 aero forces (normal, axial, side), 2 aero moments (pitching, yawing), thrust, weight]
    - Suitable for cruciform missile with small surfaces
  - 6-DOF [3 aero forces (normal, axial, side), 3 aero moments [pitching, rolling, yawing), thrust, weight]
    - Suitable for most (relatively rigid) missiles

Fig. 5.9 Conceptual Design vs Preliminary Design Models of Flight Trajectory Degrees of Freedom (DOF).

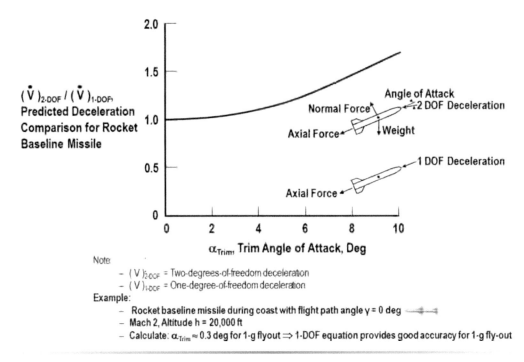

Fig. 5.10 1 Degree of Freedom (1-DOF) Equation of Motion Has Good Accuracy if the Fly-out Is at Low Angle of Attack.

5.4 Driving Parameters for Missile Flight Performance

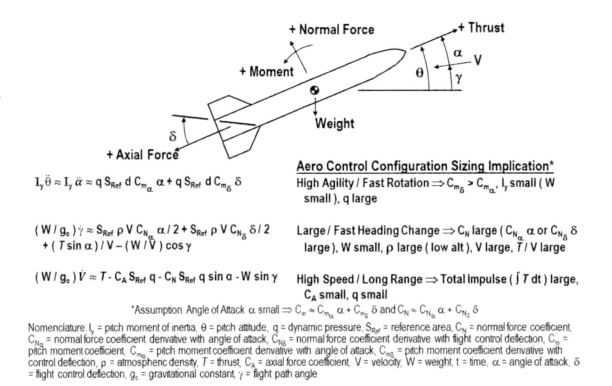

$I_y \ddot{\theta} \approx I_y \ddot{\alpha} \approx q\, S_{Ref}\, d\, C_{m_\alpha}\, \alpha + q\, S_{Ref}\, d\, C_{m_\delta}\, \delta$

$(W/g_c) \dot{\gamma} \approx S_{Ref}\, \rho\, V\, C_{N_\alpha}\, \alpha/2 + S_{Ref}\, \rho\, V\, C_{N_\delta}\, \delta/2 + (T \sin \alpha)/V - (W/V) \cos \gamma$

$(W/g_c) \dot{V} \approx T - C_A\, S_{Ref}\, q - C_N\, S_{Ref}\, q \sin \alpha - W \sin \gamma$

Aero Control Configuration Sizing Implication*

High Agility / Fast Rotation $\Rightarrow C_{m_\delta} > C_{m_\alpha}$, I_y small (W small), q large

Large / Fast Heading Change $\Rightarrow C_N$ large ($C_{N_\alpha} \alpha$ or $C_{N_\delta} \delta$ large), W small, ρ large (low alt), V large, T/V large

High Speed / Long Range $\Rightarrow$ Total Impulse ($\int T\, dt$) large, C_A small, q small

*Assumption: Angle of Attack α small $\Rightarrow C_m \approx C_{m_\alpha} \alpha + C_{m_\delta} \delta$ and $C_N \approx C_{N_\alpha} \alpha + C_{N_\delta} \delta$

Nomenclature: I_y = pitch moment of inertia, θ = pitch attitude, q = dynamic pressure, S_{Ref} = reference area, C_N = normal force coefficient, C_{N_α} = normal force coefficient derivative with angle of attack, C_{N_δ} = normal force coefficient derivative with flight control deflection, C_m = pitch moment coefficient, C_{m_α} = pitch moment coefficient derivative with angle of attack, C_{m_δ} = pitch moment coefficient derivative with control deflection, ρ = atmospheric density, T = thrust, C_A = axial force coefficient, V = velocity, W = weight, t = time, α = angle of attack, δ = flight control deflection, g_c = gravitational constant, γ = flight path angle

Fig. 5.11a 3 Degrees of Freedom (3-DOF) Equations of Motion Show Drivers for Missile Configuration Sizing.

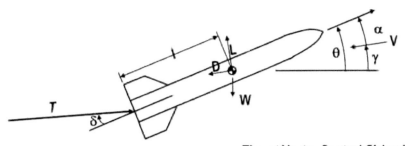

$I_y \ddot{\theta} \approx I_y \ddot{\alpha} \approx T l \sin \delta$

Thrust Vector Control Sizing Implication*
High Agility / Fast Rotation $\Rightarrow T l \delta$ large, I_y small (W small)

$(W/g_c)\dot{\gamma} \approx (T \sin \alpha)/V + S_{Ref} \rho V C_L /2 - (W/V) \cos \gamma$

Large / Fast Heading Change $\Rightarrow T \sin \alpha / V$ large

$(W/g_c)\dot{V} \approx T \cos \alpha - C_D S_{Ref} q - W \sin \gamma$

High Speed / Long Range $\Rightarrow$ Total Impulse ($\int T$ dt) large, C_D small, q small

*Assumption: Aero moment small compared to TVC moment

Nomenclature: TVC = thrust vector control, I_y = pitch moment of inertia, θ = pitch attitude, S_{Ref} = reference area, ρ = atmospheric density, γ = flight path angle, g_c = gravitational constant, C_L = lift coefficient, L = lift force, T = thrust, l = thrust moment arm to center of gravity, C_D = drag coefficient, D = drag force, V = velocity, W = weight, α = angle of attack, δ = thrust vector control deflection, q = flight dynamic pressure, t = time

Fig. 5.11b 3 Degrees of Freedom (3-DOF) Equations of Motion Show Drivers for Missile Configuration Sizing (cont).

5.5 Steady-State Flight and Constant Bearing Intercept

Steady Level Flight (Cruise)*

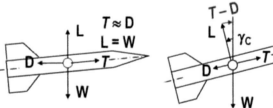

$T \approx D$
$L = W$

$T \approx W / (L/D)$

*Note: Low Required Thrust $\Rightarrow$ Lower Fuel Flow Rate $\Rightarrow$ Longer Range

Steady Climb

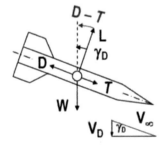

$\sin \gamma_C = (T-D)/W = V_C/V_\infty$
$V_C = (T-D) V_\infty / W$
$R_C = \Delta h / \tan \gamma_C = \Delta h (L/D)$

Steady Descent (Glide)

$\sin \gamma_D = (D-T)/W = V_D/V_\infty$
$V_D = (D-T) V_\infty / W$
$R_D = \Delta h / \tan \gamma_D = \Delta h (L/D)$

Assumptions for Equations: Small Angle of Attack, Equilibrium Flight

Symbol Definitions:
V_C = Velocity of Climb, V_D = Velocity of Descent, γ_C = Flight Path Angle During Climb, γ_D = Flight Path Angle During Descent, V_∞ = Free Stream Velocity, Δh = Incremental Altitude, R_C = Horizontal Range in Steady Climb, R_D = Horizontal Range in Steady Dive (Glide), L = Lift, W = Weight, T = Thrust, D = Drag, h = Altitude

Reference: Chin, S.S., "Missile Configuration Design."

Fig. 5.12 Steady-State Flight Range Is Enhanced by High Lift-to-Drag Ratio.

138 Missile Design Guide

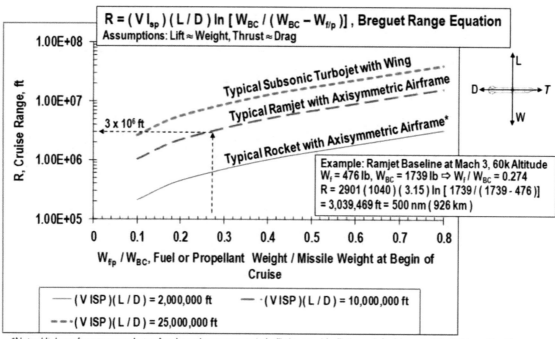

Fig. 5.13 Drivers for Cruise Range Are Velocity, Specific Impulse, Lift-to-Drag Ratio, and Weight Fraction of Fuel and Propellant.

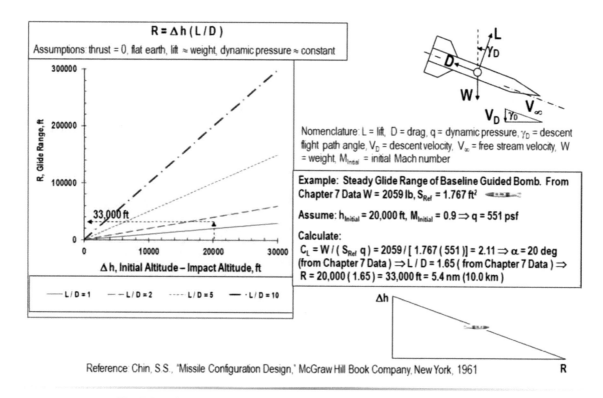

Fig. 5.14 Steady-State Glide Range Is Driven by Initial Glide Altitude and Lift-to-Drag Ratio.

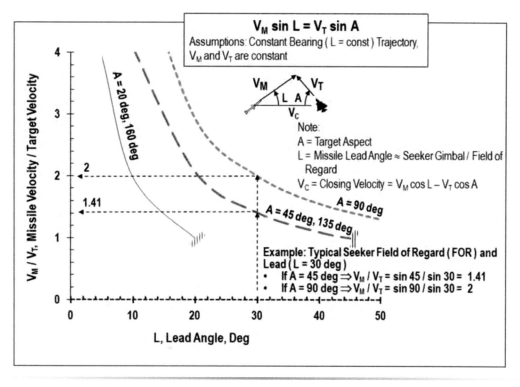

Fig. 5.15 A Constant Bearing Intercept Has Constant Lead Angle.

5.6 Boost, Glide, Coast, Ballistic, and Divert Flight

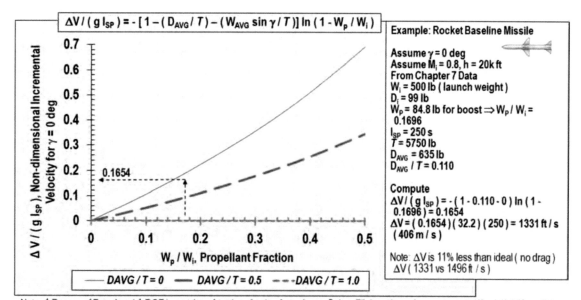

Fig. 5.16 Incremental Boost Velocity Is Driven by Propellant Weight, Launch Weight, Specific Impulse, and Drag.

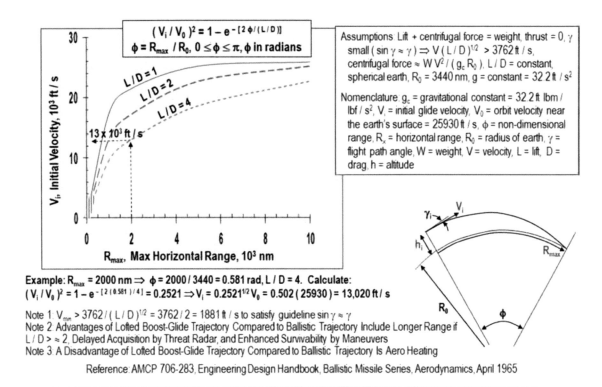

Fig. 5.17 Lofted Boost-Glide Range Is Driven by Initial Glide Velocity, Initial Glide Altitude, and Lift-to-Drag Ratio.

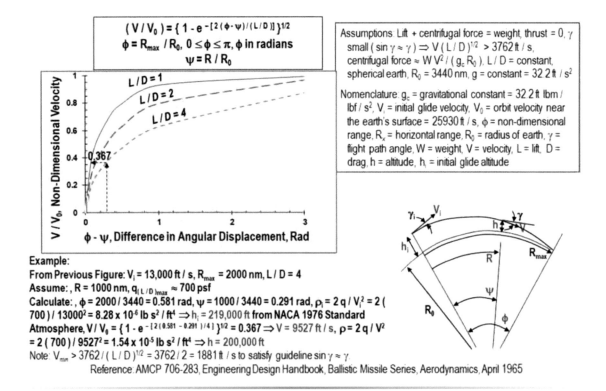

Fig. 5.18 Lofted Boost-Glide Velocity Decay Is Driven by Initial Glide Velocity, Initial Glide Altitude, Range, and Lift-to-Drag Ratio.

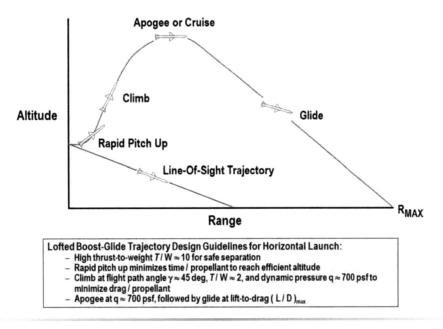

Fig. 5.19 Flight Trajectory Shaping Provides Extended Range for High Performance Missiles.

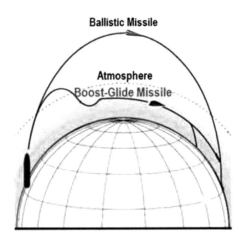

Lofted Boost-Glide Trajectory Benefits:
- Lift-to-Drag Ratio with L / D > 2 provides longer range than ballistic trajectory
- Maneuver capability provides larger footprint and less predictable trajectory to target defenses than ballistic trajectory
- Lower apogee reduces line-of-sight range and exposure time to threat surveillance ground radar than ballistic trajectory

Lofted Boost-Glide Trajectory Design Guidelines:
- Apogee at dynamic pressure q ≈ 700 psf to maximize glide range
- Glide at (L / D)$_{max}$ to maximize glide range

Fig. 5.20 Boost-Glide Trajectory May Have Larger Footprint Capability and Less Exposure to Threat Radar.
Note: See video supplement, Benefits of Lofted Hypersonic Boost-Glide Missile.

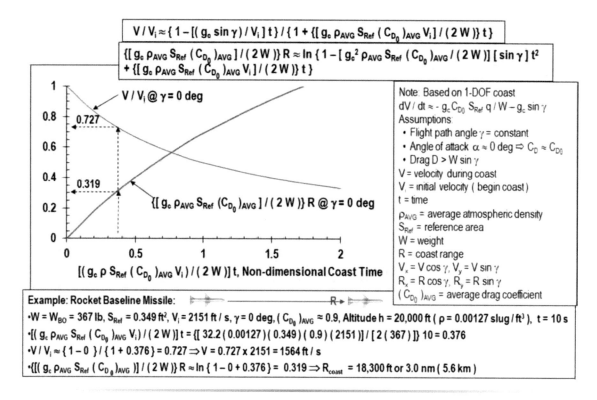

Fig. 5.21 1 Degree of Freedom Coast Range Is Driven by Initial Velocity, Altitude, Drag, and Weight.

Example for Rocket Baseline Missile Coast at Constant Altitude (h = 20k) from Burnout (Mach 2.07)										
$C_L = W / (S_{Ref} \, q)$, $\dot{V} \approx - (g_c S_{Ref} / W) C_D q$, $\Delta V = \dot{V} \, \Delta t$, $\Delta R = (1/2) \dot{V} (\Delta t)^2 + V \, \Delta t$										
t, s	Δt, s	V, fps	M	q, psf	C_L	α, deg	C_D	$\dot{V}$, ft/s²	ΔV, fps	ΔR, ft	R, ft
0		2151	2.07	2938	0.36	0.5	0.90	-81	-81	-	0
1	1	2070	1.99	2717	0.39	0.6	0.90	-75	-75	2111	2111
2	1	1995	1.92	2523	0.42	0.6	0.90	-71	-71	2033	4144
3	1	1924	1.86	2347	0.45	0.7	0.91	-65	-65	1960	6104
4	1	1859	1.79	2191	0.48	0.7	0.91	-61	-61	1892	7996
5	1	1798	1.73	2050	0.51	0.8	0.91	-57	-57	1829	9825
6	1	1741	1.68	1919	0.55	0.8	0.91	-53	-53	1770	11595
7	1	1688	1.63	1806	0.58	0.9	0.91	-50	-50	1715	13310
8	1	1638	1.58	1701	0.62	0.9	0.91	-47	-47	1663	14973
9	1	1591	1.53	1605	0.66	1.0	0.91	-45	-45	1615	16588
10	1	1546	1.49	1515	0.69	1.0	0.91	-42	-42	1569	18157

W = weight = 367 lb, S_{Ref} = reference area = 0.349 ft², h = 20k ft (atmospheric density ρ = 0.00127 slug/ft³, speed of sound a = 1037 fps), t = time, V = velocity, M = Mach number = V / a, g_c = gravitation constant, q = dynamic pressure = (1 / 2) ρ V², C_L = lift coefficient, α = angle of attack (from Chapter 7 data), C_D drag coefficient (Chapter 7 data), R = range

Note: 2-DOF end of coast velocity and range at t = 10 s are within 1% of the 1-DOF prediction (1546 vs 1564 fps, 18,157 vs 18,300 ft)

Fig. 5.22 2 Degrees of Freedom Coast Prediction Is More Accurate than 1-DOF, but Requires Numerical Solution.

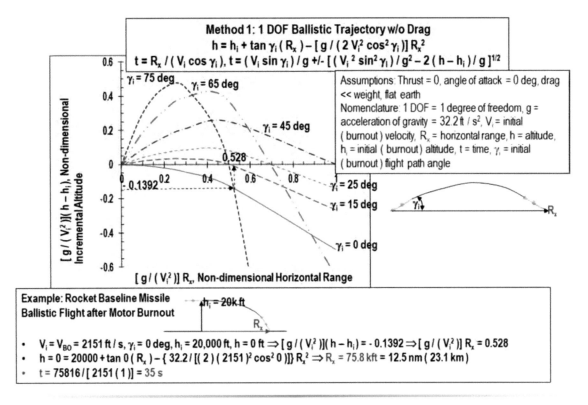

Fig. 5.23a Short Range Ballistic Flight Drivers Include Burnout Velocity, Burnout Angle, Burnout Altitude, and Drag.

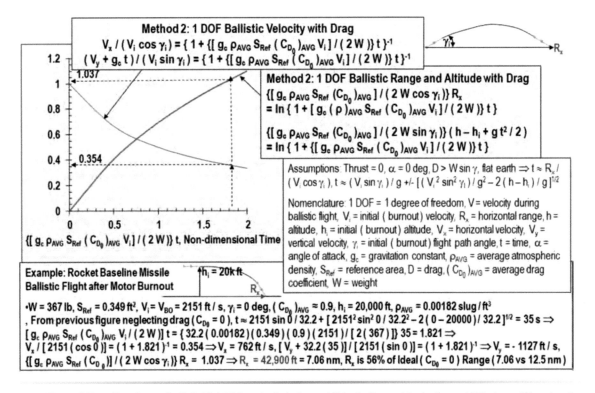

Fig. 5.23b Short Range Ballistic Flight Drivers Include Burnout Velocity, Burnout Angle, Burnout Altitude, and Drag (cont).

144 Missile Design Guide

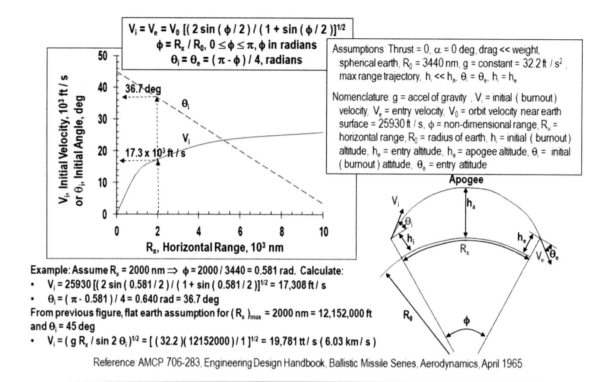

Fig. 5.24 Drivers of Long-Range Ballistic Flight Are Burnout Velocity and Burnout Attitude Angle.

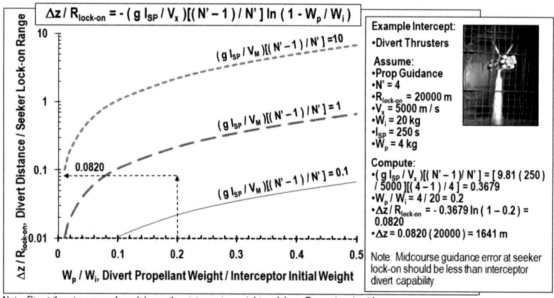

Fig. 5.25 Interceptor Divert with Thrusters is Driven by Propellant Weight, Interceptor Weight, Axial Velocity, and Seeker Lock-on Range.

5.7 Turn Performance

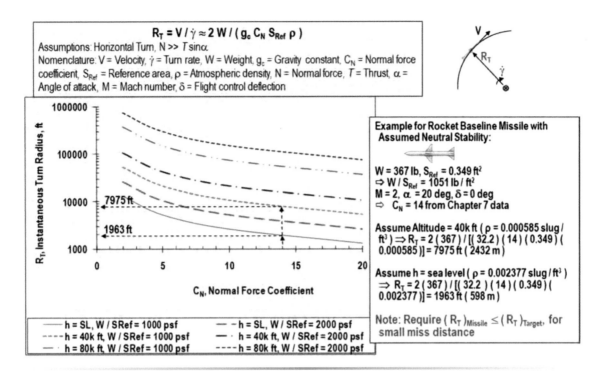

Fig. 5.26 Small Turn Radius with Aero Control Requires High Normal Force Coefficient, Light Weight, and Low Altitude.

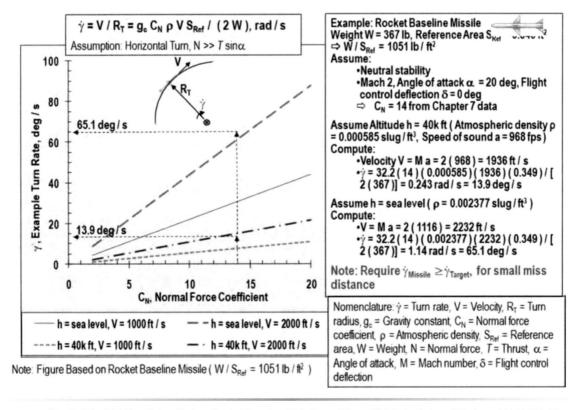

Fig. 5.27 High Turn Rate with Aero Control Requires High Normal Force, High Velocity, Low Altitude, and Light Weight.

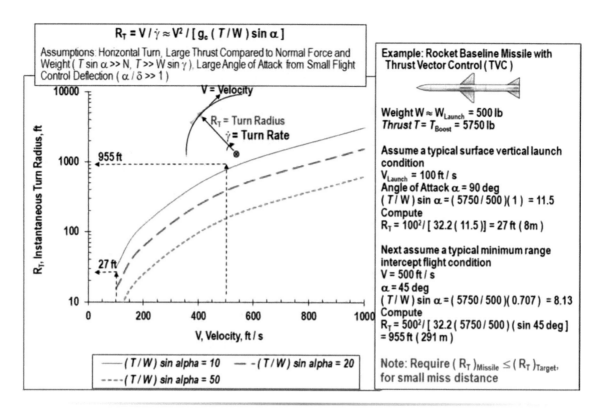

Fig. 5.28 Small Turn Radius with TVC Requires High Thrust-to-Weight Ratio, High Angle of Attack, and Low Velocity.

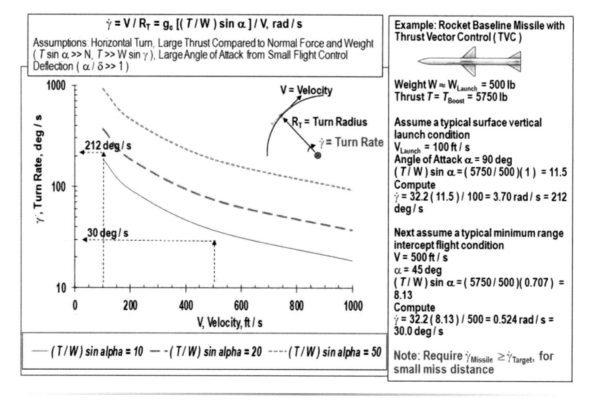

Fig. 5.29 High Turn Rate with TVC Requires High Thrust-to-Weight Ratio, High Angle of Attack, and Low Velocity.

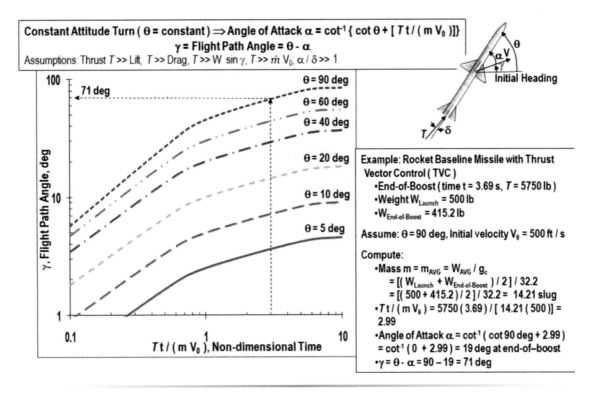

Fig. 5.30 Large or Rapid Heading Change with TVC Requires High Thrust-to-Weight, High Angle of Attack, and Low Velocity.

Chapter 6 Other Measures of Merit

6.1 Introduction

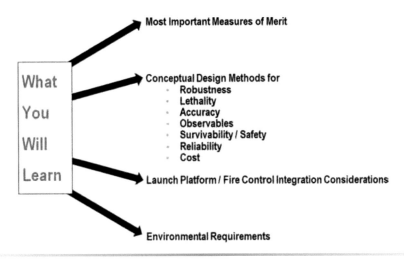

Fig. 6.1 Chapter 6 Other Measures of Merit—What You Will Learn.

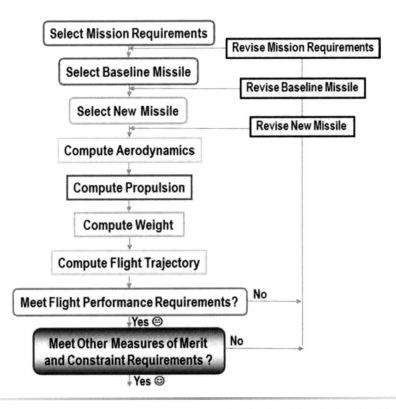

Fig. 6.2 Missile Conceptual Design and System Engineering Requires Broad, Creative, Rapid, and Iterative Evaluations.

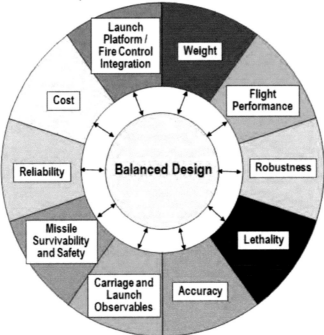

Fig. 6.3   A Balanced Missile Design Requires Harmonized Mission Requirements and Measures of Merit.

6.2   Robustness

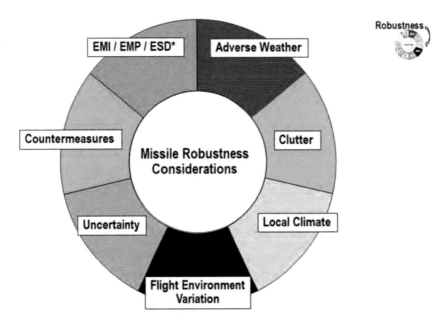

Fig. 6.4   Missiles Should Have Robust Capability.

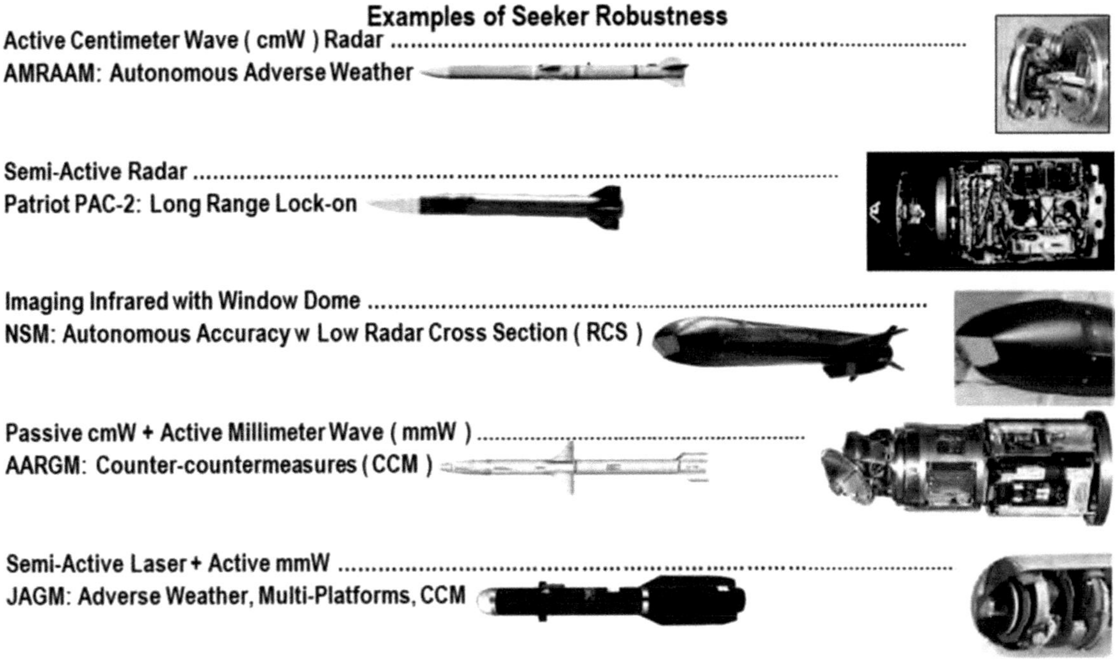

Examples of Seeker Robustness

Active Centimeter Wave (cmW) Radar ....
AMRAAM: Autonomous Adverse Weather

Semi-Active Radar ....
Patriot PAC-2: Long Range Lock-on

Imaging Infrared with Window Dome ....
NSM: Autonomous Accuracy w Low Radar Cross Section (RCS)

Passive cmW + Active Millimeter Wave (mmW) ....
AARGM: Counter-countermeasures (CCM)

Semi-Active Laser + Active mmW ....
JAGM: Adverse Weather, Multi-Platforms, CCM

Fig. 6.5 Seeker Robustness Considerations Include Weather, Autonomy, Range, Observables, and Countermeasures.

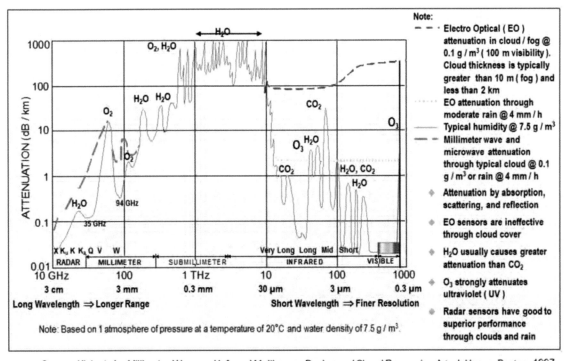

Source: Klein, L.A., *Millimeter-Wave and Infrared Multisensor Design and Signal Processing*, Artech House, Boston, 1997

Fig. 6.6 Radar Seekers Have Less Attenuation in Adverse Weather, Infrared Seekers Have Finer Resolution.

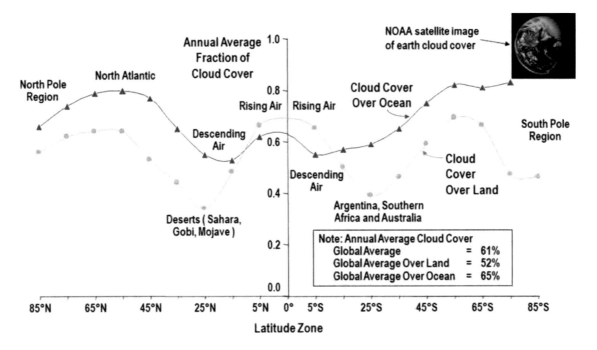

Fig. 6.7 Cloud Cover Is Pervasive.

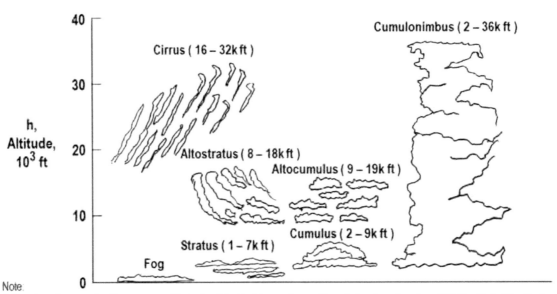

Fig. 6.8 A Radar Seeker and Sensor Has an Advantage of More Robust Operation Through Clouds.

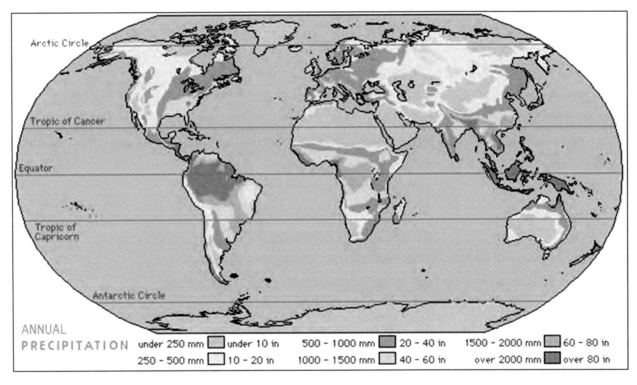

Source: Microsoft Encarta

Fig. 6.9 A Consideration in Missile Seeker and Sensor Selection Is the Probability of Precipitation.

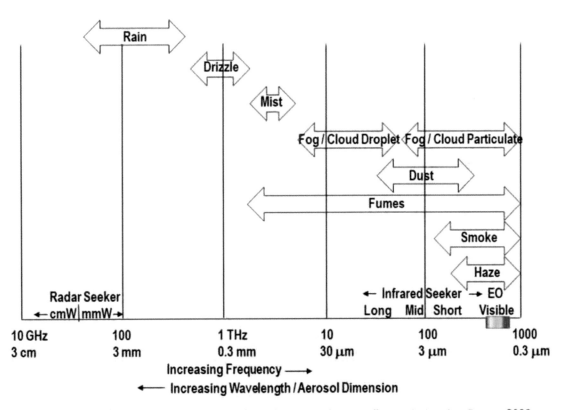

Source: Richardson, M.A., et al, *Surveillance and Target Acquisition Systems*, Brassey's, London, Boston, 2000

Fig. 6.10 Aerosols Drive Seeker and Sensor Atmospheric Attenuation if Aerosol Dimension Is Comparable to Wavelength.

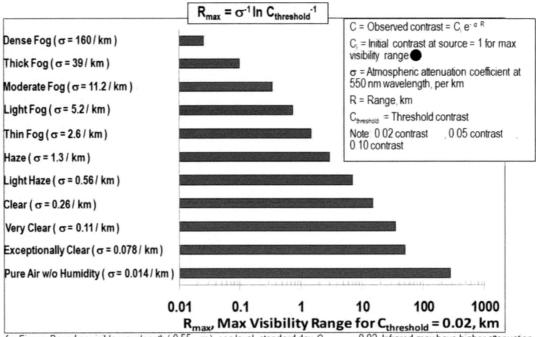

Fig. 6.11 Threshold Contrast Range of a Passive Electro-Optical (EO) Seeker Is Typically Comparable to Human Sight Range.

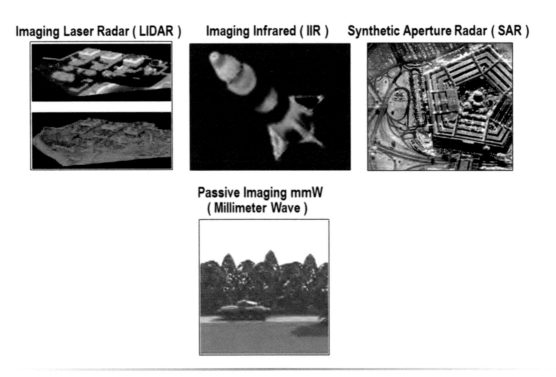

Fig. 6.12 An Imaging Seeker Has Enhanced Accuracy, Target Acquisition, and Target Discrimination.
Note: See video supplement, AGM-130 IIR; SAR Physics.

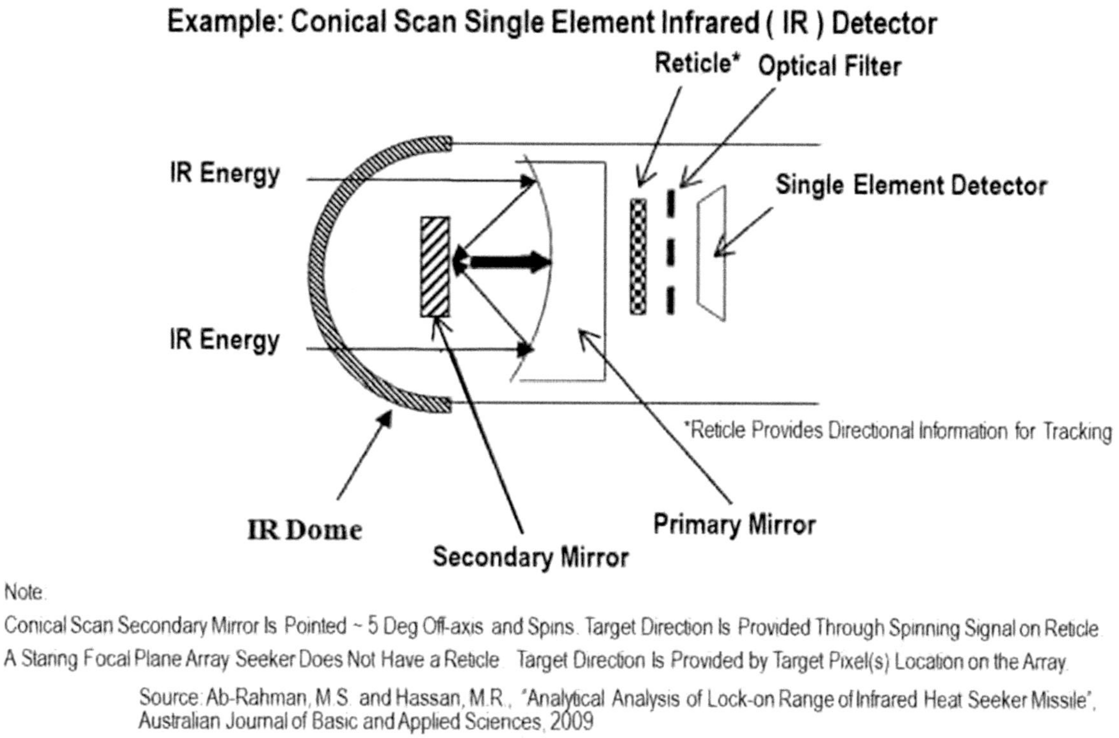
Example: Conical Scan Single Element Infrared (IR) Detector
Reticle*
Optical Filter
IR Energy
Single Element Detector
IR Energy
*Reticle Provides Directional Information for Tracking
IR Dome
Primary Mirror
Secondary Mirror
Note:
Conical Scan Secondary Mirror Is Pointed ~ 5 Deg Off-axis and Spins. Target Direction Is Provided Through Spinning Signal on Reticle
A Staring Focal Plane Array Seeker Does Not Have a Reticle. Target Direction Is Provided by Target Pixel(s) Location on the Array.
Source: Ab-Rahman, M.S. and Hassan, M.R., "Analytical Analysis of Lock-on Range of Infrared Heat Seeker Missile", Australian Journal of Basic and Applied Sciences, 2009

Fig. 6.13 An Infrared Seeker Focuses IR Energy on Detector(s).

IIR Seeker Type	Seeker Weight	Seeker Range	Clutter	Field of View	Counter measure	Accuracy	Cost
Staring FPA (e.g., AIM-9X)	●	●	●	◒	●	●	○ ◒
Linear Scanning Array (e.g., IRIS-T)	◒	◒	◒	●	●	◒	◒
Pseudo Imaging Rosette Scan (e.g., AA-11)	◒	○	○	◡	○	○	●

● Superior ◒ Good ○ Average ◡ Poor

Fig. 6.14 Imaging Infrared Seeker Trades Include Staring Focal Plane Array vs. Scanning Array vs. Pseudo Image.

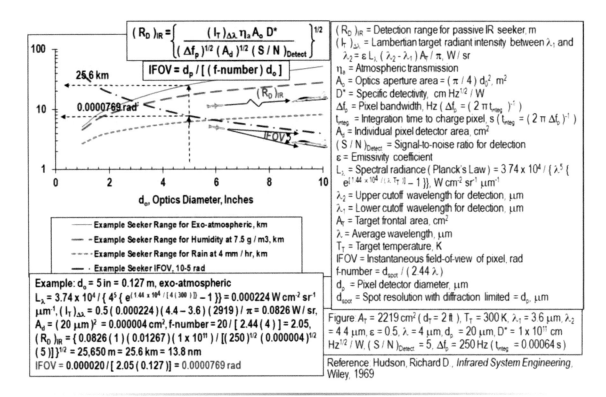

Fig. 6.15 A Large Diameter Infrared Seeker Has Longer Detection Range and Better Resolution.

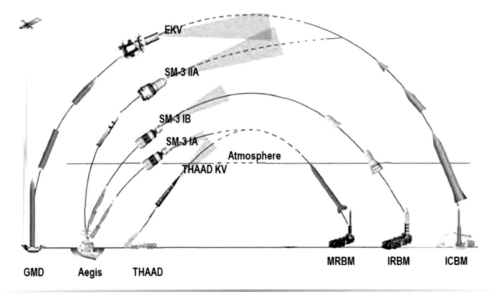

Fig. 6.16 Exo-atmospheric Hit-to-Kill Missiles Use High Resolution Imaging Infrared Seekers.

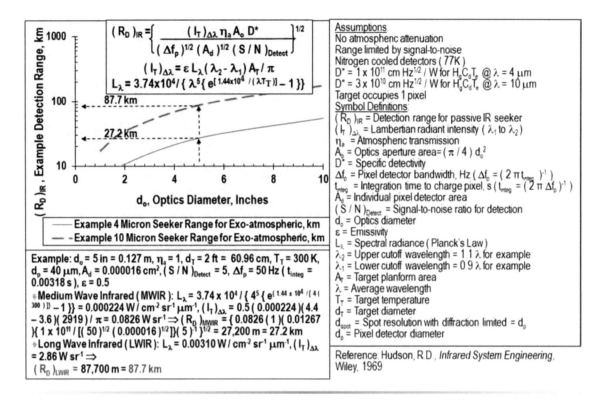

Fig. 6.17 Long Wave Infrared Detection Range for a Cold Target Is Typically Greater Than Mid-Wave Infrared Detection Range.

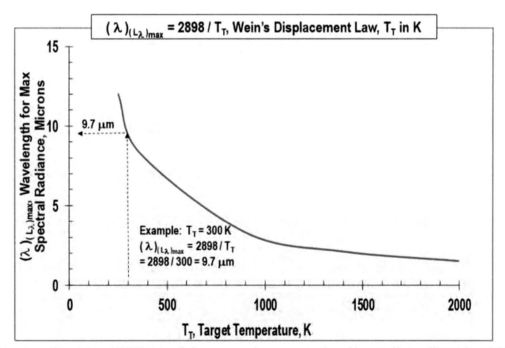

Note: Long Wave Infrared (LWIR) typically has superior detection range for cold target. Medium Wave Infrared (MWIR) is typically superior for contrast/clutter performance, focal plane array (FPA) producibility, resolution, focal length, and cost.

Fig. 6.18 Most Cold Target Energy Is at Long Wavelength—Most Warm Target Energy at Shorter Wavelength.

Example: • ATR of Tank • IIR Seeker (focal length = 100 mm, Pixel diameter (d_p) = 50 μm)	IFOV N x IFOV • N is number of line pairs or resolution cell IFOVs • IFOV is instantaneous field of view of IIR seeker pixel
Identify Threat Tank	N = 8 to 22 Line Pairs Range ~ 1 km 1 Pixel ~ 0.5 m
Recognize Threat Tank from Another Tracked Vehicle	N = 6 to 14 Line Pairs R ~ 2 km 1 Pixel ~ 1 m
Classify Tracked Vehicle from Object	N = 3 to 6 Line Pairs R ~ 4 km 1 Pixel ~ 2 m
Detect Object with Sufficient S / N, S / C, and Pixels	N = 2 to 4 Line Pairs R ~ 8 km 1 Pixel ~ 4 m
Detect Object with Sufficient S / N, S / C, but ≤ 1 Pixel	N ≤ 1 Line Pair R ≥ ~ 16 km 1 Pixel ≥ ~ 8 m

Note: Imaging infrared (IIR) seeker detection requires sufficient signal-to-noise (S / N), signal-to-clutter (S / C), contrast, and pixels

Fig. 6.19 ATR with an Imaging Seeker Requires Signal-to-Noise, Signal-to-Clutter, Contrast, and Pixels on Target.
Note: See video supplement, SPICE Automatic Target Recognition(ATR).

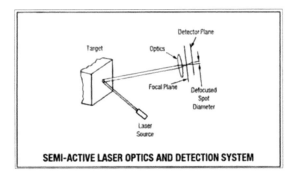
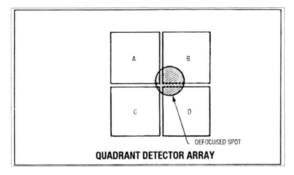
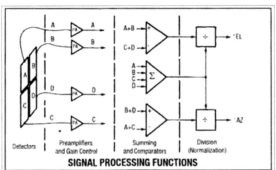

Reference: Wolfe, W.I. and Zissis, G.J., The Infrared Handbook, Revised Edition, Office of Naval Research, 1985

Fig. 6.20 Semi-active Laser Seeker with Quadrant Detector Has High Accuracy for a Small Laser Spot.
Note: See video supplement, SAL JDAM.

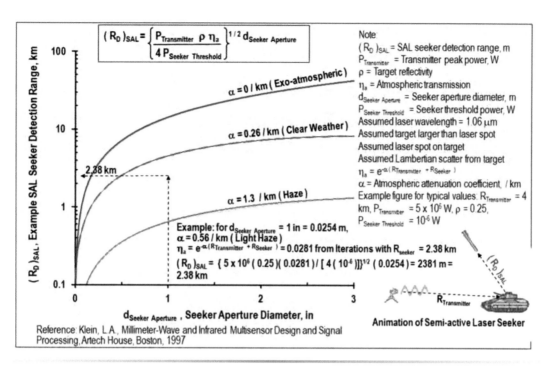

Fig. 6.21 A Semi-active Laser Seeker with Large Aperture Has Longer Detection Range.
Note: Animation of Semi-active Laser Seeker.

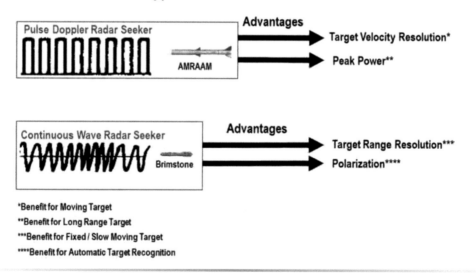

Fig. 6.22 Radar Seekers Are Typically Either Pulse Doppler or Continuous Wave.

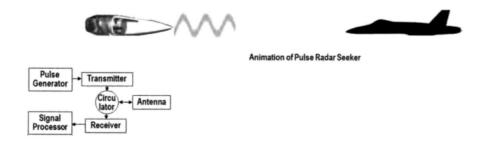

Fig. 6.23 A Pulse Active Radar Seeker Transmits and Receives Reflected Energy from the Target.
Note: Animation of Pulse Radar Seeker.

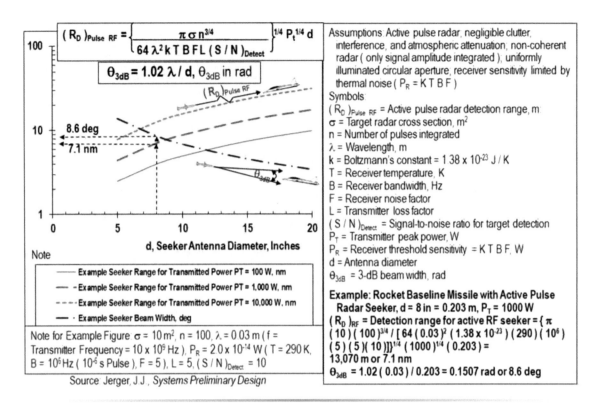

Fig. 6.24 An Active Radar Seeker with Large Diameter Has Longer Detection Range and Better Resolution.

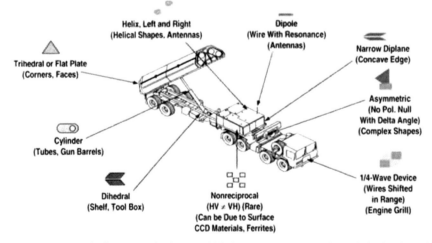

Fig. 6.25 A Radar Seeker with Polarization Can Provide Automatic Target Recognition (ATR).

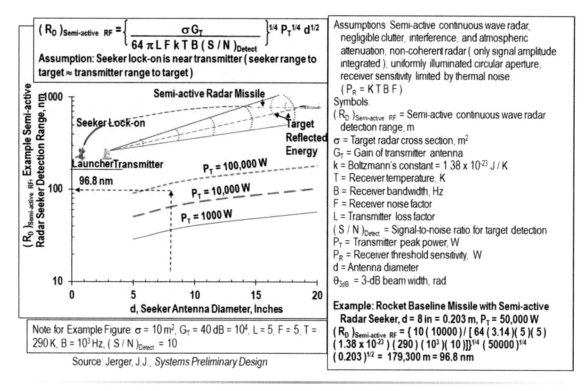

Fig. 6.26 A Semi-active Radar Seeker with Large Diameter Has Longer Detection Range.

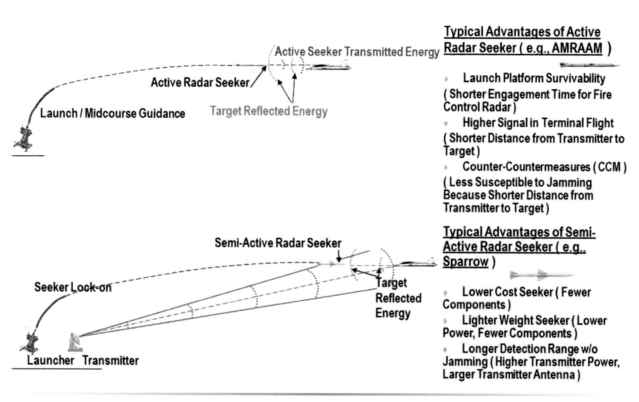

Fig. 6.27 Typical Active Radar Seeker Benefit Is Launch Platform Survivability, Semi-Active Radar Seeker Is Lower Cost.

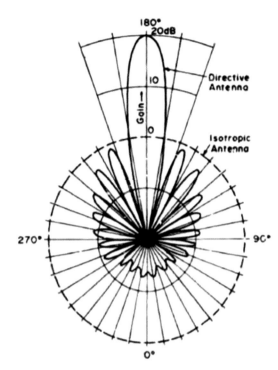

Equations for ideal gain and resolution of circular RF antenna:
- $G = (\pi d / \lambda)^2$
- $\theta_{3dB} = 1.02 \lambda / d$

Example for rocket baseline missile antenna (circular RF antenna with d ≈ 8 in = 0.203 m, f = 10 GHz ⇒ λ = 0.03 m)
- $G = (\pi d / \lambda)^2 = [\pi (0.203) / 0.03]^2 = 452 = 27$ dB
- $\theta_{3dB} = 1.02 \lambda / d = 1.02 (0.03) / 0.203 = 0.1307$ rad = 8.6 deg

Note: G = Ideal gain at center of main lobe, θ_{3dB} = Ideal half-power beamwidth for uniformly illuminated circular aperture, d = diameter, f = frequency, λ = wavelength

Aperture gain for typical planar array ≈ 0.7 of ideal gain, aperture gain for typical parabolic reflector ≈ 0.45 of ideal gain, actual θ_{3dB} is larger than ideal θ_{3dB} (for a typical tapered illumination θ_{3dB} = 1.15 λ / d)

Sources:
- Stimson, G. W., *Introduction to Airborne Radar*, SciTech Publishing, 1998
- Heaston, R.J. and Smoots, C.W., "Precision Guided Munitions," GACIAC Report HB-83-01, May 1983

Fig. 6.28 Directional Antenna Advantages Are Gain, Range, Resolution, and Less Sensitive to Countermeasures.

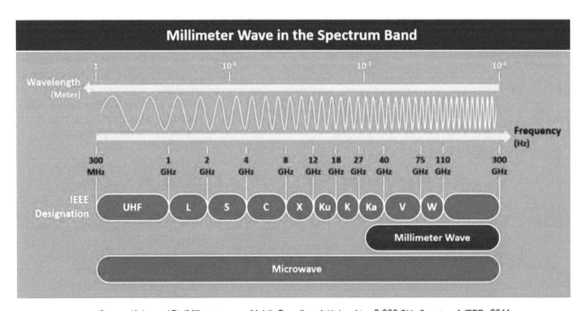

Fig. 6.29 Millimeter Wave Seeker Has Shorter Wavelength.

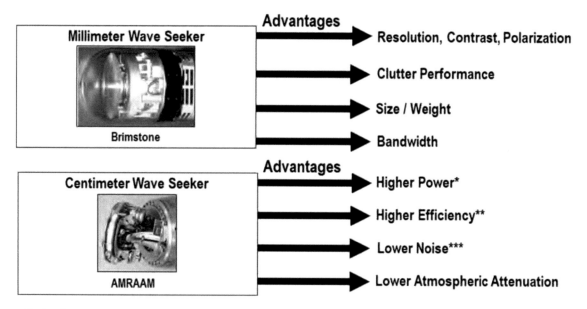

*Higher Power Limit Compared to mmW Seeker (Voltage Discharge Breakdown from Smaller Gap for mmW Seeker)
**Higher Efficiency Transmission Feed Compared to mmW Seeker (Higher Loss in Transmission Feed for mmW Seeker)
***Usually Lower Noise Receiver Compared to mmW Seeker Receiver

Note: mmW Seeker Is Higher Frequency (e.g., 35 GHz, 94 GHZ) than a cmW Seeker (e.g., 10 GHz)

Fig. 6.30 A mmW Seeker Has Better Resolution, but a cmW Seeker Typically Has Higher Power and Longer Range
Note: mmW Seeker Is Typically Less Sensitive to Countermeasures (Threat Typically at Lower Frequency).

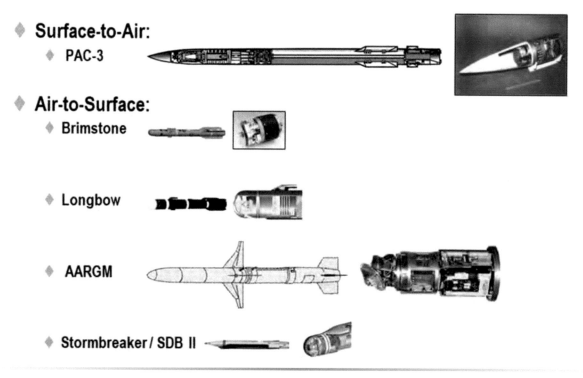

Fig. 6.31 Examples of Millimeter Wave Seekers.

Transmitter Amplifier	Max Power @ 10 GHz	Max Power @ 94 Gz	Max Voltage	Electromagnetic Pulse (EMP) Susceptibility	Weight, Volume	Bandwidth	Maturity
TWT	50,000 W ●	1000 W ●	●	●	⌣	15% ○	●
Solid State (GaN)	200 W ⌣	5 W ⌣	⌣	⌣	●	50% ●	⌣

Note:

- TWT high voltage limit (~ 100,000 V @ 10 GHz) ⇒ low current ⇒ low heating ⇒ high available transmitted power
- TWT vacuum ⇒ lower heating ⇒ higher power
- TWT high voltage power supply ⇒ heavier weight
- Solid state transmitter breakdown / arcing voltage limit (~ 100 V) from small gap ⇒ low available transmitted power
- Solid state heating ⇒ lower transmitted power
- Gallium Nitride (GaN) solid state breakdown / arcing voltage limit > Gallium Arsenide (GaAs) ⇒ GaN higher available power than GaAs
- GaN less mature than GaAs

● Superior   ◐ Good   ○ Average   ⌣ Poor

Fig. 6.32 Active RF Seeker Transmitter Options Are Traveling Wave Tube Versus Solid State Power Amplifier.

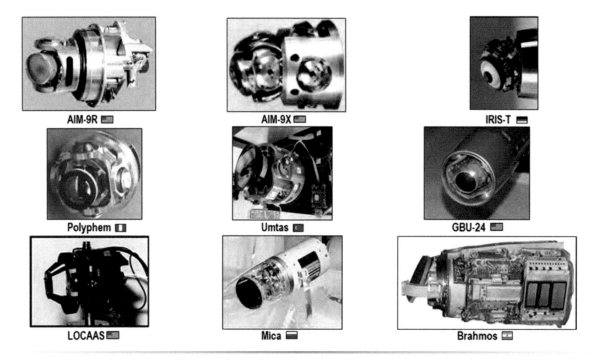

Fig. 6.33a Most Seekers Have a Gimbaled Platform.

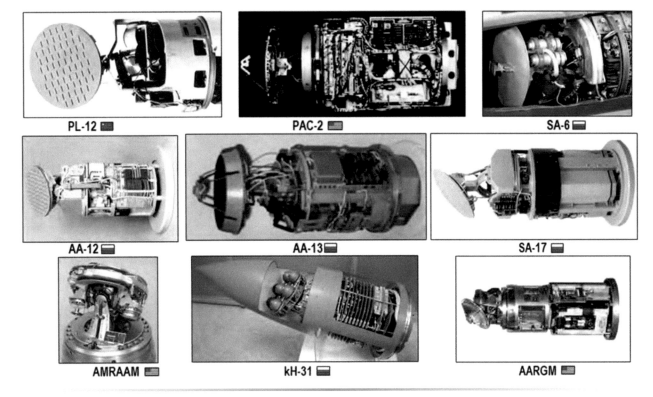

Fig. 6.33b Most Seekers Have a Gimbaled Platform (cont).

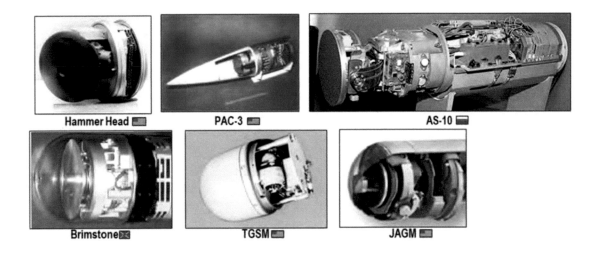

Note:
- Seeker gimbal system mechanically tracks the target and stabilizes the seeker. Seeker gimbal stabilization facilitates target tracking by minimizing interference from the motion of the missile body
- Most missile seekers have two gimbals (yaw, pitch). High maneuvering missiles (e.g., IRIS-T) may have three gimbals (yaw, pitch, roll)

Fig. 6.33c Most Seekers Have a Gimbaled Platform (cont).

166   Missile Design Guide

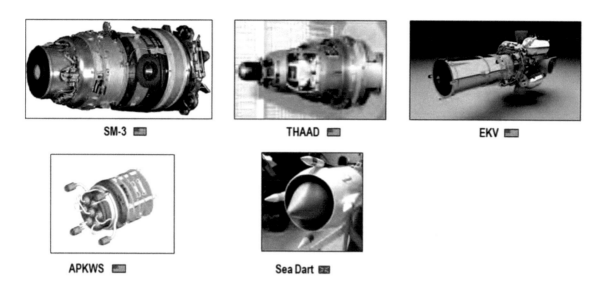

Note:
- A strapdown seeker electronically tracks the target within its field of view
- Most exo-atmospheric missile defense interceptors have strapdown imaging infrared (IIR) seekers
- An IIR seeker with larger size focal plane array has larger field of view
- Field of view of an IIR seeker is typically 1-to-5 degrees, depending upon the size of its focal plane array (e.g., 64x64, 640x480) and resolution

Fig. 6.34 Examples of Strapdown Seekers.

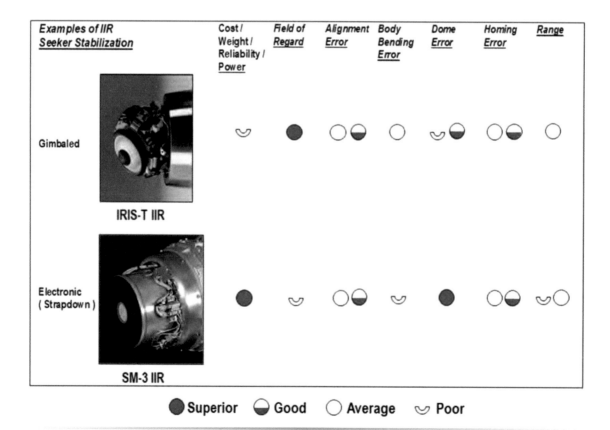

Fig. 6.35 An Imaging Infrared (IIR) Seeker Tradeoff Is Gimbaled Stabilization vs Strapdown Electronic Stabilization.

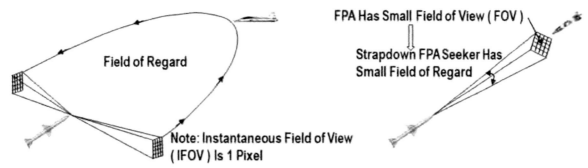

Fig. 6.36 Field of Regard of a Gimbaled Imaging IR Seeker Is Usually Much Larger than Field of View of a Strapdown Seeker. *Note*: See video supplement, Motion of Large FOR Gimbaled Seeker (AIM-9X Seeker).

Seeker / Sensor	Angular Resolution	Adverse Weather Impact	Automatic Target Recognition	Range	Moving Target	Volume Search Time	Hypersonic Dome Compat.	Diameter Required	Weight and Cost	Maturity
• Synthetic Aperture Radar (SAR)	●◐	●	●◐	●	○ -	●	●	ᴗ -	-	-
• Active Imaging millimeter wave (mmW)	◐○	◐○	◐○	○ -	●○	○	●	○	-	-
• Passive Imaging mmW	◐	◐○	●	-	○ᴗ	○	●	- ᴗ	ᴗ○	-
• Active Imaging Infrared (LIDAR)	●	-	●	○	●	-	ᴗ○	●	◐	●○
• Semi-active Laser / Active Non-image Infrared (LADAR)	●	-	○	○	●○	◐	○ᴗ	◐	●	●
• Active / Semi-active Radar	ᴗ	●	○ -	●●	●	●	●	○●	○ -	●
• Passive Imaging IR	●◐	-	●○	◐○	○	●	-	●	○◐	◐○
• Passive Non-image Infrared	○	-	-	○	○	●	-	●	●	◐
• Acoustic	-	●◐	◐	-	○ -	●	-	-	●	◐
• GPS / INS / Data Link	●ᴗ	●	●	●	ᴗ -	●	●	●	●	◐

Note: ● Superior ◐ Good ○ Average ᴗ Below Average - Poor

Fig. 6.37 Seekers Are Improved by Global Positioning System (GPS), Inertial Navigation System (INS), and Data Link Sensors.

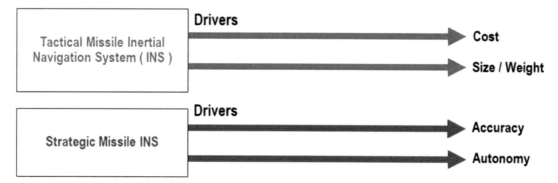

Note:
- Strategic Missile INS
  - INS Requires Large Kalman Filtering (e.g., > 100 States), Small Alignment Error, and Low Drift Rate - to Minimize INS Error in Long Duration Flight with Autonomy (No / Few Updates)
  - Requires INS and Update Insensitivity to Threat Countermeasures
  - Examples of Gimbaled INS: Trident (with Stellar Update), Minuteman, ALCM (with Terrain Contour Matching (TERCOM) Update, Tomahawk (with TERCOM, Digital Scene Matching Area Correlation (DSMAC) Update

- Tactical Missile INS
  - INS Driven by Low Cost and Small Size
  - Small Size Strapdown INS Usually Has Relatively Large Alignment Error & Large Drift Rate. This May Be Corrected by Frequent Updates to INS (e.g., Satellite Global Positioning System (GPS) Updates)
  - Examples of Strapdown INS: Excalibur Micro-machined Electro-Mechanical Systems (MEMS) with GPS Updates, JDAM ring laser gyro (RLG) with GPS updates, SLAM-ER RLG with GPS updates, Polyphem fiber optic gyro (FOG) with GPS updates

Fig. 6.38 Tactical Inertial Navigation System Drivers Are Cost & Weight, Strategic INS Drivers Are Accuracy & Autonomy.

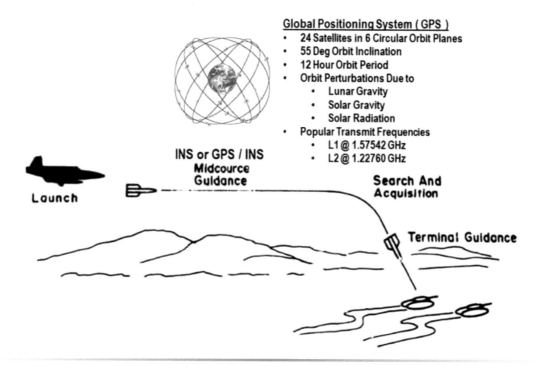

Fig. 6.39 Most US Missile Midcourse Guidance Is Provided by Either Inertial Navigation System (INS) or GPS-INS.

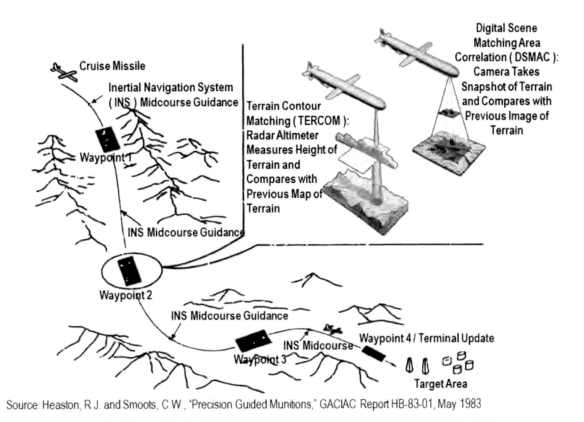

Fig. 6.40 Waypoint Guidance Can Provide Midcourse Guidance Updates from Stored Terrain Features.

Strapdown Gyro	Example Global Positioning System (GPS) / Inertial Navigation System (INS) Application	Gyro Cost	Gyro Size / Weight	Accuracy if GPS not Jammed	Accuracy if no GPS or GPS Jammed
Ring Laser	JDAM	○	○	●	◐
Fiber Optic	Polyphem	◐	◐	●	◐
Micro-Machined Electro-Mechanical Systems (MEMS)	Excalibur	●	●	●	◐ ○

● Superior　◐ Good　○ Average　⌣ Poor
Note: Relative accuracy compared to earth rate (○ 15 deg / h)

Fig. 6.41 A GPS-INS Based on Strapdown MEMS Gyros Is Low Cost and Light Weight, with Good Accuracy.

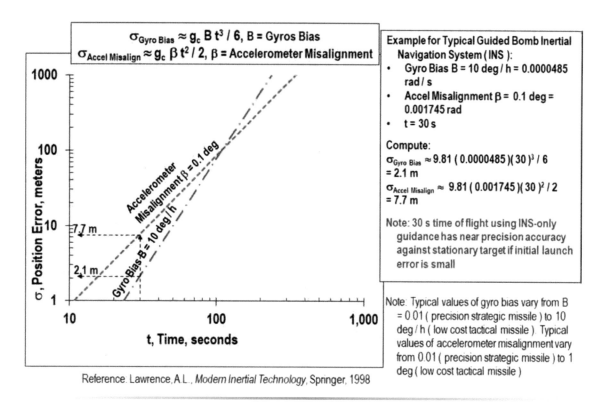

Fig. 6.42　Drivers for Strapdown INS Accuracy without GPS Are Accelerometer Misalignment, Gyro Bias, and Time.

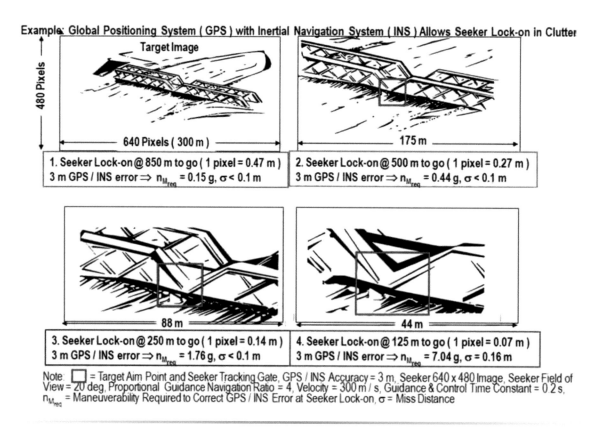

Fig. 6.43　GPS-INS Allows Robust Seeker Lock-on in Adverse Weather and Clutter.

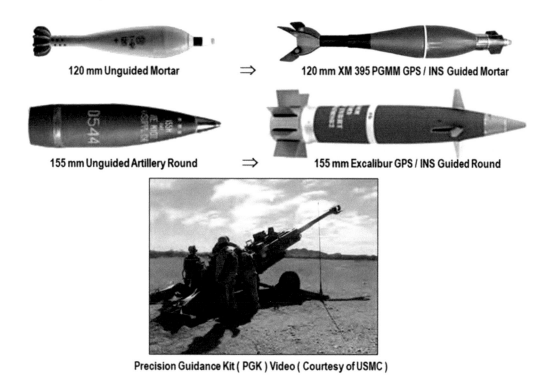

Fig. 6.44 Global Positioning System with Inertial Navigation System Allows Precision, Extended Range, and Vertical Impact. *Note:* See video supplement, Excalibur; Precision Guidance Kit (PGK).

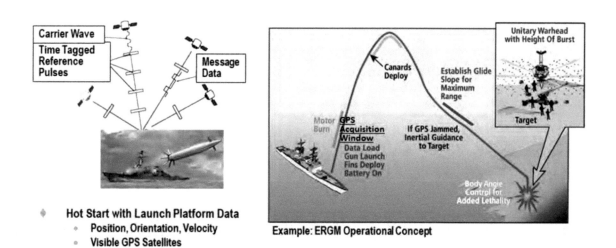

- **Hot Start with Launch Platform Data**
  - Position, Orientation, Velocity
  - Visible GPS Satellites
  - Precise Time (~ 10^{-6} s), Satellite Almanac, Correct Ephemeris Error, Correct Atmosphere Time Delay
  - GPS Code and Doppler Bins
- **Missile GPS Requirements for Fast Acquisition**
  - GPS Receiver with Many (e.g., 4 to 12) Channels
  - Parallel Processing During GPS Signal Search
  - Search for Visible Satellites Only
  - First Acquire Satellite with Highest Strength
  - Coupled GPS / Inertial Navigation System (INS) Kalman Filter of Acceleration and Rate Dynamics

Fig. 6.45 A Global Positioning System Guided Short Range Weapon Requires Fast (~ 5s) GPS Acquisition.

Pseudolites Outside Threat Jamming Area Transmit High Power GPS Signal to Missile

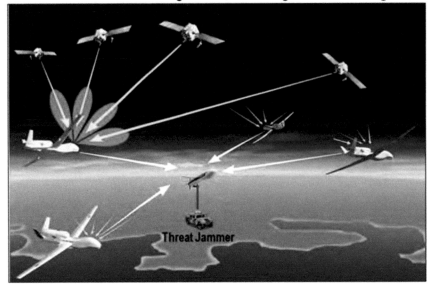

Example of Unmanned Air Vehicle (UAV) Pseudolites

Note: Typical Differences in Pseudolite Range Require Larger Dynamic Range for Missile GPS Receiver.
Note: Ground Pseudolites Have Better Accuracy. Airborne Pseudolites Have Longer Line-of-Sight (LOS) Range and Flexible Basing

Source: Sklar, J.R., "Interference Mitigation Approaches for the Global Positioning System", Lincoln Lab Journal, 2003

Fig. 6.46 Pseudolites Provide Resistance to Global Positioning System (GPS) Jammers.

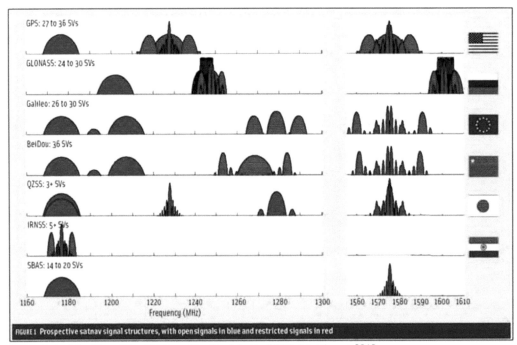

Source: Betz, John, "Something Old, Something New", Inside GNSS, July / August 2013
Note: Satellite Based Augmentation System (SBAS) Provides Wide Area Differential Corrections Using Ground Reference Stations.
Note: QZSS (Japan) and IRNSS (India) Are Regional Systems.

Fig. 6.47 USA, Russia, European Union, China, Japan, and India Have Navigation Satellite Systems.

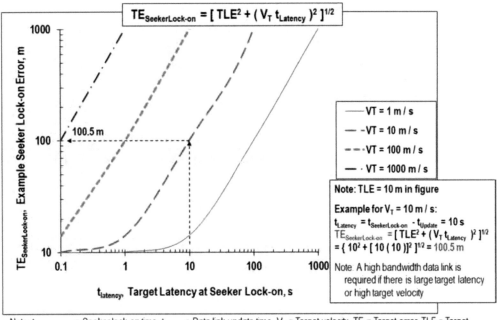

Fig. 6.48 A High Bandwidth Data Link Reduces Moving Target Location Error.

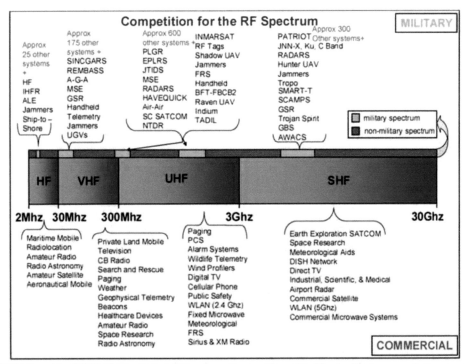

Fig. 6.49 Military Radar Frequency (RF) Spectrum Requirements Must Compete with Expanding Commercial Requirements.

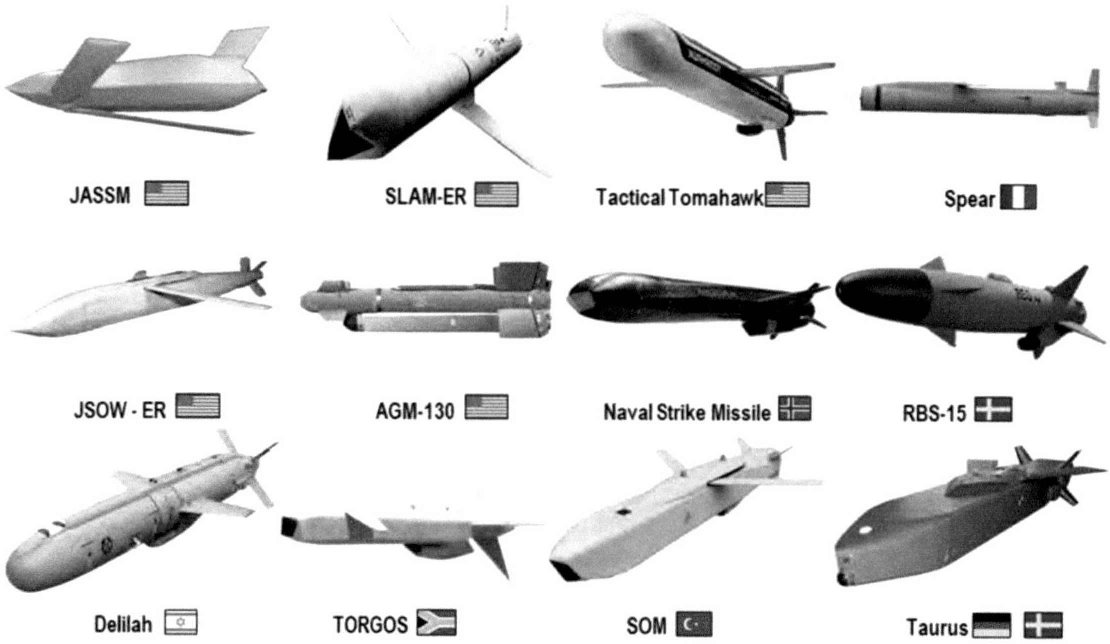

174 Missile Design Guide

JASSM 🇺🇸	SLAM-ER 🇺🇸	Tactical Tomahawk 🇺🇸	Spear 🇮🇹
JSOW - ER 🇺🇸	AGM-130 🇺🇸	Naval Strike Missile 🇳🇴	RBS-15 🇩🇰
Delilah 🇮🇱	TORGOS 🇿🇦	SOM 🇹🇷	Taurus 🇩🇪🇩🇰

Fig. 6.50 Modern Precision Strike Missiles Often Combine Seeker, Inertial Navigation, Global Positioning System, and Data Link Guidance.

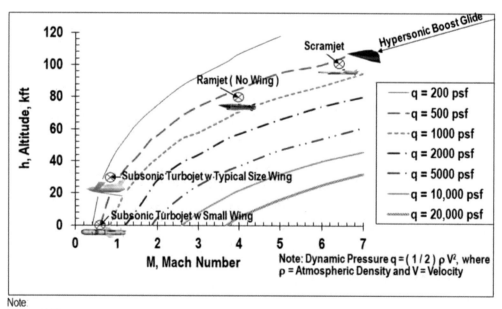

Note:
- U.S. 1976 Standard Atmosphere
- For a Lifting Body, Efficient Cruise / Glide Occurs at Maximum Lift / Drag (L/D)$_{Max}$ and $500 < q < 1000$ psf.
- (L/D)$_{Max}$ for Cruise Missile with Typical Size Wing Occurs at about $200 < q < 500$ psf.
- $q \approx 200$ psf Lower Limit for Aero Control.
- Subsonic Cruise at Medium Altitude (e.g., $h = 30$k ft) Usually Requires Low Radar Cross Section for Survivability

Fig. 6.51 Optimum Cruise and Glide Are a Function of Mach Number, Type Propulsion, Altitude, and Planform.

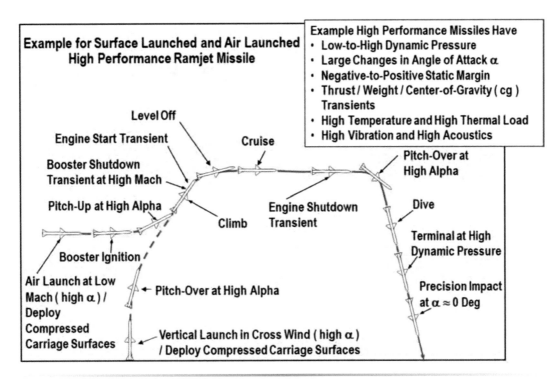

Fig. 6.52 Missile Guidance and Control Must Be Robust for Changing Events, Environment, and Uncertainty.

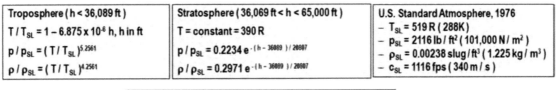

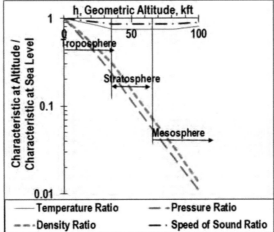

Fig. 6.53 Design Robustness Requires Consideration of Large Variation of Atmospheric Properties with Altitude.
Note: See video supplement, Illustrating Minuteman Flight through Atmosphere.

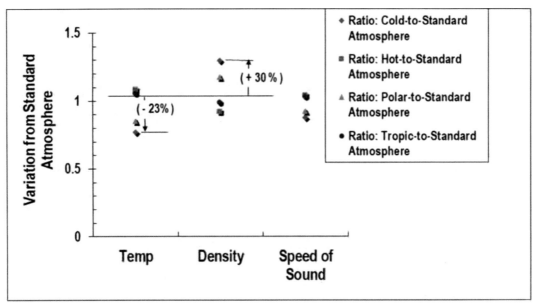

Fig. 6.54 Design Robustness Requires Consideration of Standard Atmospheric Modeling Differences.

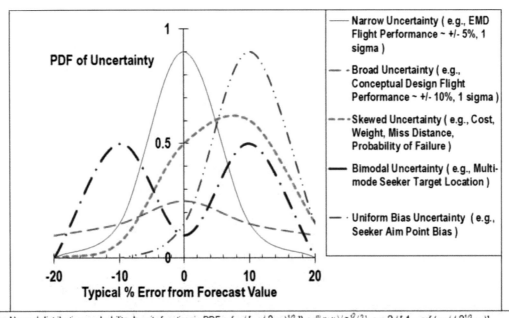

Fig. 6.55 Design Robustness Requires Consideration of Uncertainty.

Examples of Countermeasures (CM) by
Target / Threat
- Electro-optical Countermeasures
  - Directed Laser
  - Flare
  - Smoke / Oil Fog
  - IR Blinkers
- Radar Frequency (RF) Countermeasures
  - Active Deception (e.g., Gate Stealer, Angle Track Error) and Phase Cancel
  - Noise Jammer
  - Chaff
  - Phased Array (Rapid Acquire)
- Decoy (e.g., MALD, Nulka)
- Low Observables
  - Radar Cross Section (RCS), Infrared (IR) Signature
  - Fire Control
- Speed
- Altitude
- Maneuverability
- Lethal Defense
  - Kinetic Energy Missiles and Projectiles
  - High Power Laser
  - High Power Microwave

Examples of Counter-Countermeasures
(CCM) by Missile
- Seeker
  - Imaging
  - Multi-spectral / Mode / Band
  - Temporal Processing
  - Home-on-Jam
  - Burn Through After Inertial Navigation System (INS) Flight
  - Frequency Agility
  - Polarization
  - Frequency Modulation Continuous Wave (FMCW) Radar (Lower Observables)
- Global Positioning System
  - Standoff Acquisition
  - Integrated with INS
  - Directional Antenna
  - Pseudolite / Differential
  - High Altitude Flight
- Automatic Target Recognition
- High Speed
- Altitude (Terrain Following or Very High Altitude Flight)
- Maneuverability (Mid-course, Terminal)
- Low Observables
- Laser Hardness, Electro-Magnetic Interference (EMI) Hardness
- Saturation by Multiple Missiles

Fig. 6.56 Missile Robustness Requires Consideration of Countermeasures and Counter-Countermeasures.
Note: See video supplement, Example Flare, RF (Cross-Eye), and Decoy (Nulka) Countermeasures.

Imaging Infrared (IIR): AGM-130 ....

Two Color IIR (Python 5)

Multi-mode Acoustic – IIR: BAT..

Multi-mode IIR – Light Detection and Ranging (LIDAR): LOCAAS)

Multi-mode Anti-Radiation Homing (ARH) – Millimeter Wave (mmW): AARGM ....

Multi-mode IIR - Semi-Active Laser (SAL) - mmW: Stormbreaker / SDB II

Multi-mode ARH - IIR – TV: Harop Video................................

Fig. 6.57 Examples of Countermeasure Resistant Seekers.
Note: See video supplement, Multi-mode ARH - IIR – TV: Harop Video.

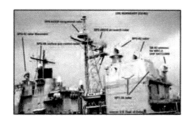  

Ship Radar Electromagnetic Interference (EMI) Electromagnetic (EM) Rail Gun   Electrostatic Discharge (ESD)*

 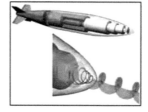 

Electromagnetic Pulse (EMP)**   Non-Nuclear EMP Weapon   Nuclear Artillery Shell EMP

*Post-explosion photograph of Pershing II missile rocket motor. Rocket motor accumulated a static charge in cold dry weather. ESD between kevlar composite motor case and plastic shipping container caused rocket motor to explode when pulled from shipping container.
**Lightning strike of Sidewinder missile on F-18 aircraft. Direct lightning strike current is typically about 100,000 amperes.

Note: Missile components that require in-depth EMC analysis include energetic components (rocket motor, warhead, power supply) and electronics (especially commercial-off-the-shelf)

Note: EMC specifications include MIL-STD-461G, 464C, 469A, 1512, 1857, 1795A, and MIL-A-17161D

Fig. 6.58 Electromagnetic Compatibility (EMC) Considers EMI, ESD, and EMP Hardening.

6.3 Lethality

Examples of Standoff Precision Strike Targets That Drive Warhead Design / Technology

- Small Size, Hard Target: Tank ⇒ Shaped Charge, Explosively Formed Projectile (EFP), Kinetic Energy (KE) Warhead
- Large Size, Hard Target: Ship ⇒ KE Penetrator + Blast Frag Warhead
- Deeply Buried Hard Target: Underground Bunker ⇒ Large KE Penetrator + Blast Frag Warhead
- Large Size, Soft Urban Target: Building ⇒ Low Collateral (Thin Case, Composite Case, Consumable Fragments, "Dial-a-Power") Blast Frag Warhead
- Soft Target Grouped Over Large Area: Surface Air-to-Missile (SAM) Site ⇒ Sensor Fuzed Weapons (SFWs), Submunitions Blast Frag Warhead, Thermobaric Warhead

Fig. 6.59 Target Size, Hardness, and Collateral Drive Missile Warhead Design and Technology.
Note: See video supplement, Standoff Precision Strike Targets/Missiles.

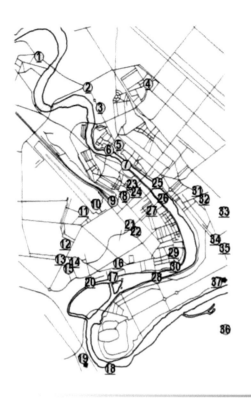

Targets: 1. Directorate of Military Intelligence; **2, 5, 8, 13, 34.** Telephone switching stations; **3.** Ministry of Defense national computer complex; **4.** Electrical transfer station; **6.** Ministry of Defense headquarters; 7. Ashudad highway bridge; 9. Railroad yard; 10. Muthena airfield (military section); **11.** Air Force headquarters; **12.** Iraqi Intelligence Service; **14.** Secret Police complex; 15. Army storage depot; **16.** Republican Guard headquarters; **17.** New presidential palace; **18.** Electrical power station; 19. SRBM assembly factory (Scud); **20.** Baath Party headquarters; **21.** Government conference center; 22. Ministry of Industry and Military Production; **23.** Ministry of Propaganda; **24.** TV transmitter; **25, 31.** Communications relay stations; 26. Jumhuriya highway bridge; **27.** Government Control Center South; 28. Karada highway bridge (14th July Bridge); **29.** Presidential palace command center; **30.** Presidential palace command bunker; **32.** Secret Police headquarters; **33.** Iraqi Intelligence Service regional headquarters; **35.** National Air Defense Operations Center; 36. Ad Dawrah oil refinery; **37.** Electrical power plant
Source: AIR FORCE Magazine, 1 April 1998

Fig. 6.60   76% of Baghdad Targets Struck First Night of Desert Storm Were Time Critical Targets.

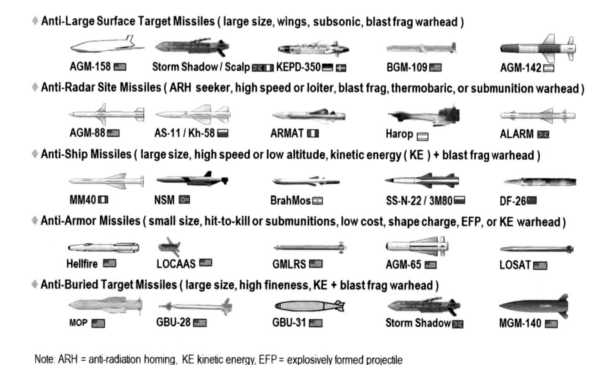

Note: ARH = anti-radiation homing, KE kinetic energy, EFP = explosively formed projectile

Fig. 6.61   Type of Target Drives Precision Strike Missile Size, Weight, Speed, Cost, Seeker, and Warhead.

180 Missile Design Guide

Weapon (Guidance)	Fixed Surface Targets[1]	Moving Targets[2]	Time Critical Targets[3]	Buried Targets[4]	Adverse Weather[5]	Firepower[6]
Griffin (Semi-Active Laser (SAL), GPS/INS)	Average	Average	Average	Poor	Average	Superior
Maverick (SAL, INS)	Superior	Average	Average	Average	Poor	Poor
SDB (SAL, mmW, Imaging Infrared (IIR), GPS/INS)	Superior	Average	Average/Poor	Average	Superior	Average
TOW (Wire Command)	Good	Average	Poor	Average	Average	Good
Hellfire (SAL, INS)	Good	Average	Average	Average	Poor	Good
LOCAAS (IIR, GPS/INS)	Average	Good	Average	Poor	Average	Superior
APKWS (SAL, INS)	Average	Average	Average	Poor	Poor	Superior

Note: ● Superior, ◒ Good, ○ Average, – Poor

(1) – Multi-mode and larger warhead desired. Global Positioning System (GPS) / inertial navigation system (INS) provides precision (3 m) accuracy
(2) – Long range seeker, high bandwidth data link, and duration desired
(3) – High speed with duration desired ⇒ High speed to target with loiter or high speed with powered submunition
(4) – Kinetic energy penetration warhead desired ⇒ High impact speed, low drag, high density, long length
(5) – GPS / INS, Synthetic Aperture Radar (SAR) seeker, imaging millimeter wave (mmW) seeker, radar seeker, and data link have high payoff
(6) – Light weight and small size desired for high firepower. Light weight also provides low cost

Fig. 6.62 Lightweight Multi-Purpose Precision Strike Weapons Are Based on Many Tradeoffs.
Note: See video supplement, Lightweight Multi-Purpose Precision Strike Weapon: Griffin.

MQ-1 Predator: Hellfire (~100 lb)

MQ-9 Reaper: GBU-12 Laser Guided Bomb (~500 lb)

X-45: GBU-39 SDB (~250 lb)

MQ-5B Hunter: Viper Strike (~40 lb)

Note: UCAV sensor suites may consist of forward looking infrared (FLIR), television (TV) charge coupled device (CCD) camera, data link, synthetic aperture radar (SAR), laser detection and ranging (LADAR), and electronic warfare (EW) sensors

Fig. 6.63 Unmanned Combat Air Vehicles (UCAVs) Require Lightweight Weapons and Sophisticated Avionics.
Note: See video supplement, X-45 Ejection Launch SDB, Predator Rail Launch Hellfire.

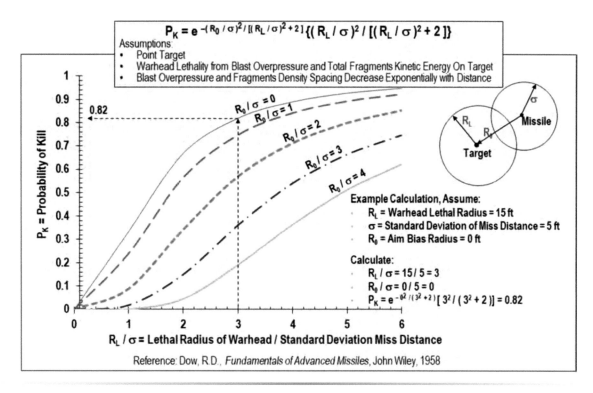

Fig. 6.64 A Typical Probability of Kill Criteria for a Blast Fragmentation Warhead Is Lethal Radius Greater Than 3 Sigma Miss Distance.

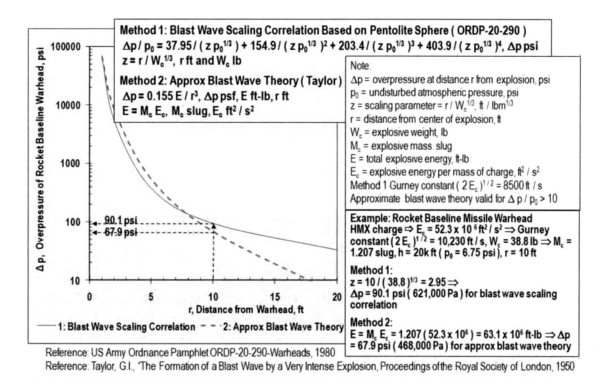

Fig. 6.65 Warhead Blast Kill Requires Small Miss Distance.

- **Central Core Detonation Blast**
  - Similar to Conventional Blast / Frag Warhead
- **Secondary Combustion Blast of Excess Fuel with Air**
  - Sustained Overpressure Over Longer Distance
- **Blast Most Effective Against**
  - Soft Outside Targets
  - Soft Interior Targets (Buildings, Caves, Tunnels)

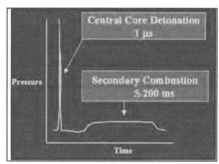

Type	Example	Main Charge	Burster / Dispensing Charge
Unitary Thermobaric	Steel Case, Fuze, Unitary Thermobaric Fill	HTPB / HMX / Al	None
Burster Thermobaric	High Explosive Burster, Fuze, Steel Case, Metal / IPN Slurry	IPN / Mg Slurry	RDX
Metal Augmented Charge (MAC)	Steel Case, Fuze, Explosive, Metal Augment	Al / Viton Elastomer	PBXN-112

Fig. 6.66 Excess Fuel in Thermobaric (Metal Augmented Charge) Warhead Provides Secondary Blast Combustion.

$$A_{FragmentSpray} = \int_0^{2\pi} (l + r \sin \theta) \, d\phi = 2\pi r (l + r \sin \theta)$$

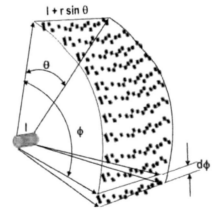

Note
$A_{FragmentSpray}$ = Cylindrical warhead fragmentation spray surface area, l = Warhead case length, r = Distance from center of explosion, θ = Vertical spray angle, ϕ = Radial spray angle
Vertical spray angle typically varies from $2 < \theta$, 50 deg
Types of Kill:
- Catastrophic Kill (k-kill) destroys target
- Fire power kill (F-kill) destroys target fire control system. For long repair time, target may be inoperative for duration of war
- Mobility kill (M-kill) destroys target propulsion system. For long repair time, target may be inoperative for duration of war

Fig. 6.67 The Primary Kill Mechanism of a Typical Blast Frag Cylindrical Warhead Is the Warhead Fragments.

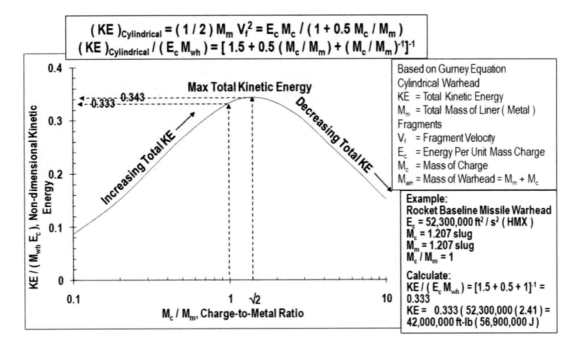

Fig. 6.68 A High Warhead Total Kinetic Energy Requires Charge Weight Be Approximately Equal to Fragments Weight.

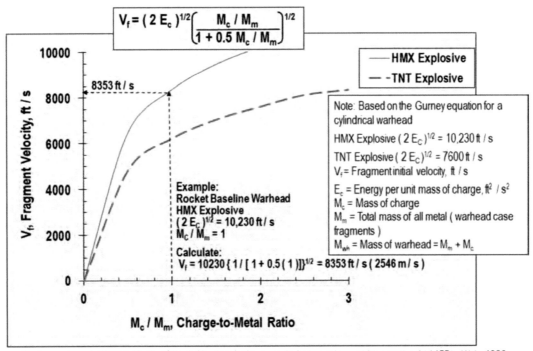

Fig. 6.69 Warhead High Fragment Velocity Requires High Charge-to-Metal Ratio.

184 Missile Design Guide

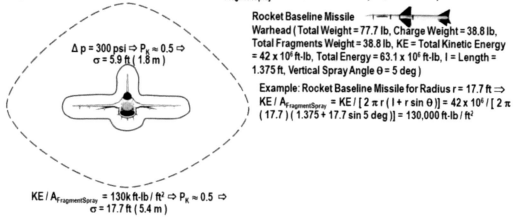

Fig. 6.70 Warhead Fragment Kinetic Energy Usually Has a Larger Lethal Radius Than Blast Overpressure. *Note*: See video supplement, AIM-7 Sparrow Warhead (Aircraft Targets).

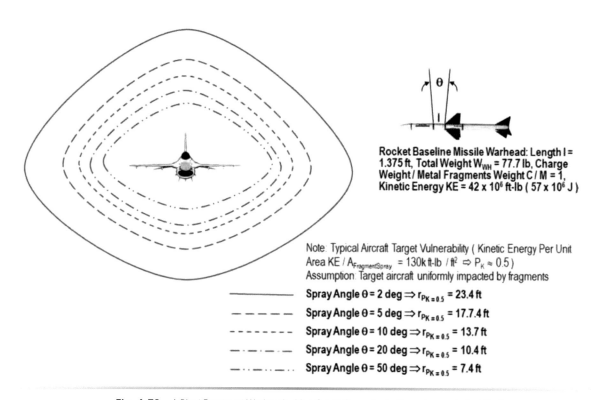

Fig. 6.71 A Blast Fragment Warhead with a Small Spray Angle Has a Larger Lethal Radius.

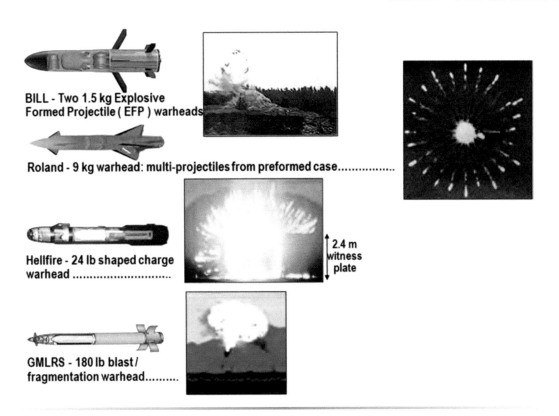

Fig. 6.72 Accurate Guidance Provides Higher Lethality, Lighter Warhead Weight, and Lower Collateral Damage.
Note: See video supplement, BILL, Roland, Hellfire, and GMLRS, Warheads.

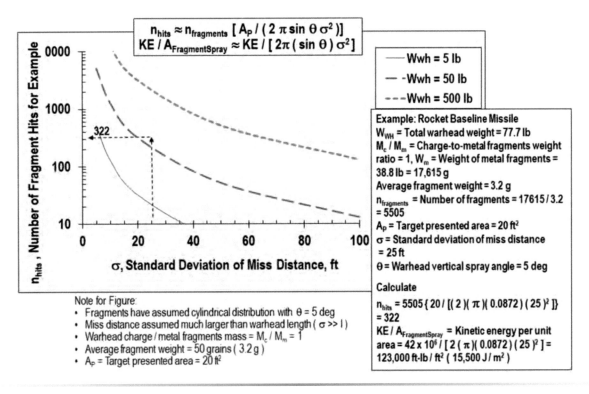

Fig. 6.73 Small Miss Distance Improves the Number of Warhead Fragment Hits.

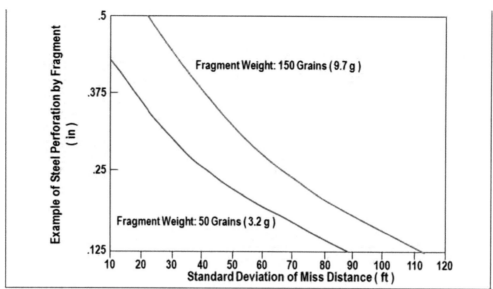

Note for Figure:
- Fragment initial velocity 5000 ft / s
- Sea level
- Average fragment weight 3.2 g
- Fewer than 0.3% of the fragments weigh more than 9.7 g for nominal 3.2 g preformed warhead fragments
- Small miss distance gives less reduction in fragment velocity, enhancing penetration

Fig. 6.74 Small Miss Distance Improves Warhead Fragment Penetration.

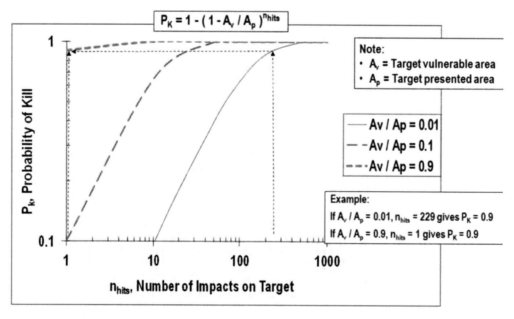

Assumptions: Each fragment has same kinetic energy. No interactions between fragment impacts

Fig. 6.75 Blast Frag Warhead Lethality Drivers—Total Fragment Impacts, Target Vulnerable Area, and Target Size.

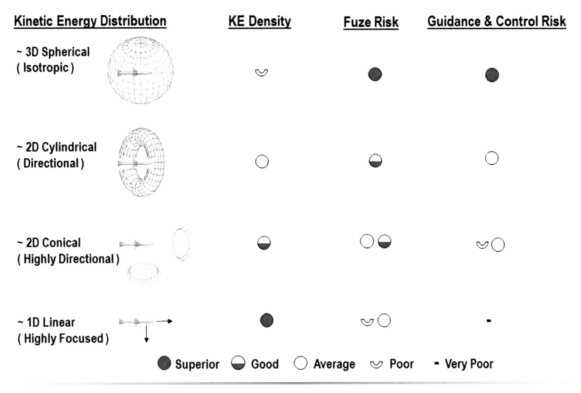

Fig. 6.76 A Directional Warhead Has High Kinetic Energy Density but It Also Has Fuzing and Guidance & Control Risk.

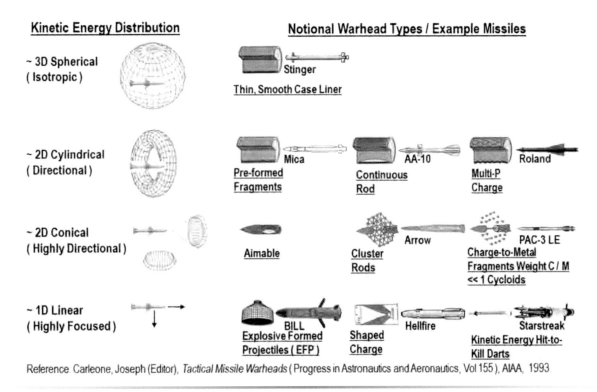

Fig. 6.77 Desired Kinetic Energy Directional Distribution Drives the Type of Warhead Design.

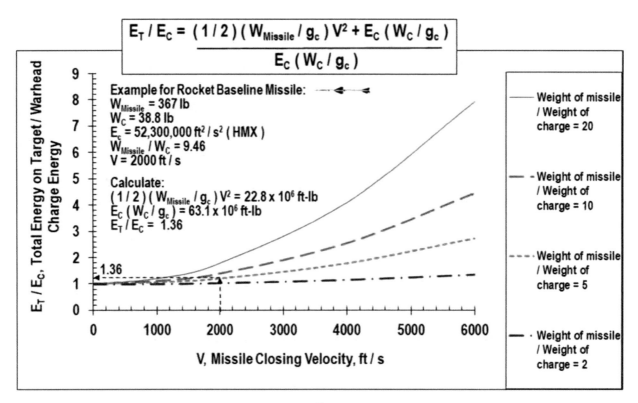

Note: Warhead explosive charge energy based on HMX, $(2 E_c)^{1/2}$ = 10,230 ft / s. (3118 m / s)
1 kg weight at Mach 3 closing velocity has kinetic energy of 391,000 J $\Rightarrow$ equivalent chemical energy of 0.4 lb TNT

Fig. 6.78 Hypersonic Hit-to-Kill Enhances Energy on Target, Especially for a Missile with a Small Warhead.

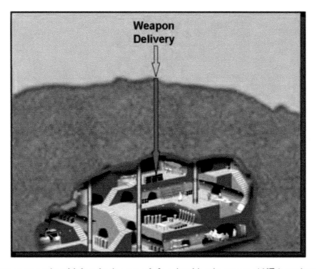

Note: Defeating a deeply buried target requires high velocity non-deforming kinetic energy (KE) warhead penetration, followed by smart fuzing of blast frag warhead.

Fig. 6.79 Defeating a Deeply Buried Target Requires High Speed Impact, Kinetic Energy Warhead, and Smart Fuzing.
Note: See video supplement, Animation of Warhead and Fuzing Against Deeply Buried Target.

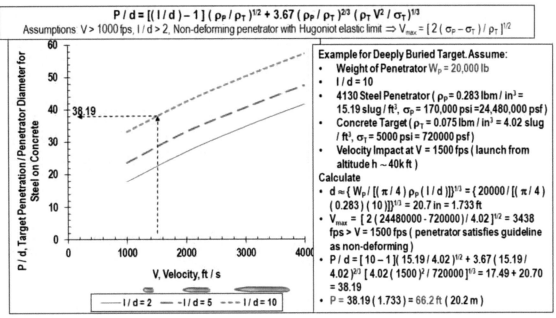

Nomenclature: l = Penetrator length, d = Penetrator diameter, V = Impact velocity(fps), ρ_P = Penetrator density (lbm / in^3), ρ_T = Target density (lbm / in^3), σ_T = Target yield stress (psf), σ_P = Penetrator yield stress (psf)

Source: Christman, D.R., et al, "Analysis of High-Velocity Projectile Penetration," Journal of Applied Physics, Vol 37, 1966

Fig. 6.80 Kinetic Energy Penetrator Weight, Density, Length, and Velocity Increase Target Penetration.

Fig. 6.81 Kinetic-Kill Targets Include Ballistic Missiles, Aircraft, Projectiles, Buried Targets, and Armored Vehicles. *Note*: See video supplement, LOSAT.

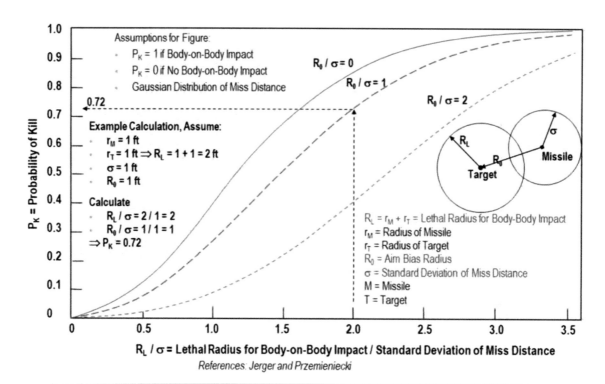

Fig. 6.82 A Kinetic Kill Missile Requires Hit-to-Kill Accuracy.

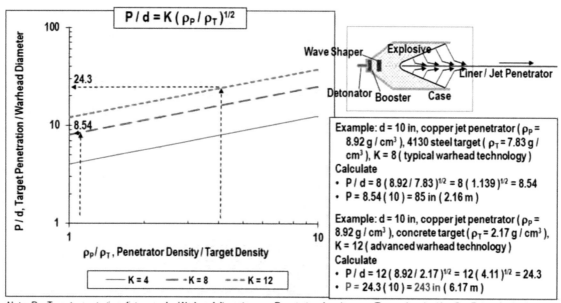

Note: P = Target penetration distance, d = Warhead diameter, ρ_P = Penetrator density, ρ_T = Target density, K = Coefficient for jet length
- K = 8 is a typical warhead
- K > 8 for warhead with high ductility jet penetrator, optimum standoff distance, and low strength target
- K < 8 for warhead w/o standoff, compact length, less precise manufacturing tolerances, low ductility jet penetrator, missile flight dynamics at impact, high strength target, and non-homogeneous target

Fig. 6.83 Shaped Charge Warhead Jet Penetration Is Driven by Diameter, Penetrator Density, and Target Density.

Liner Design Source: Carleone	Cost	Jet Length w/o Stretching	Jet Stretching	Hole Diameter
Conical 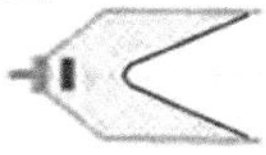	●	○	○	○
Tulip 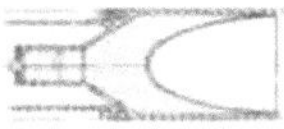	○	●	●	◐
Biconic 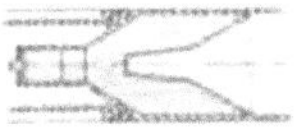	○	◐	◐	○
Trumpet 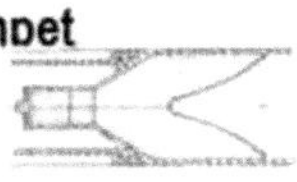	○	◐	●	○
Hemisphere 	◐	◐	●	○

Note: ● Superior   ◐ Good   ○ Average   — Below Average

Fig. 6.84 Shaped Charge Warhead Liner Design Geometry Has Many Trade-offs.

Liner Material / Jet Penetrator	Density lbm / in^3	Cost	Jet Length w/o Stretching	Jet Stretching	Hole Diameter
Copper	0.32	◐	○	◐	○
Aluminum	0.10	◐	—	◐	●
Tungsten	0.70	—	●	—	—
Depleted Uranium	0.69	—	●	—	—
Molybdenum	0.37	◐	◐	—	○
Tantalum	0.60	○	●	◐	—

Note:
- High density liner ⇒ long jet length ⇒ high penetration
- High ductility liner ⇒ high jet stretching ⇒ longer jet length ⇒ higher penetration
- Low density liner ⇒ large hole diameter ⇒ robustness in manufacturing, flight dynamics, and counter-countermeasures (e.g., reactive armor).

Note: ● Superior   ◐ Good   ○ Average   — Below Average

Fig. 6.85 Shaped Charge Warhead Liner Material Has Many Trade-offs.

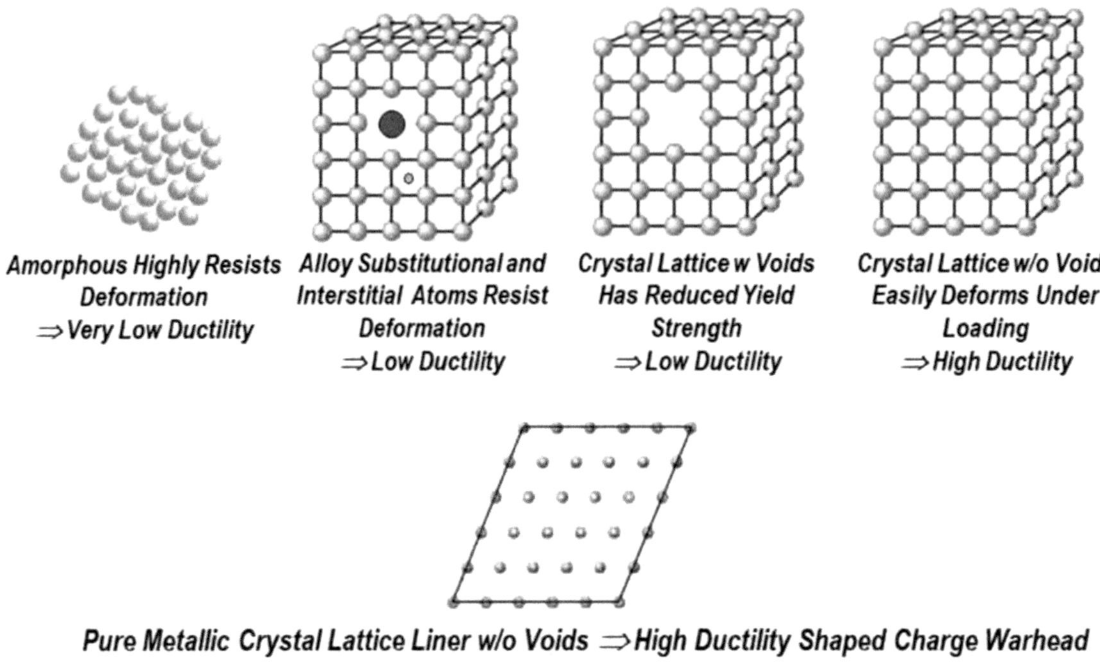

192 Missile Design Guide

![Amorphous Highly Resists Deformation ⇒ Very Low Ductility] ![Alloy Substitutional and Interstitial Atoms Resist Deformation ⇒ Low Ductility] ![Crystal Lattice w Voids Has Reduced Yield Strength ⇒ Low Ductility] ![Crystal Lattice w/o Voids Easily Deforms Under Loading ⇒ High Ductility]

Pure Metallic Crystal Lattice Liner w/o Voids ⇒ High Ductility Shaped Charge Warhead

Fig. 6.86 Shaped Charge Warhead with High Ductility Requires a Pure Metallic Crystal Lattice Liner without Voids.

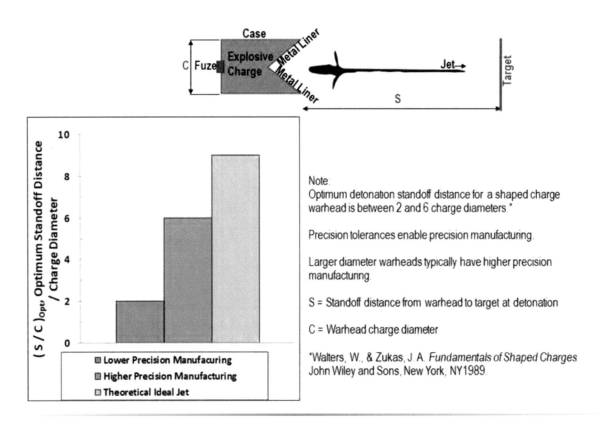

Fig. 6.87 Optimum Detonation Standoff Distance for a Shaped Charge Warhead is Typically About 4 Charge Diameters from the Target.

- **Precision Targeting, Precision Guidance, and Data Link**
  - Accuracy
  - Optimum Impact Angle (e.g., Vertical)
  - Regrets Avoidance
- **High Reliability**
  - Electronic Safe, Arm, and Fuzing
- **Non-Lethal Warhead**
  - Non-Nuclear Electro-Magnetic Pulse (EMP)
- **Focused Energy to Target Area**
  - Laser
  - Hit-to-Kill Warhead
    - Kinetic Energy (KE)
    - Explosively Formed Projectile (EFP)
    - Shaped Charge
- **Blast / Frag Warhead Tailored to Minimize Collateral Damage**

BLU-129/B Lower Collateral Damage Bomb (Carbon Fiber Case)

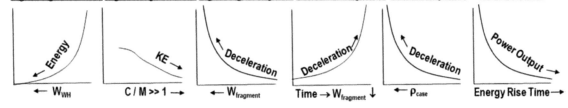

Fig. 6.88 There Are Multiple Approaches to Minimize Collateral Damage.

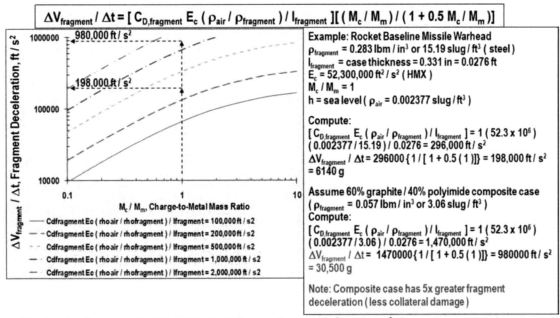

Fig. 6.89 Low Density Warhead Case Material Has Rapid Deceleration of Fragments, Resulting in Lower Collateral Damage.

Fig. 6.90 A Lightweight Warhead with Precision Guidance Accuracy Has Lower Collateral Damage.
Note: See video supplement, Low Collateral Damage Rocket Video: BAEAPKWS, Northrop Grumman GATR, and Raytheon Talon.

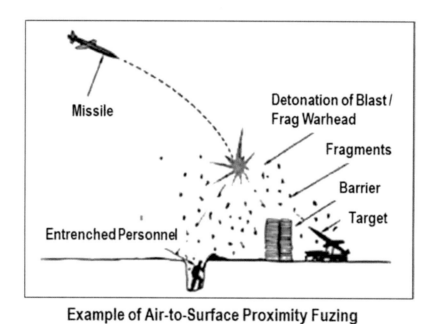

Fig. 6.91 A Proximity Fuze Selects the Standoff Distance at Warhead Detonation.
Note: See video supplement, Proximity Fuze Historical Development.

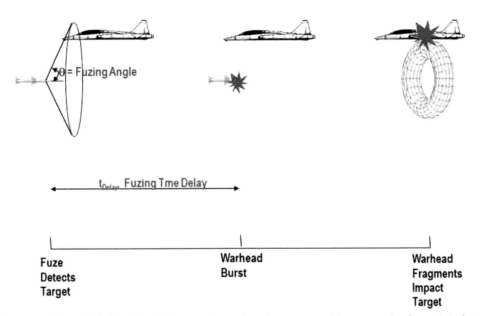

Note: The most popular proximity fuzes for missiles use either radar or laser sensors. Other approaches for proximity fuzing include optical, acoustic, pressure, and electrostatic sensors.

Fig. 6.92 A Proximity Fuze for a Blast-Frag Warhead Against an Air Target Requires Selecting Fuze Angle and Time Delay.

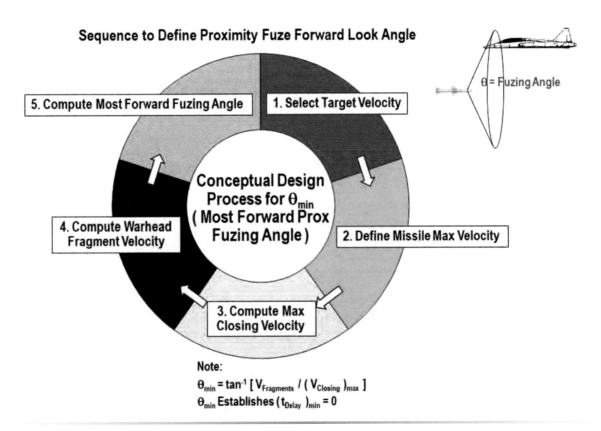

Fig. 6.93a Air Target Proximity Fuze Min Angle Is Driven by Maximum Closing Velocity and Warhead Fragments Velocity.

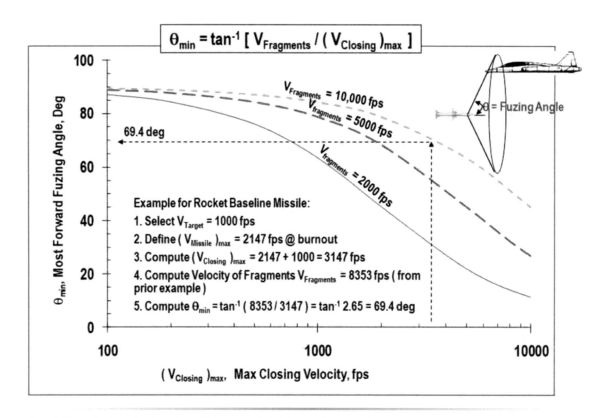

Fig. 6.93b Air Target Proximity Fuze Minimum Angle Is Driven by Maximum Closing Velocity and Warhead Fragments Velocity (cont).

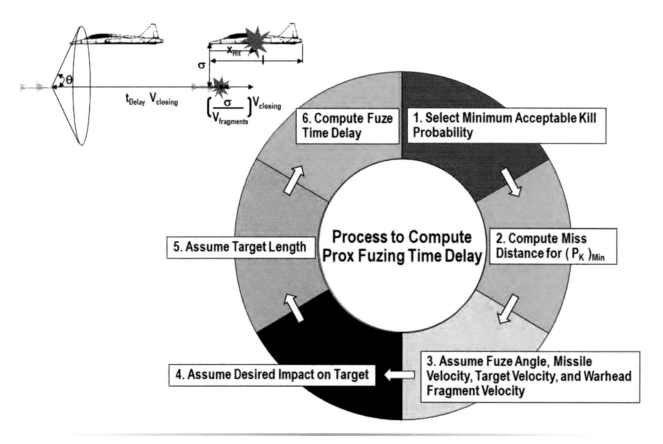

Fig. 6.94a Air Target Proximity Fuze Time Delay Is Driven by Fuze Angle, Miss Distance, Closing Velocity, and Target Length.

CHAPTER 6 Other Measures of Merit   197

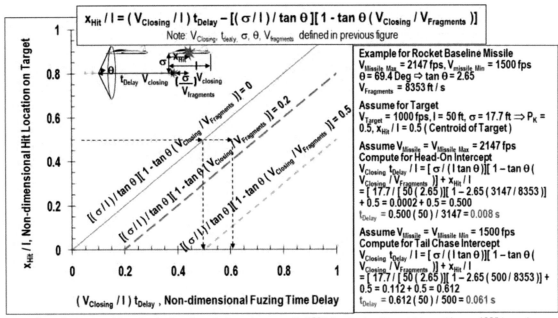

Fig. 6.94b Air Target Proximity Fuze Time Delay Is Driven by Fuze Angle, Miss Distance, Closing Velocity, and Target Length (cont).

6.4 Accuracy

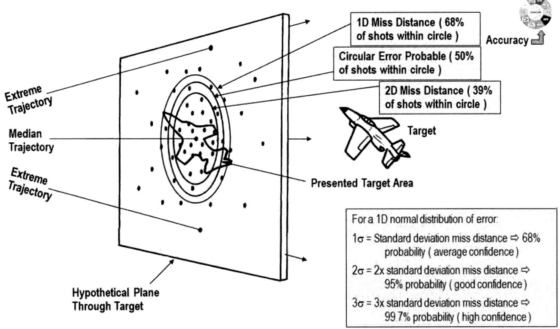

Fig. 6.95 CEP Is Approximately Equal to Miss Distance.

Type of Guidance	Max Range for 3 m Accuracy Against Fixed Target	Max Range for 3 m Accuracy Against Moving Target	Collateral Damage	Unit Cost	Counter-Counter Measures (CCM)	Reliability & Maintainability (R&M)
Seeker Terminal Guided w GPS / INS / Data Link Midcourse Guided	~ No Limit ●	●	◐	–	○	○
Global Positioning System (GPS) / INS / Data Link Guided	~ No Limit ●	○ –	◐	○	◐	◐
Inertial Navigation System (INS) Guided	R ~ 5 km ○	–	○	– ◐	●	◐
Unguided Projectile Unguided Bomb Unguided Rocket	R ~ 600 m R ~ 300 m – R ~ 150 m	–	–	●	●	●

Note: ● Superior   ◐ Good   ○ Average   – Poor

Fig. 6.96 Guided Advantages—Accuracy and Low Collateral Damage, Unguided Advantages—Low Cost, CCM, R&M.

Type of Terminal Guidance	Seeker / Fire Control System (FCS) Sensor Correction	Fixed Target Accuracy Short / Long Range	Moving Target Accuracy Short / Long Range
Seeker Proportional Homing 	Seeker Line-of-Sight (LOS) Angle Rate Error $\Rightarrow 0$	●	●
FCS Command LOS / 3 Point 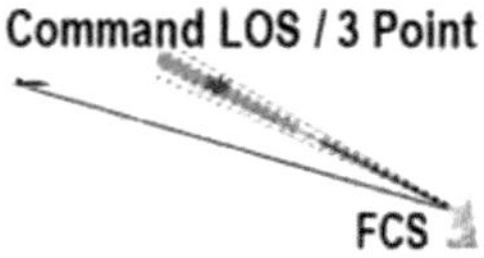	Fire Control System Commands Missile to Lead FCS Line-of-Sight to Target	● ◐	◐ ○
Seeker Pursuit Homing 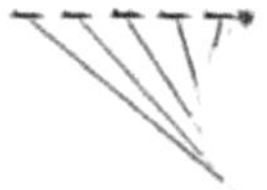	Seeker Line-of-Sight Angle Error $\Rightarrow 0$	●	–
FCS Command Beam Rider 	Fire Control System Commands Line-of-Sight Angle / Position Error $\Rightarrow 0$	◐ ○	○ –

Note: ● *Superior*   ◐ *Good*   ○ *Average*   – *Poor*

Fig. 6.97 Seeker Proportional Homing Is the Most Common Terminal Guidance Law Against High-Speed Targets.

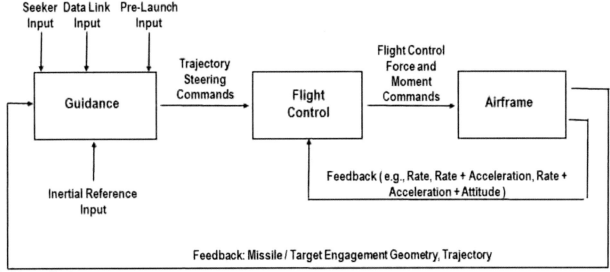

Fig. 6.98 Missile Guidance & Control Fundamentals Can Be Illustrated with a G&C Block Diagram.

Guidance System
1. Compares Missile / Target Engagement Geometry with Desired Engagement Geometry
2. Compares Missile Current Inertial Reference Position with Desired Position
3. Issues Steering Commands to Flight Control System to Correct Errors in Engagement Geometry and Inertial Position

Flight Control System
1. Quickly Compares Flight Trajectory Parameters (e.g., Attitude, Altitude, Rate, Acceleration) with Guidance Commands
2. Quickly Issues Commands to Flight Control Devices to Provide Forces and Moments to Meet Guidance Requirements

Note: Flight control system feedback (inner loop) bandwidth should be 5-to-10x guidance system (outer loop) feedback bandwidth
Note: A statically unstable missile must have rate stabilization with the flight control feedback rate gain $K_{Rate} \leq C_{N_\alpha} q S_{Ref} d (\alpha / \delta)$, where C_{N_α} = normal force coefficient derivative with angle of attack, q = dynamic pressure, S_{Ref} = reference area = body cross sectional area, d = body diameter, α = angle of attack, δ = flight control deflection

Reference: Eichblatt

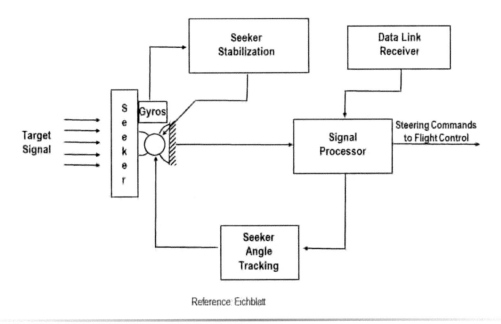

Reference: Eichblatt

Fig. 6.99 Example of Gimbaled Seeker Simplified Block Diagram.

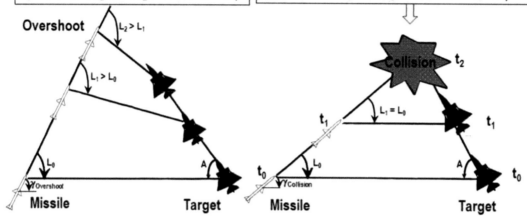

Note:
- $t_0, t_1, t_2,$ = Time sequence of events
- L_0, L_1, L_2 = Line-of-sight (LOS) angle from missile seeker to target
- A = Target aspect angle relative to missile
- γ = Flight path angle

Note:
Proportional guidance: $\dot{\gamma} = N \dot{L}$, N = gain

Fig. 6.100 Proportional Guidance Provides a Constant Bearing Flight Path.

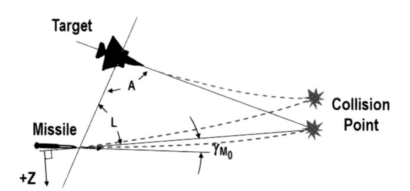

Proportional Guidance:
Step Maneuver Target
$$\tau \frac{d^2Z}{dt^2} + \frac{dZ}{dt} + N' \frac{Z}{t_o - t} = \left(\frac{-N'}{t_o - t}\right)\left(\frac{\cos A}{\cos L}\right)\left(\frac{1}{2} a_T t^2\right)$$

Proportional Guidance:
Initial Heading Error
$$\tau \frac{d^2Z}{dt^2} + \frac{dZ}{dt} + N' \frac{Z}{t_o - t} = -V_M \gamma_{M_0}$$

Note: $t_o - t = 0$ at intercept, causing discontinuity in above equations. N' = Effective navigation ratio = $N [V_M \cos L / (V_M \cos L + V_T \cos A)]$, N = Navigation ratio = $(d\gamma/dt)/(dL/dt)$, τ = Missile time constant, V_M = Velocity of missile, γ_{M_0} = Initial flight path angle error of missile, t_o = Total time of flight, a_T = Acceleration of target, V_T = Velocity of target, A = Target aspect angle, L = Missile lead angle, z = Distance normal to trajectory, t = time

Reference: Jerger, J.J., *Systems Preliminary Design Principles of Guided Missile Design*, D. Van Nostrand Company, Inc., Princeton, New Jersey, 1960

Fig. 6.101 A Maneuvering Target and An Initial Heading Error Result in Missile Miss Distance.

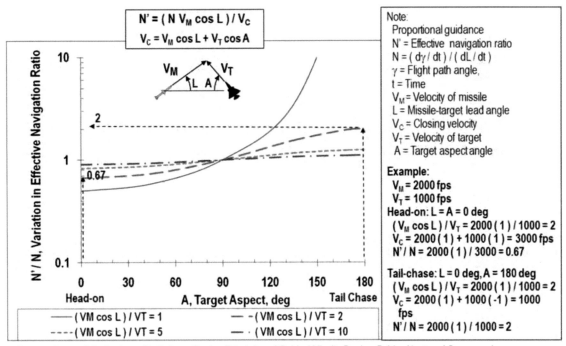

Source: Jerger, J.J., *Systems Preliminary Design Principles of Guided Missile Design*, D. Van Nostrand Company, Inc., Princeton, New Jersey, 1960

Fig. 6.102 Effective Navigation Ratio Is a Function of Missile Velocity, Target Velocity, and Engagement Geometry.

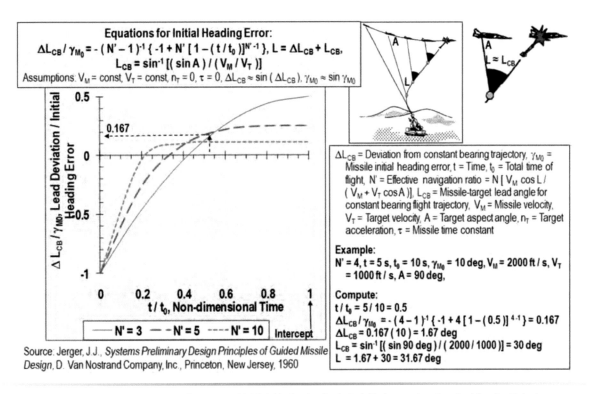

Fig. 6.103 Proportional Guidance with High Navigation Ratio Quickly Approaches Constant Bearing Trajectory.

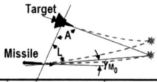

Implementation	Countermeasure and Noise Robustness	Required Sensors / Target Data	Accuracy Against Off-Boresight Target	Accuracy Against Maneuvering Target
N = Navigation Ratio = $(d\gamma/dt)/(dL/dt)$ N = Navigation Ratio γ = Flight Path Angle L = Missile-Target Lead Angle A = Target Aspect Angle V_M = Missile Velocity V_T = Target Velocity t = Time	●	●	−	−
$N' = N[V_M \cos L / (V_M \cos L + V_T \cos A)]$ N' = Effective Navigation Ratio	○	○	◐	○
"Optimal" Guidance (Includes Measurement of Target Acceleration, Kalman Filtering)	−	−	− ◐	− ◐

Note: ● Superior   ◐ Good   ○ Average   − Poor

Fig. 6.104 Terminal Homing Guidance Law Tradeoffs Include Robustness, Required Sensors, Required Target Data, and Accuracy.

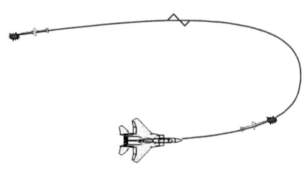

Conceptual design typically uses a single time constant τ for entire missile (optimistic)
- τ is a measure of missile agility to respond to target condition changes
- τ equals elapsed time from input command maneuver until missile has completed 63% or $(1 - e^{-1})$ of the commanded maneuver ($t = \tau$)
- τ also called "rise time"

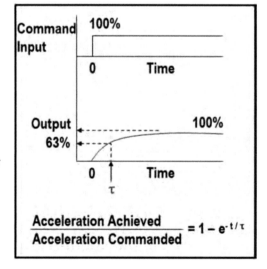

$$\frac{\text{Acceleration Achieved}}{\text{Acceleration Commanded}} = 1 - e^{-t/\tau}$$

Note: A highly maneuverable missile typically has a small time constant

Fig. 6.105a Missile Guidance Time Constant Is an Indication of Maneuver Response Time.

Contributors to Missile Time Constant (τ)

1. Flight Control Effectiveness (τ_δ)
2. Flight Control Actuator Rate ($\tau_{\dot\delta}$)
3. Error Slope Filter for Seeker Dome ($\tau_{Dome\ Filter}$)
4. Other Guidance & Control (G&C) Filters ($\tau_{G\&C\ Filter}$)
5. Other Guidance & Control Dynamics*
6. Seeker Errors**

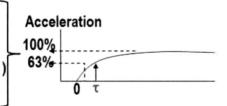

$$\frac{\text{Acceleration Achieved}}{\text{Acceleration Commanded}} = 1 - e^{-t/\tau}$$

Primary contributors to missile time constant are usually
- Flight control effectiveness (τ_δ)
- Flight control actuator rate ($\tau_{\dot\delta}$)
- Error slope filter for seeker dome ($\tau_{Dome\ Filter}$)

*Other G&C dynamics include gyro dynamics, accelerometer dynamics, processor latency, etc
**Seeker errors include resolution, latency, blind range, tracking error, glint / angular noise, and amplitude noise

A first-order (usually optimistic) conceptual design expression of missile guidance time constant is
$\tau \approx \tau_\delta + \tau_{\dot\delta} + \tau_{Dome\ Filter}$

> **Example for Rocket Baseline Missile:**
> Mach Number M = 2, Altitude h = 20k ft, Coast
> $\tau \approx \tau_\delta + \tau_{\dot\delta} + \tau_{Dome\ Filter}$
> $= 0.096 + 0.070 + 0.043 = 0.209\ s$

Fig. 6.105b Missile Guidance Time Constant Is an Indication of Maneuver Response Time (cont).

- **Assumptions for calculating conceptual design time constant due to aero flight control effectiveness (τ_δ)**
  - Command maximum angle of attack (α_{Max})
  - Flight control operating in race-break "bang–bang" +/- δ_{Max} deflection mode (minimum time to α_{Max})
  - Fast flight control actuator deflection
    ($\delta_{Max}/\dot\delta \ll \tau_\delta/2$)
  - Near neutral missile static stability ($\alpha/\delta \gg 1$)
    - Note: If α/δ is not $\gg 1 \Rightarrow$ larger value of τ_δ
- **Equation of motion is**
  - $\ddot\alpha \approx [\rho V^2 S_{Ref}\ d\ C_{m_\delta}/(2\ I_y)]\ \delta_{Max}$
- **Integrate to solve for α_{Max}**
  - $\alpha_{Max} = [\rho V^2 S_{Ref}\ d\ C_{m_\delta}/(8\ I_y)]\ \delta_{Max}\ \tau_\delta^2$
- **τ_δ is given by**
  - $\tau_\delta = [8\ I_y\ (\alpha_{Max}/\delta_{Max})/(\rho V^2 S_{Ref}\ d\ C_{m_\delta})]^{1/2}$

Note: Typical drivers to small τ_δ for aero control missile
- Low fineness (small $I_y/(S_{Ref}\ d)$)
- High dynamic pressure (low altitude / high speed)
- High flight control effectiveness (high C_{m_δ})

Note: Reaction jet thruster typically has a smaller time constant for flight control than the time constant for aero flight control

> **Example for Rocket Baseline Missile:**
> W = 367 lb, d = 0.667 ft, S_{Ref} = 0.349 ft², I_y = 94.0 slug-ft²,
> M = 2, h = 20k ft (ρ = 0.001267 slug/ft³), α_{Max} = 9.4 deg,
> δ_{Max} = 12.6 deg, C_{m_δ} = 51.6 per rad
> Calculate
> $\tau_\delta = \{8(94.0)(9.4/12.6)/[0.001267(2074)^2(0.349)(0.667)(51.6)]\}^{1/2} = 0.096\ s$

Fig. 6.106 Missile Maneuver Time Constant Is Often Driven by Flight Control Effectiveness.

- ♦ **Assumptions for calculating conceptual design time constant due to aero flight control dynamics ($\tau_{\dot{\delta}}$)**
  - "Bang – bang" flight control
  - Flight control deflection dynamics $\dot{\delta} = +/- \dot{\delta}_{Max}$ driven by actuator rate limit
  - Near neutral missile static stability ($\alpha / \delta \gg 1$)
- ♦ **Equation of Motion for $\dot{\delta} = +/- \dot{\delta}_{Max}$**
  - $\ddot{\alpha} = [\rho V^2 S_{Ref} \, d \, C_{m\delta} / (2 \, I_y)] \dot{\delta}_{Max}$
- ♦ **Equation of Motion for "Perfect" Response $\dot{\delta} = \infty$, $\delta = \delta_{Max}$**
  - $\ddot{\alpha} = [\rho V^2 S_{Ref} \, d \, C_{m\delta} / (2 \, I_y)] \delta_{Max}$
- ♦ $\tau_{\dot{\delta}}$ is Difference Between Actual Response to α_{Max} and "Perfect" (τ_δ) Response
- ♦ **Then**
  - $\tau_{\dot{\delta}} = 2 \, \delta_{Max} / \dot{\delta}_{Max}$

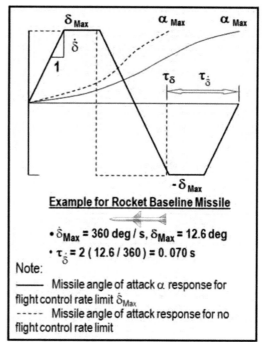

Example for Rocket Baseline Missile

- $\dot{\delta}_{Max} = 360 \text{ deg/s}$, $\delta_{Max} = 12.6 \text{ deg}$
- $\tau_{\dot{\delta}} = 2(12.6/360) = 0.070 \text{ s}$

Note:
—— Missile angle of attack α response for flight control rate limit $\dot{\delta}_{Max}$
----- Missile angle of attack response for no flight control rate limit

Note: Reaction jet thruster dynamics time constant is typically smaller than aero flight control dynamics time constant

Fig. 6.107 Missile Time Constant Contributor from Flight Control Dynamics Is Usually Driven by Actuator Dynamics.

- ♦ **Radome Error Slope** $|R| = |\Delta\varepsilon / \Delta\delta_{seeker}| \approx 0.05 (l_N/d - 0.5)[1 + 15(\Delta f/f)]/(d/\lambda)$
  - Valid for $l_N/d \geq 0.5$
- ♦ Based on Routh's Stability Requirement for Guidance & Control Feedback Loop
  - $\tau_{Dome \, Filter} = N'(V_C/V_M)|R|(\alpha/\dot{\gamma})$
- ♦ Angle of Attack Sensitivity to Missile Flight Path Turn Rate Is
  - $\alpha/\dot{\gamma} = \alpha(W/g_c)V_M / \{q \, S_{Ref}[C_{N_\alpha} + C_{N_\delta}/(\alpha/\delta)]\}$
- ♦ **Substituting** ⇨ $\tau_{Dome \, Filter} = N' W V_C |R| / \{g_c \, q \, S_{Ref}[C_{N_\alpha} + C_{N_\delta}/(\alpha/\delta)]\}$

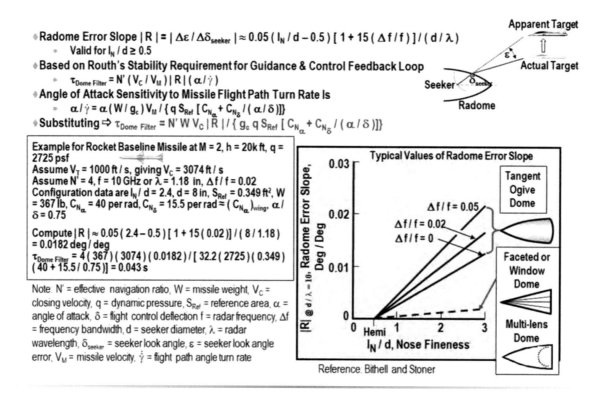

Example for Rocket Baseline Missile at M = 2, h = 20k ft, q = 2725 psf
Assume V_T = 1000 ft/s, giving V_C = 3074 ft/s
Assume N' = 4, f = 10 GHz or λ = 1.18 in, $\Delta f/f$ = 0.02
Configuration data are l_N/d = 2.4, d = 8 in, S_{Ref} = 0.349 ft², W = 367 lb, C_{N_α} = 40 per rad, C_{N_δ} = 15.5 per rad ≈ $(C_{N_\alpha})_{wing}$, α/δ = 0.75

Compute $|R| \approx 0.05(2.4 - 0.5)[1 + 15(0.02)]/(8/1.18)$ = 0.0182 deg/deg
$\tau_{Dome \, Filter}$ = 4(367)(3074)(0.0182)/[32.2(2725)(0.349)(40 + 15.5/0.75)] = 0.043 s

Note: N' = effective navigation ratio, W = missile weight, V_C = closing velocity, q = dynamic pressure, S_{Ref} = reference area, α = angle of attack, δ = flight control deflection, f = radar frequency, Δf = frequency bandwidth, d = seeker diameter, λ = radar wavelength, δ_{seeker} = seeker look angle, ε = seeker look angle error, V_M = missile velocity, $\dot{\gamma}$ = flight path angle turn rate

Reference: Bithell and Stoner

Fig. 6.108 Drivers for Radome Time Constant Include Radome Error Slope and Closing Velocity.

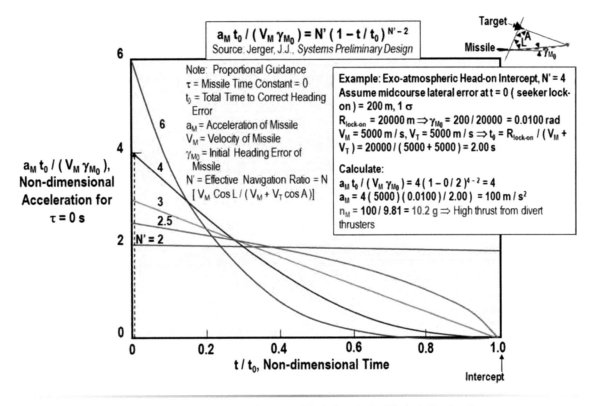

Fig. 6.109 High Initial Maneuverability Is Required to Eliminate Heading Error.

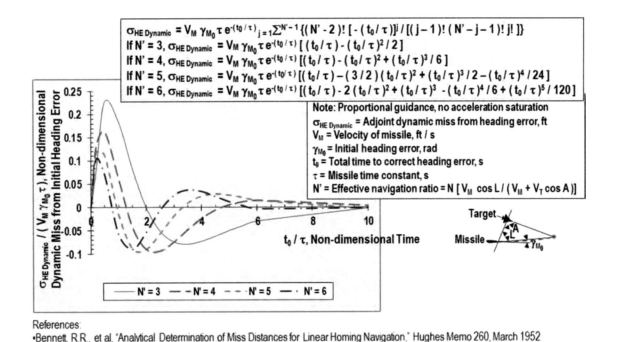

References:
- Bennett, R.R., et al, "Analytical Determination of Miss Distances for Linear Homing Navigation," Hughes Memo 260, March 1952
- Zarchan, P., *Tactical and Strategic Missile Guidance*, Vol. 5, American Institute of Aeronautics and Astronautics, 2007

Fig. 6.110a Maximum Miss Distance from Heading Error Is Driven by Time to Intercept, Navigation Ratio, and Time Constant.

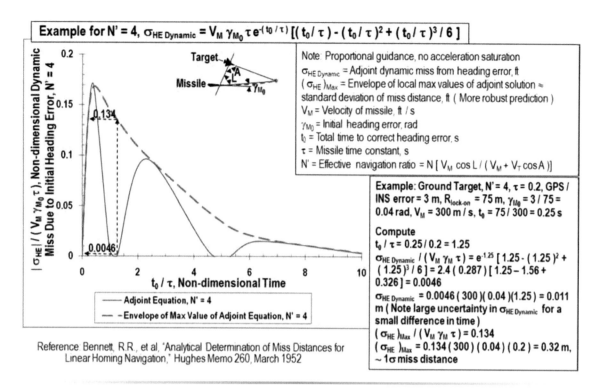

Fig. 6.110b Maximum Miss Distance from Heading Error Driven by Time to Intercept, Navigation Ratio, and Time Constant (cont).

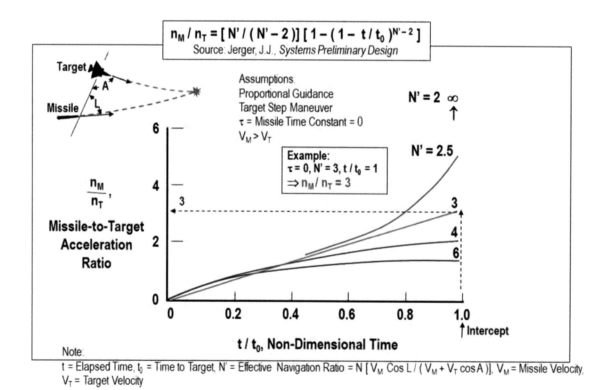

Fig. 6.111 Required Missile Maneuverability for Maneuvering Target Is about 3x the Target Maneuverability.

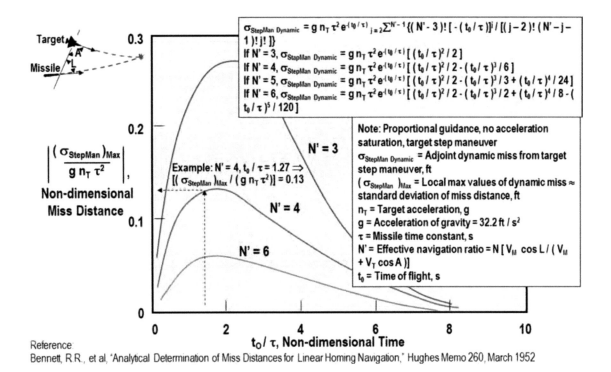

Fig. 6.112 A Target Step Maneuver Requires 6 to 10 Time Constants to Settle Out Miss Distance.

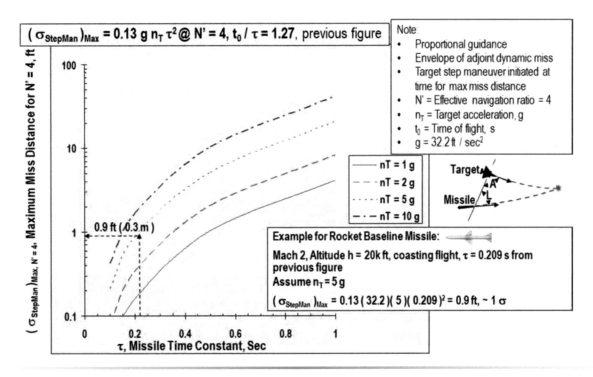

Fig. 6.113 A Small Time Constant Is Required for Small Miss Distance Against a High Maneuvering Target.

208 Missile Design Guide

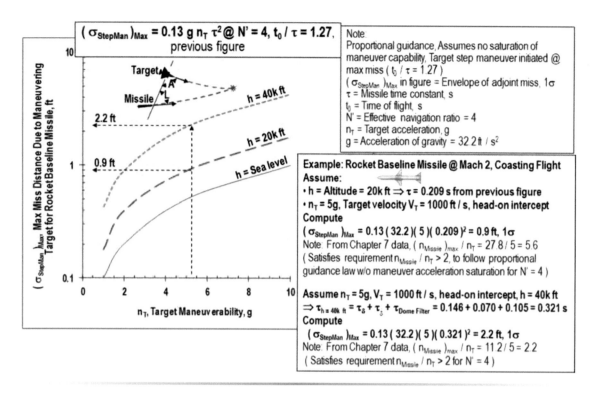

Fig. 6.114 An Aero Control Missile Has Smaller Miss Distance at Low Altitude and High Dynamic Pressure.

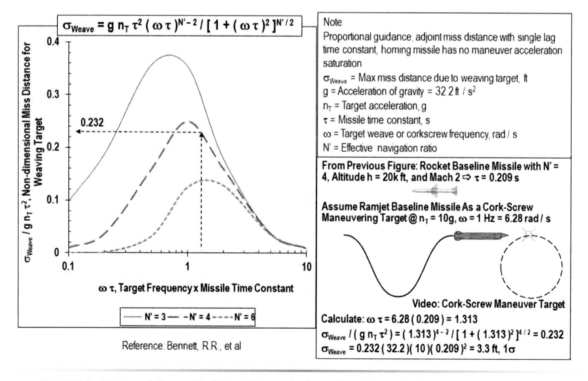

Fig. 6.115 Weaving, Jinking, and Cork-Screw Maneuvering Targets Require Large Navigation Ratio for Small Miss Distance.
Note: See video supplement, Cork–Screw Maneuver Target.

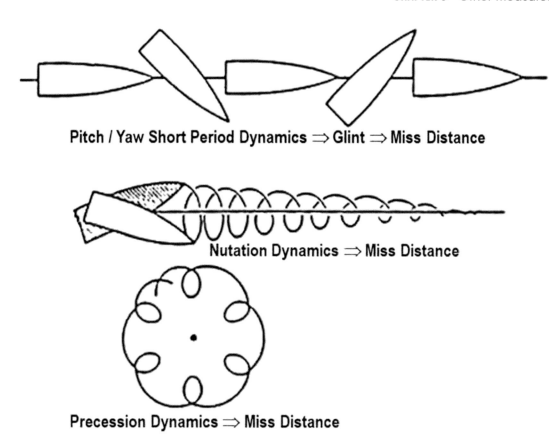

Fig. 6.116 Target Flight Trajectory Dynamics Result in Missile Miss Distance.

Animation of Radar Glint from Head-on Scatter Centers of Fighter Aircraft

Average Glint Angle ε:

$\varepsilon = 1/3\,(b_{Target}/R)$ for Head-on Intercept, ε in rad

$\varepsilon = 1/3\,(L_{Target}/R)$ for Side Intercept, ε in rad

b_{Target} = Target Span

L_{Target} = Target Length

R = Range to Target

Animation of Radar Glint from Side Scatter Centers of Destroyer Ship

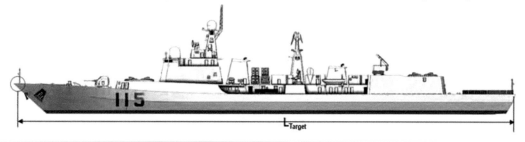

Fig. 6.117 Radar Glint Error Occurs from the Angular Flashes of Target Scatter Centers
Note: Animation of Radar Glint from Head-on Scatter Centers of Flighter Aircraft and Side Scatter Centers of Destroyer Ship. .

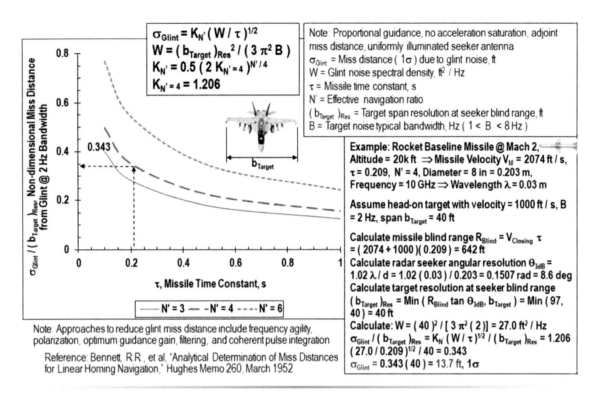

Fig. 6.118 Glint Miss Distance Is Driven by Seeker Resolution, Missile Time Constant, Navigation Ratio, and Target Size.

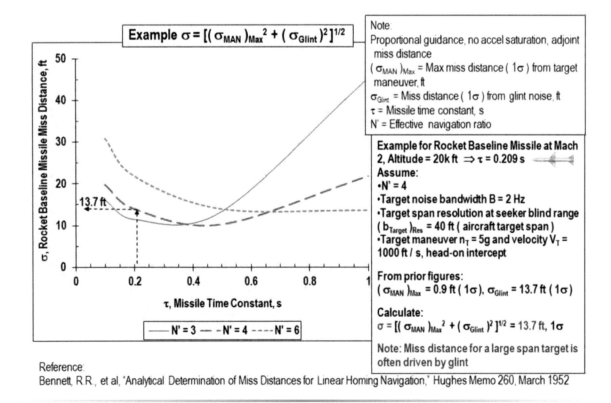

Fig. 6.119 Minimum Miss Distance Requires Optimum Time Constant and Optimum Navigation Ratio.

6.5 Carriage and Launch Observables

Conventional External Pylon / Rail Carriage*: High RCS

Conformal Carriage**: Reduced RCS

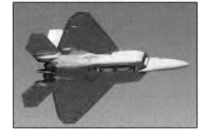
Internal Carriage***: Low RCS

*External Pylon / Rail Carriage
- Advantage: High firepower
- Disadvantages: High RCS (requires longer range missiles for self defense), high drag, aeroelasticity

**Conformal (Semi-Submerged) Carriage
- Advantage: Reduced RCS, reduced drag
- Disadvantages: Specialized store and aircraft attachments

***Internal Carriage
- Advantage: Low RCS, low drag
- Disadvantages: Low firepower, weapon span and length constraints

Fig. 6.120 Internal Weapon Carriage Is Required for a Low Radar Cross Section Aircraft Launch Platform.

Center Weapon Bay Best for Ejection Launch of Large / Heavy Missiles from Aircraft

F-22 Semi-Bay Load-out: 2 SDB I, 1 AIM-120C

F-35 Semi-Bay Load-out: 4 SDB II

B-1 Single Bay Load-out: 8 GBU-31

Side Weapon Bay Best for Rail Launch of Small / Light Missiles

F-22 Side Bay: 1 AIM-9 in Each Side Bay

RAH-66 Side Bay: 1 AGM-114, 2 FIM-92, 4 Hydra 70 in Each Side Bay

Fig. 6.121 Aircraft Center Bay Is Best for Missile Ejection—Aircraft Side Bay Is Best for Missile Rail Launch.
Note: See video supplement, F-22(AMRAAM/JDAM/AIM-9).

212 Missile Design Guide

High Smoke Example: AIM-7
Particles (e.g., metal fuel oxide) at all atmosphere temperature.

Reduced Smoke Example: AIM-120
Contrail (HCl from AP oxidizer) at T < -10° F atmospheric temperature.

Minimum Smoke Example: Javelin
Contrail (H_2O) at T < -35° F atmospheric temperature.

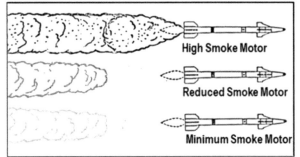
Sketches of High Smoke, Reduced Smoke, and Min Smoke Contrails.

Fig. 6.122 Minimum Smoke Propellant Reduces Launch Observables.
Note: See video supplement, Sparrow, Archer, AMRAAM, Javelin, and Hellfire Smoke Contrails.

6.6 Missile Survivability and Safety

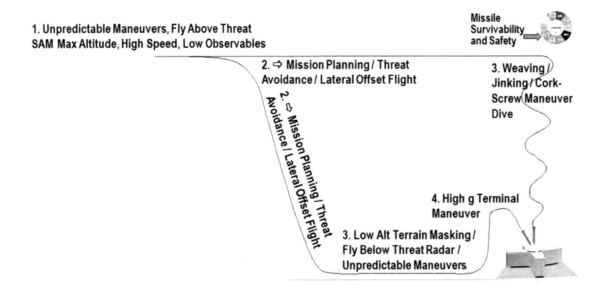

Fig. 6.123 Options for Survivability Include Stealth, Altitude, Speed, Threat Avoidance, Terrain Masking, and Maneuverability.
Note: See video supplement, Tomahawk Using Terrain Following and Jinking.

High Speed, High Altitude, and Maneuverability

HTV-2 (Hypersonic Lofted Boost-Glide)

SS-N-22 Sunburn (Supersonic w Ramjet Propulsion)

Terminal High Speed with Low Altitude

3M-54E Sizzler (Rocket Supersonic Penetrator / Turbojet Subsonic Fly-out)

Low Radar Cross Section (RCS) with Terminal Low Altitude Flight

NSM (Faceted Dome, Serpentine Inlet, Decoupled Airframe, Body Chines, Graphite Composite Structure)*

*Note: Decoupled airframe allows seeker tracking with inlet on bottom for efficient flyout and inlet on top for lower RCS in terminal flight.

JASSM (Window Dome, Flush Serpentine Inlet, Trapezoidal Body, Single Vertical Tail Shielded by Body from Forward Radar, Canted Nozzle, Graphite Composite Structure)

Fig. 6.124 Long Range Strike Missiles Use Speed, Altitude, Maneuverability, and RCS for Survivability.

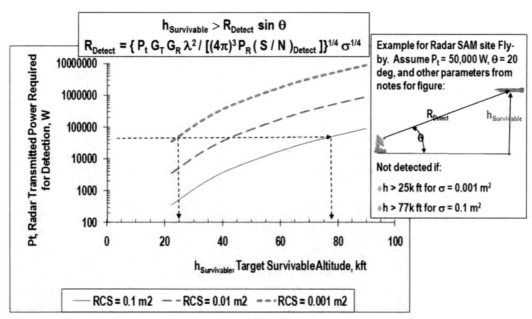

Notes for Figure:
R_{Detect} = Surface threat radar detection range, $(S/N)_{Detect}$ = Signal-to-Noise required for detection = 1, Unobstructed line-of-sight, θ = Surface threat radar elevation angle = 20 deg, G_T = Transmitter gain = 40 dB, G_R = Receiver gain = 40 dB, λ = Wavelength = 0.03 m, P_R = Receiver sensitivity = 10^{-14} W, σ = Target radar cross section

Fig. 6.125 High Altitude Flight and Low Radar Cross Section Enhance Survivability by Reducing Detection Range.

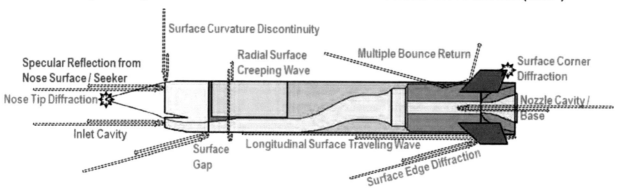

Note:
- Frontal RCS typically affects survivability more than rear RCS
- Specular reflection from the nose / seeker is typically the major contributor to missile frontal RCS
- Inlet cavity is typically the second largest contributor to missile frontal RCS
- Other contributors shown above are typically tertiary contributors to missile frontal RCS
- Nozzle cavity is typically a tertiary contributor to missile frontal RCS, but is typically a major contributor to rear RCS
- Specular reflection from base is typically a tertiary contributor to missile frontal RCS, but is typically a major contributor to rear RCS

Fig. 6.126 There Are Many Contributors to Frontal Radar Cross Section.

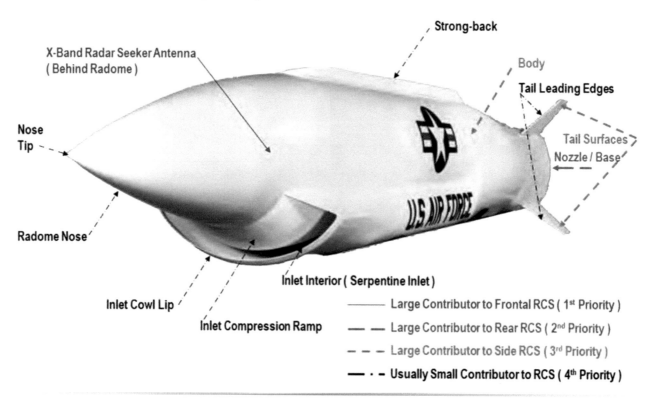

Fig. 6.127 There Are Many Geometry Contributors to Radar Cross Section.

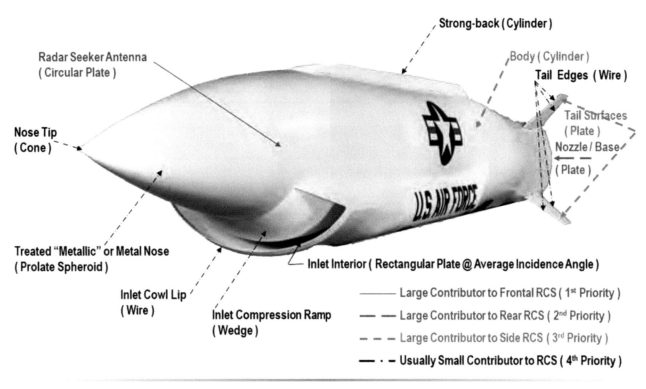

Fig. 6.128 Conceptual Design Radar Cross Section May Be Computed from Simple Shapes Scattering.

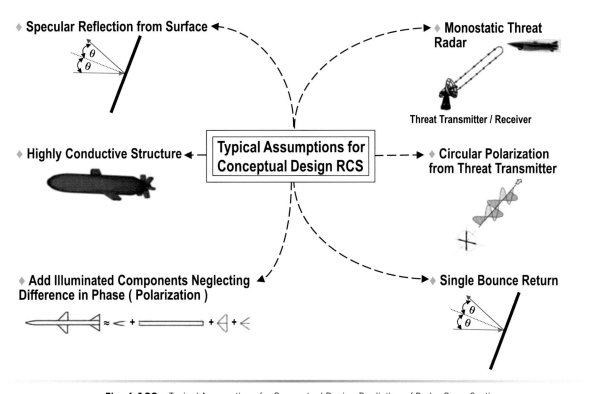

Fig. 6.129 Typical Assumptions for Conceptual Design Prediction of Radar Cross Section.

216 Missile Design Guide

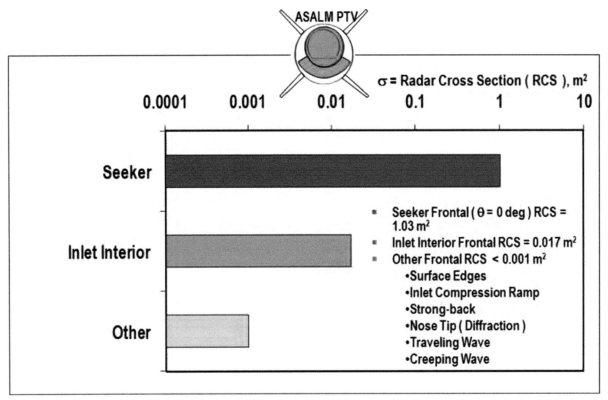

Source: Crispin, J. W., Jr., Goodrich, R. F., and Siegel, K. M., "A Theoretical Method for the Calculation of the Radar Cross Section of Aircraft and Missiles," University of Michigan Report 2591-1-H, July 1959

Fig. 6.130 Pareto of Ramjet Baseline Missile Frontal RCS Shows Most of the Frontal RCS Is from Seeker Antenna.

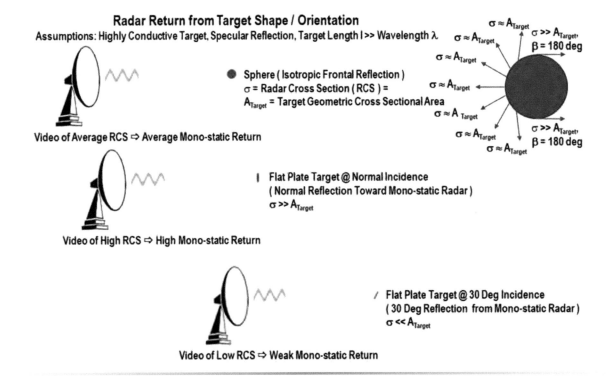

Fig. 6.131 Evaluate RCS Reduction from Shape and Orientation before Considering Radar Absorbing Material.
Note: Video of Average RCS (Arrow symbol) Average Mono-static Return, Video of High RCS (Arrow symbol) High Mono-static Return, Video of Low RCS (Arrow symbol) Weak Mono-static Return.

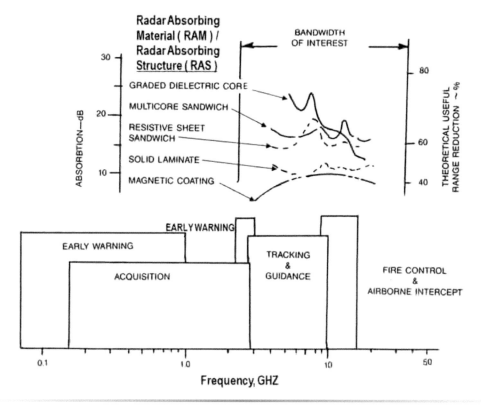

Fig. 6.132 Missile Radar Absorbing Material (RAM) Is Usually Sized for a High Frequency Radar Threat.

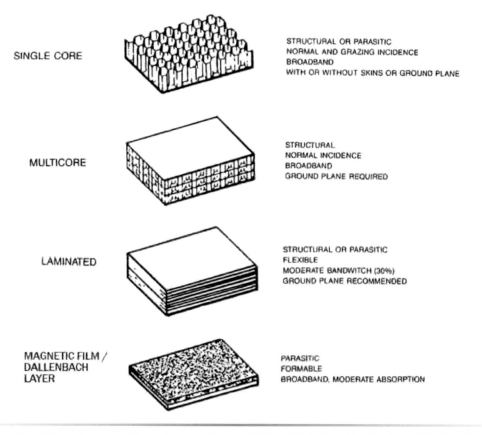

Fig. 6.133 Examples of Radar Absorbing Material (RAM) and Radar Absorbing Structure (RAS) Alternatives.

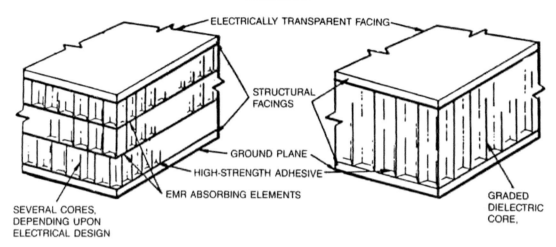

Fig. 6.134 Example of Radar Absorbing Structure (RAS).

- **Dallenbach Magnetic Radar Absorbing Material (RAM) Consists of**
    - Elastomeric Polymer (e.g., Neoprene, Silicone, Urethane, Nitrile – Rubber, Fluro-elastomers)
    - Magnetic Absorber (i.e., Ferrites)
    - Metal Reflective Backing for Reflected Wave λ
- **"Quarter Wave" Dallenbach Magnetic RAM Reduces Radar Cross Section (RCS) ~ 10 db**
    - Incident Wave λ_0 Reflected on Front Face Out of Phase with Reflected Wave λ from Back Face
    - Thickness of Dallenbach RAM $t_{RAM} = (\lambda / 4) \sin \theta$
    - Example for Incidence Angle $\theta = 90$ deg:

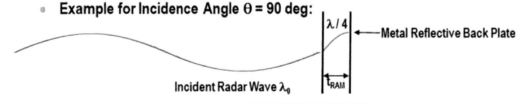

Fig. 6.135 Dallenbach Magnetic RAM Is Popular for Tactical Missiles Because It Has Smaller Thickness.

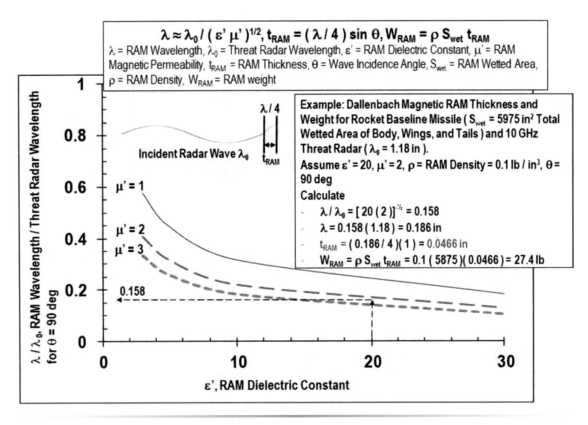

Fig. 6.136 Dallenbach Radar Absorbing Material Thickness Is Driven by Radar Frequency, Dielectric Constant, and Permeability.

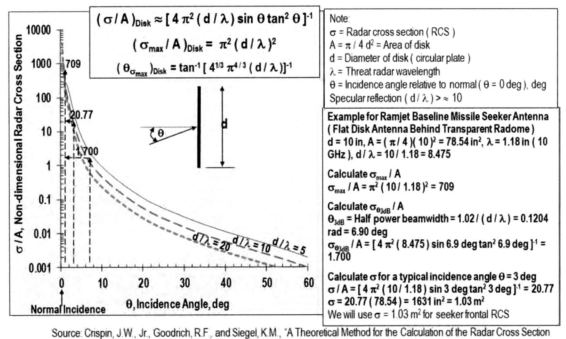

Fig. 6.137 Flat Disk Seeker Antenna Has Large RCS at Near Normal Incidence and Low RCS at Off Boresight.

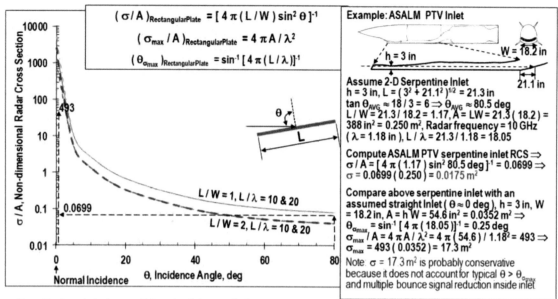

Fig. 6.138 A Rectangular Flat Surface Has Large RCS at a Normal Incidence Angle and Low RCS at an Inclination Angle.

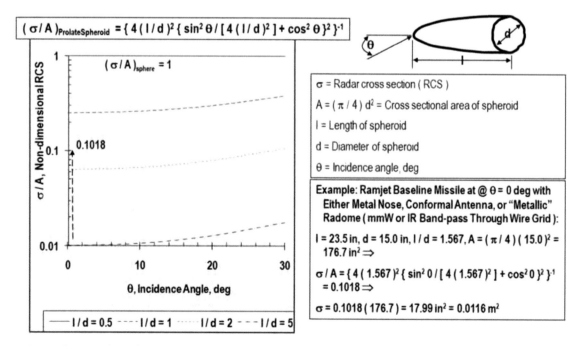

Fig. 6.139 A High Fineness Metallic Nose or a High Fineness Conformal Antenna Has a Low Frontal RCS.

- **Ramjet Baseline Missile Frontal RCS (Untreated Radome, X-Band Radar Seeker)** $\Rightarrow$

$$\sigma \approx \sigma_{Seeker} + \sigma_{Inlet} = 1.03 + 0.0175 = 1.05\ m^2 = 0.2\ dB$$

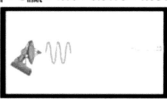

Animation of Signal Return for Monostatic Threat Radar Receiver and Baseline RCS Missile

- **Ramjet Baseline Missile with Reduced Frontal RCS** $\Rightarrow$

$$\sigma \approx \sigma_{Nose} + \sigma_{Inlet} = 0.0116 + 0.0175 = 0.03\ m^2 = -15\ dB$$

Animation of Signal Return for Monostatic Threat Radar Receiver and Reduced RCS Missile

Note

Ramjet baseline missile frontal geometric area A = 2.264 ft² = 0.210 m²
Above values of σ_{Seeker} = 1.03 m², σ_{Inlet} = 0.0175 m², and σ_{Nose} = 0.0116 m² obtained from previous figures
Approaches to reduce missile frontal RCS by reflecting radar return away from monostatic threat receiver include

- High fineness metal nose. Missile has no seeker (uses inertial navigation, Global Positioning System, data link guidance)
- High fineness conformal metal antenna. Missile has phased array radar seeker.
- Metal wire grid over high fineness radome with spacing to reflect typical cmW radar threat. Missile radome has band-pass for a short wavelength seeker (e.g., mmW or IR)

Fig. 6.140 Example of Radar Cross Section Reduction for Ramjet Baseline Missile.
Note: Animation of Signal Return for Monostatic Threat Radar Receiver and Baseline RCS Missile, Animation of Signal Return for Monostatic Threat Radar Receiver and Reduced RCS Missile.

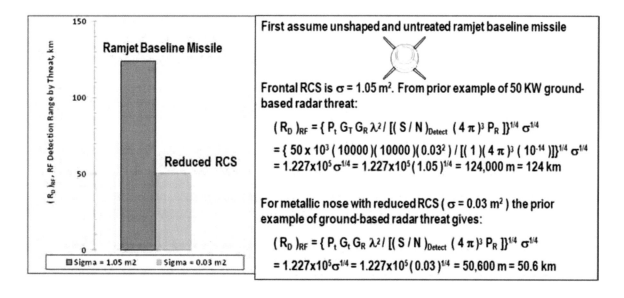

First assume unshaped and untreated ramjet baseline missile

Frontal RCS is $\sigma = 1.05\ m^2$. From prior example of 50 KW ground-based radar threat:

$(R_D)_{RF} = \{ P_t\ G_T\ G_R\ \lambda^2 / [(S/N)_{Detect}\ (4\pi)^3\ P_R] \}^{1/4}\ \sigma^{1/4}$

$= \{ 50 \times 10^3\ (10000)(10000)(0.03^2) / [(1)(4\pi)^3 (10^{-14})] \}^{1/4}\ \sigma^{1/4}$

$= 1.227 \times 10^5\ \sigma^{1/4} = 1.227 \times 10^5 (1.05)^{1/4} = 124{,}000\ m = 124\ km$

For metallic nose with reduced RCS ($\sigma = 0.03\ m^2$) the prior example of ground-based radar threat gives:

$(R_D)_{RF} = \{ P_t\ G_t\ G_R\ \lambda^2 / [(S/N)_{Detect}\ (4\pi)^3\ P_R] \}^{1/4}\ \sigma^{1/4}$

$= 1.227 \times 10^5\ \sigma^{1/4} = 1.227 \times 10^5 (0.03)^{1/4} = 50{,}600\ m = 50.6\ km$

Symbols: $(R_D)_{RF}$ = Detection range for radar threat, P_t = Transmitter power = 50×10^3 W, G_T = Transmitter gain = 40 dB = 10000, G_R = Receiver gain = 40 dB = 10000, λ = Wavelength = 0.03 m, P_R = Receiver sensitivity = 10^{-14} W, σ = Radar cross section (RCS), $(S/N)_{Detect}$ = Signal / noise required for detection = 1

Fig. 6.141 Reduced RCS Reduces Radar Detection Range.

Example: Contributors to Infrared (IR) Signature for Ramjet Baseline Missile / ASALM PTV

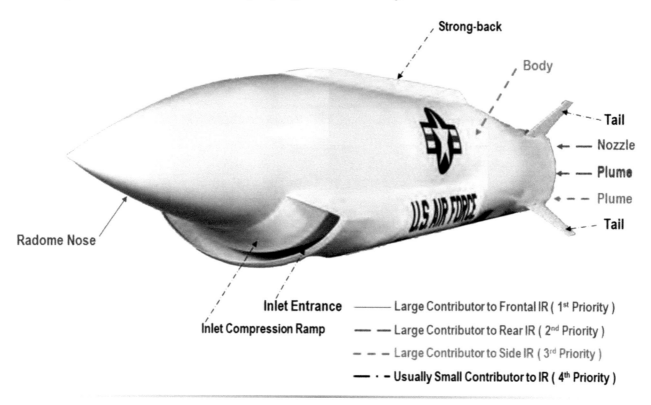

Fig. 6.142 There Are Many Contributors to Infrared Signature.

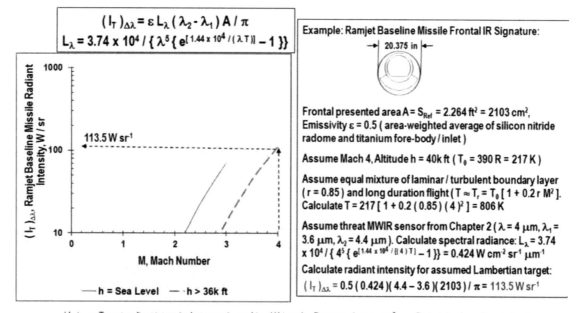

Fig. 6.143 A High-Speed Missile Has High Infrared Signature.

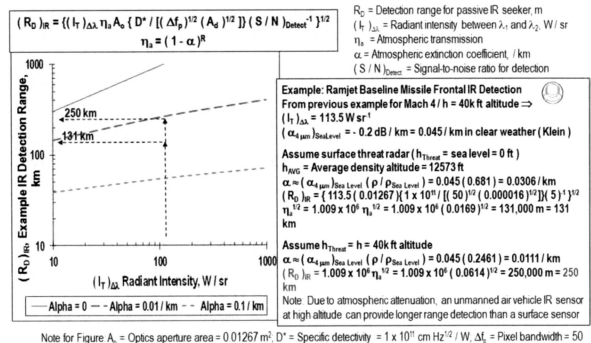

Fig. 6.144 Infrared (IR) Detection Range Is Reduced by Atmospheric Attenuation, Especially at Low Altitude.

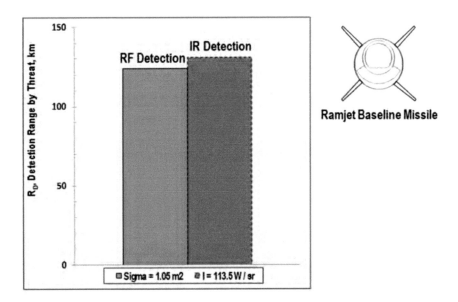

Note:
For ramjet baseline missile, frontal radar cross section (RCS) is $\sigma = 1.05\ m^2$. From prior example, with threat radar frequency (RF) sensor at sea level, $(R_D)_{RF} = 124$ km

From prior example, frontal radiant intensity of ramjet baseline missile at Mach 4, altitude h = 40 kft $\Rightarrow (I_T)_{\Delta\lambda} = 113.5$ W / sr and threat medium wave infrared (MWIR) sensor at h = sea level (0 ft) altitude with clear weather, which provides infrared (IR) detection range $(R_D)_{IR} = 131$ km

Fig. 6.145 IR Detection Range of Mach 4 Ramjet Baseline Missile Can Be Larger Than RCS Detection Range.

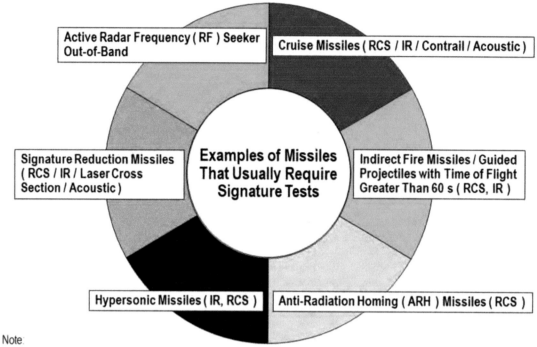

Fig. 6.146a Signature Tests Requirements Are Driven by Effectiveness and Survivability Requirements.

- ## Radar Cross Section (RCS) Ground Tests
    1. Sub-scale Metallic Model in Compact Range
    2. Full Scale Dielectric Missile in Large Anechoic Chamber
    3. Full Scale Dielectric Missile in Outdoor Range

RCS Test of MQM8-400 Vandal in US Navy Point Mugu Anechoic Chamber

- ## Infrared (IR) Ground Tests
    1. Low Temperature Wind Tunnel (Temperature Sensitive Paint with Thermographic Image)
    2. High Temperature Wind Tunnel (Spectrometer / Radiometer / Imagery)
    3. Propulsion Firing (Spectrometer / Radiometer)

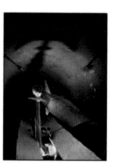

Thermal Image of Scramjet Missile in AEDC Hypervelocity Wind Tunnel Number 9

Fig. 6.146b Signature Tests Requirements Are Driven by Effectiveness and Survivability Requirements (cont).

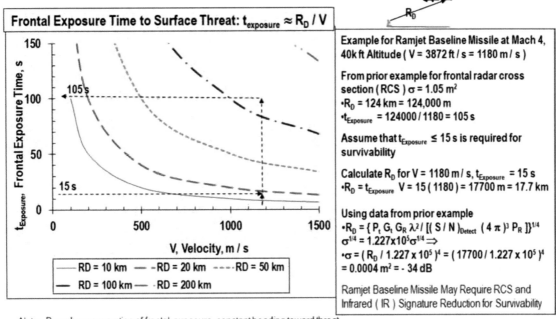

Fig. 6.147 Short Detection Range and High Speed Reduce Threat Exposure Time, Providing Enhanced Survivability.

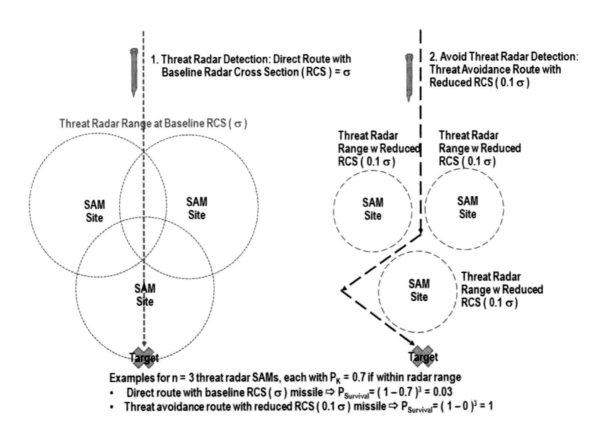

Fig. 6.148 Mission Planning with Threat Avoidance and Reduced Observables Provide Enhanced Survivability.

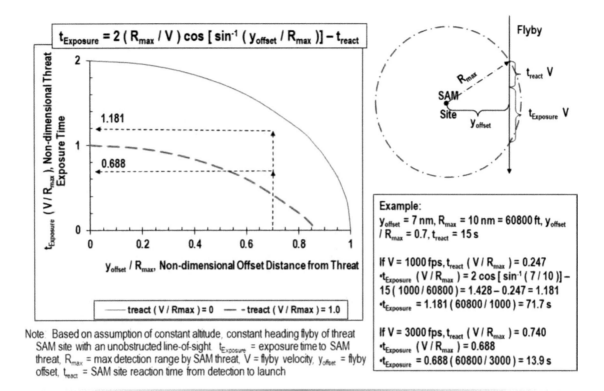

Fig. 6.149 Mission Planning with Threat Avoidance and High Speed Provide Enhanced Survivability.

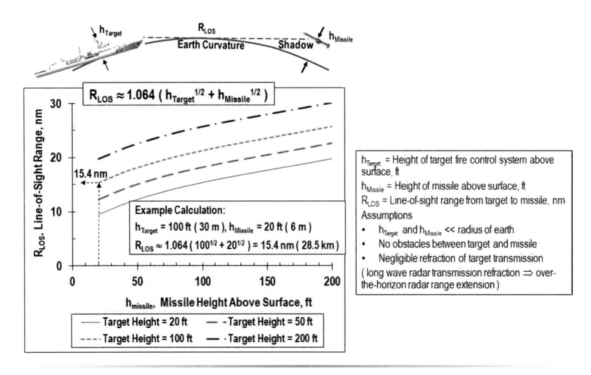

Fig. 6.150 Low Altitude Flight "Under the Radar" Reduces Detection Range.
Note: See video supplement, Exocet Versus Sheffield, Courtesy of Military.com.

- **Critical Subsystems**
  - Propellant / Fuel
  - Warhead .... W/H 
- **Severity Concerns Ranking of Power Output - Type**
  1. Detonation (~ 2 x 10^{-6} s rise time)
  2. Partial detonation (~ 10^{-4} s rise time)
  3. Explosion (~ 10^{-3} s rise time)
  4. Deflagration or propulsion rise time (~ 10^{-1} s rise time)
  5. Burning (> 1 s time) preferred if power released
- **Design and Test Conditions (MIL STD 2105C)***
  - Fragment / bullet / shaped charge impact or blast
  - Sympathetic detonation of adjacent weapons
  - Fast / slow cook-off fire release of energy by burning
  - Drop shock
  - Temperature extremes
  - Vibration during carriage
  - Carrier landing (e.g., 18 ft / s sink, 12 g impact)

* Note: NATO IM Requirements Discussed in STANAG-4439

Fig. 6.151 Insensitive Munitions Improve Launch Platform Safety and Survivability. *Note:* See video supplement, Forrestal Aircraft Carrier Fire (Year 1967).

6.7 Reliability

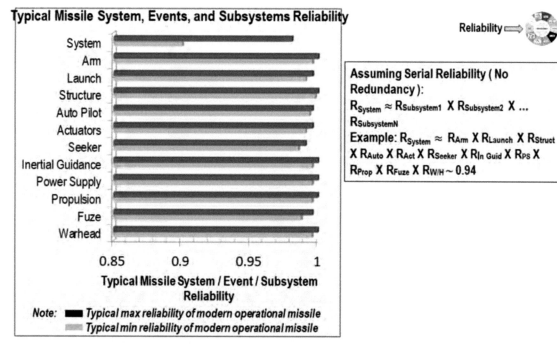

Assuming Serial Reliability (No Redundancy):

$R_{System} \approx R_{Subsystem1} \times R_{Subsystem2} \times ... R_{SubsystemN}$

Example: $R_{System} \approx R_{Arm} \times R_{Launch} \times R_{Struct} \times R_{Auto} \times R_{Act} \times R_{Seeker} \times R_{In\,Guid} \times R_{PS} \times R_{Prop} \times R_{Fuze} \times R_{W/H} \sim 0.94$

Note: Subsystems with moving parts (e.g., gimballed seeker) are typically less reliable than subsystems with non-moving parts (e.g., strapdown seeker)

Fig. 6.152 Reliability Is Provided by Few Events, Reliable Parts, Few Parts, Short Flight Time, and Benign Flight..

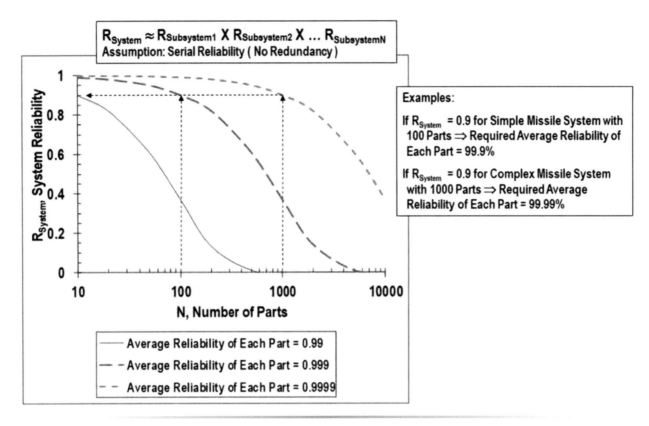

Fig. 6.153 Few Parts and Reliable Parts Provide Higher System Reliability.

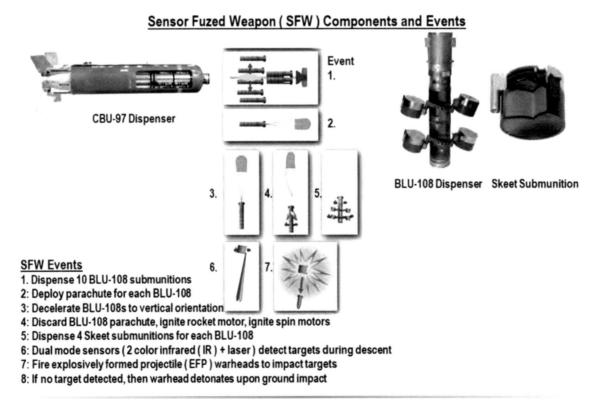

Fig. 6.154 Example of a Multi-Event Weapon That Requires High Event Reliability for Good System Reliability.
Note: See video supplement, Sensor Fuzed Weapon (SFW).

Infant Mortality	Hardware	Software
Design / Modeling Error	X	X
Manufacturing / Coding Error	X	X
Environment Overload		
Storage	X	
Transportation	X	
Launch Platform / Fire Control Integration	X	X
Flight	X	X
Reliability		
Random Wear-out vs Time	X	

Note:
- For constant failure rate, reliability follows an exponential distribution $R = e^{-\lambda t}$
- A short time of flight missile (e.g., Hellfire) typically has higher reliability than a long time of flight missile (e.g., Harpoon)
- Software failures are driven by number of lines of code / complexity
- Software failures often occur after missile deployment

Fig. 6.155 Missile Failures Can Be Characterized by Infant Mortality, Environment Overload, and Reliability.

- **Trident D-5 Infant Mortality Failure**
    - Load on Thrust Vector Control (TVC) from Launch Water Plume Exceeded TVC Design Strength Capability
        - Unanticipated load not considered in TVC design, ground test, and ground launch
    - First Submarine Launch Flight Failure March 21, 1989

Trident D-5 Flight Failure

- **After Mesh Cover Fix, Follow-on Flights Demonstrated High Reliability**
    - E.G., 184 Consecutive Successful Flights (1989 to Sep 2021)

Trident Submarine

Trident D-5 Launch

Fig. 6.156 Example of Infant Mortality Failure Design Error—Trident II D-5 TVC Failure from Water Plume Load. *Note*: See video supplement, First Under Water Launch Trident D-5; Trident D-5 Commander Evaluation Test (CET).

- **Missile Warhead Safe and Arm**
  - Operates Only Under Normal Flight Conditions (e.g., Acceleration, Time Duration of Acceleration)
  - Prevents Accidental Operation of Warhead
- **Traditional Mechanical Safe and Arm for Warhead**
  - Mechanisms Initially Out of Alignment and Align Explosive Train Only Under Normal Flight Conditions
  - Multiple Moving Parts $\Rightarrow$ Reduced Reliability
- **Electronic Safe and Arm for Warhead**
  - High Capacitor Voltage to Explosive Foil Initiator Provided Only Under Normal Flight Measurement
  - No Moving Parts (e.g., MEMS Accelerometers) and Fewer Parts $\Rightarrow$ Higher Reliability

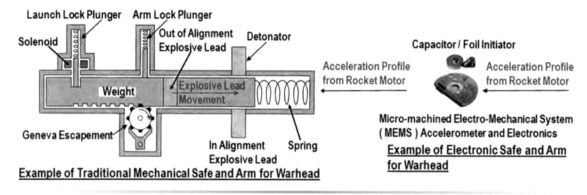

Fig. 6.157 An Example of Higher Reliability and Safety Technology Is Electronic Safe and Arm.

6.8 Cost

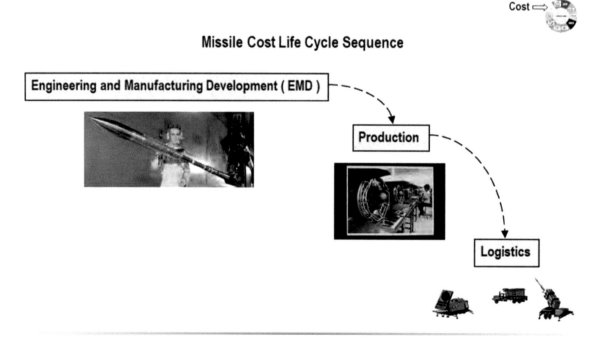

Fig. 6.158 Missile Cost Has a Life Cycle.

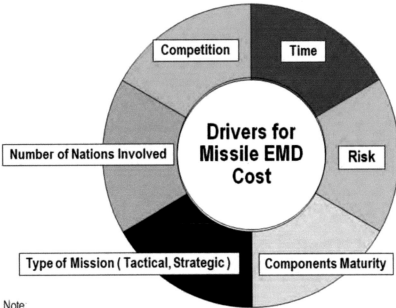

Note:
- EMD duration is typically most important driver for EMD cost (e.g., EMD cost ~ time^2)
- EMD required schedule duration depends upon risk (low risk EMD typically < 4 years, high risk EMD typically > 6 years)
- Missiles with mature components typically require less development cost / duration
- Strategic missile EMD is typically more expensive and typically requires longer duration.
- Multi-national EMD is usually more expensive and requires longer duration.
- Competition / allowable profit usually reduces EMD cost and risk.

Fig. 6.159 Missile Engineering and Manufacturing Development (EMD) Cost Is Driven by Many Factors.

Note:
- Missile weight is typically most important driver for production cost (e.g., production cost ~ Weight$^{0.5}$).
- A guided bomb or guided projectile typically costs less than tactical missile. Strategic missile typically costs more than tactical missile
- Sensors and electronics typically cost more than other components
- Airbreathing propulsion typically costs more than rocket propulsion
- Relatively immature technology components usually cost more than mature technology components.
- Complex missile with high parts count usually costs more than a simple missile with fewer parts.
- A missile with a low production rate typically costs more than a missile with a high production rate.
- Missile produced with hard tooling (e.g., high production rate) usually costs less than a missile using soft tooling (e.g., prototype)
- Single contractor (monopoly) typically is higher cost than competing contractors
- Optimum ratio of make vs buy is typically about 50%-50%. Missile system primes typically make sensors and electronics.
- Government Furnished Property (GFP) / Government Furnished Equipment (GFE) usually reduces production cost

Fig. 6.160 Missile Production Cost Is Driven by Many Factors.

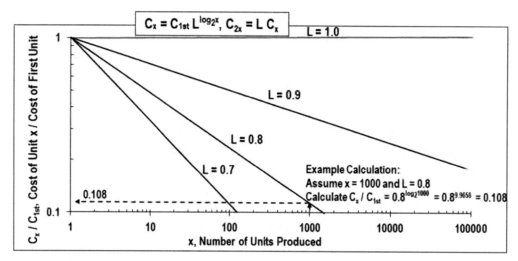

Note: Typically, there is labor intensive learning curve if $L < 0.8$ and machine intensive learning curve if $L > 0.8$
Contributors to the learning curve include:
- Labor force motivation and efficiency
- Tooling, manufacturing process, and production rate
- Components maturity and complexity

First missile may be a developmental missile that is labor intensive and produced in the laboratory using soft tooling. The 1000th missile is probably a rate production missile produced in the factory using hard tooling.
Missile production processes are typically changed about every 2-to-5 years, to incorporate new technologies and new subsystems. For a changed production process, the initial unit production cost is typically higher.

Fig. 6.161 Learning Curve and Total Production Are Drivers for Reducing Missile Unit Production Cost.

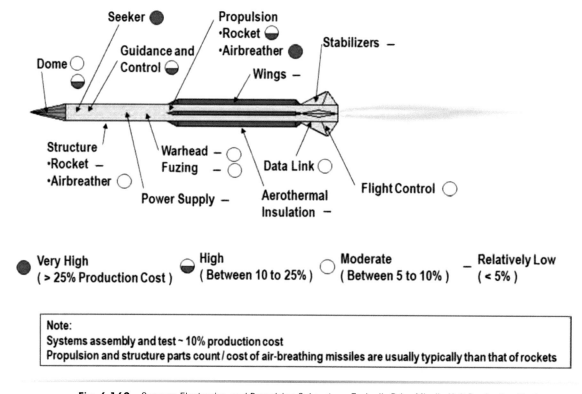

Note:
Systems assembly and test ~ 10% production cost
Propulsion and structure parts count / cost of air-breathing missiles are usually typically than that of rockets

Fig. 6.162 Sensors, Electronics, and Propulsion Subsystems Typically Drive Missile Unit Production Cost.

CHAPTER 6 Other Measures of Merit 233

Example of Relatively Expensive Missile with Many Sensors: Derby 🇮🇱 / R-Darter 🇿🇦

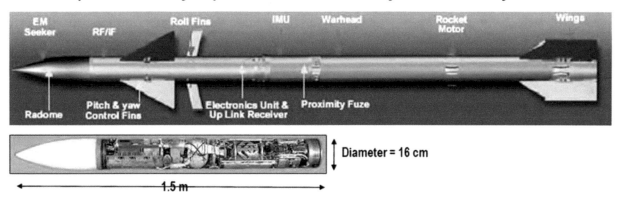

Example of Relatively Inexpensive Missile with Few Sensors: RBS-70/90 Bolide 🇸🇪

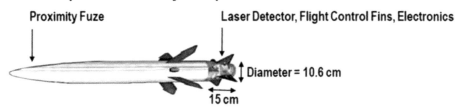

Fig. 6.163 Missiles with a Small Weight Fraction of Sensors and Electronics Are Usually Less Expensive.

Fig. 6.164 Missile Production Culture Is Driven by Rate Production of Sensors and Electronics.
Note: See video supplement, Hellfire Seeker/Electronics Production (Year 1990).

234 Missile Design Guide

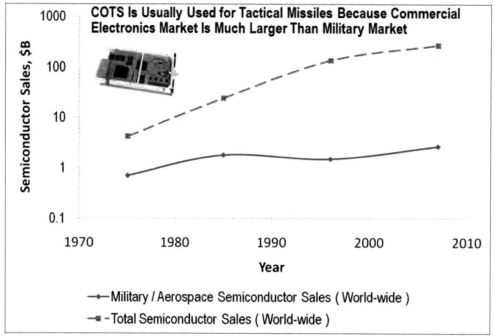

Note: COTS electronics usually require additional protection (e.g., cocooning) against harsher military environment (temperature, vibration, shock, acoustics, moisture, dust, salt, electromagnetics). COTS electronics are frequently updated (e.g., every 18 months)
Source: The McClean Report 2007, *A Complete Analysis and Forecast of the Integrated Circuit Industry*

Fig. 6.165 Missile Electronics Are Usually Based Upon Commercial-Off-The-Shelf (COTS) Electronics.

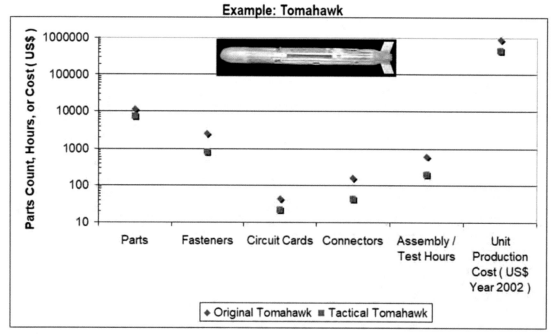

Note: Tactical Tomahawk has superior flexibility (e.g., shorter mission planning, in-flight retargeting, battle damage indication (BDI) / battle damage assessment (BDA), modular payload) at ~ 50% parts count / cost and higher reliability. Enabling technologies for low parts count include casting, pultrusion / extrusion, centralized electronics, and commercial-off-the-shelf (COTS) electronics

Fig. 6.166 A Low Parts Count Missile Has a Lower Unit Production Cost.

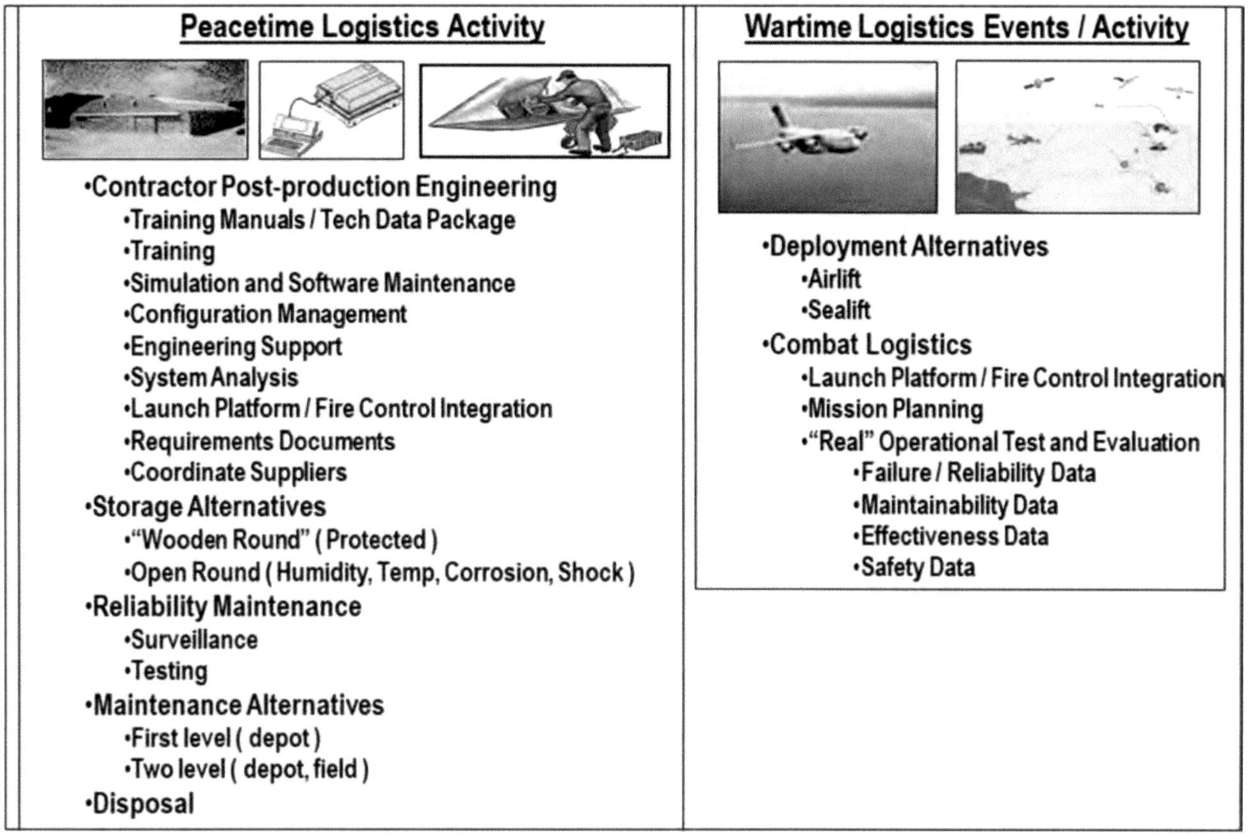

Peacetime Logistics Activity
•Contractor Post-production Engineering
•Training Manuals / Tech Data Package
•Training
•Simulation and Software Maintenance
•Configuration Management
•Engineering Support
•System Analysis
•Launch Platform / Fire Control Integration
•Requirements Documents
•Coordinate Suppliers
•Storage Alternatives
•"Wooden Round" (Protected)
•Open Round (Humidity, Temp, Corrosion, Shock)
•Reliability Maintenance
•Surveillance
•Testing
•Maintenance Alternatives
•First level (depot)
•Two level (depot, field)
•Disposal
Wartime Logistics Events / Activity
•Deployment Alternatives
•Airlift
•Sealift
•Combat Logistics
•Launch Platform / Fire Control Integration
•Mission Planning
•"Real" Operational Test and Evaluation
•Failure / Reliability Data
•Maintainability Data
•Effectiveness Data
•Safety Data

Fig. 6.167 Missile Logistics Emphasis Is Different in War and Peace.

"Wooden Round" Container
(Sealed, Insulated, Shock Isolated)
Port Access
(Desiccant, Logger)
Plug
Logger
Missile Cradle
Air Conditioned Bunker
(Temperature, Humidity)
Air Conditioner
Door

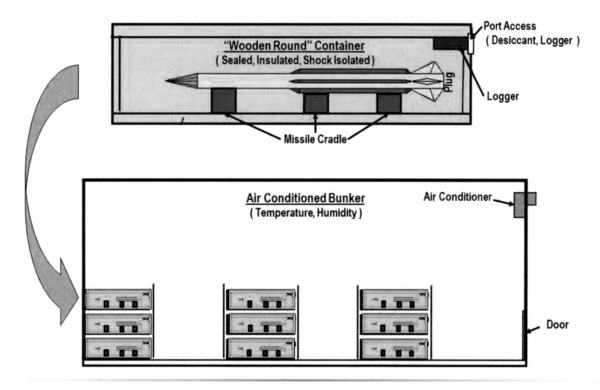

Fig. 6.168 Missile Storage in a Protected Container Reduces Aging.

Fig. 6.169 Logistics Cost Is Lower for a Simple Missile System.
Note: See video supplement, SAM Logistics Alternatives.

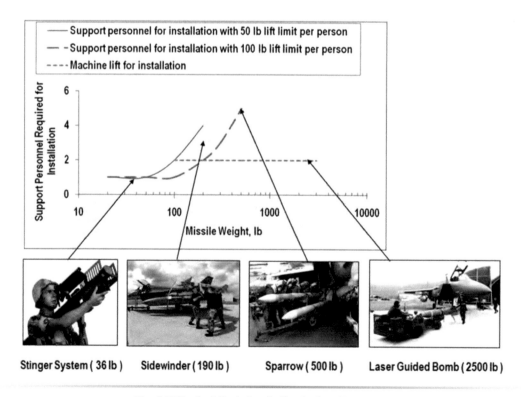

Fig. 6.170 Logistics Is Usually Simpler for a Lightweight Missile.
Note: See video supplement, Simple Logistics of Predator Lightweight Missile (21 lb).

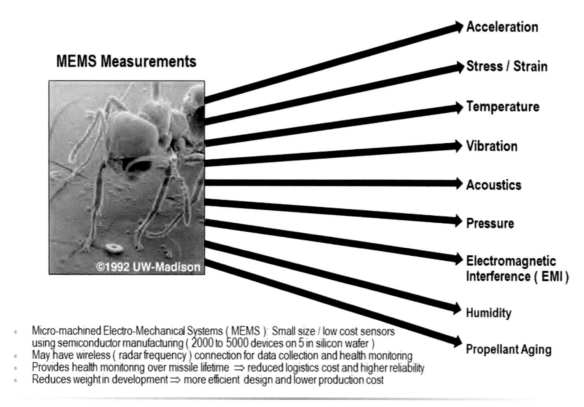

- Micro-machined Electro-Mechanical Systems (MEMS): Small size / low cost sensors using semiconductor manufacturing (2000 to 5000 devices on 5 in silicon wafer)
- May have wireless (radio frequency) connection for data collection and health monitoring
- Provides health monitoring over missile lifetime ⇒ reduced logistics cost and higher reliability
- Reduces weight in development ⇒ more efficient design and lower production cost

Fig. 6.171 Small MEMS Sensors Provide Logistics Health Monitoring, Reduce Cost, and Reduce Weight.

6.9 Launch Platform/Fire Control Integration

Launch Platform Integration / Fire Control Integration

US Launch Platform	Example Launcher	Carriage Span / Shape	Length	Weight
Surface Ships	Vertical Launch System (VLS)	22" / 22"	263"	3400 lb
Submarines	Canister Launch System (CLS) Torpedo / Capsule	22" Tomahawk, 83" Trident D-5, 13.5" Harpoon	263" / 534" / 186"	3400 lb / 130,000 lb / 1525 lb
Fighters / Bombers / Large Unmanned Combat Air Vehicles (UCAVs)	Rail / Ejection	~24" x ~24"	~168"	~500 lb to ~3000 lb
Ground Vehicles for Surface-to-Surface (STS) Missiles	Launch Pods	28" x 28" ATACMS / 28" x 28" MLRS	158" / 156"	3690 lb / 677 lb
Helicopter / Small UCAVs	Rail / Ejection	~13" x ~13"	~70"	~120 lb
Tanks / Artillery / Ship Guns	Gun Barrel	120 mm Gun Barrel	~40"	~60 lb

Fig. 6.172 Missile Carriage Size, Shape, and Weight Limits Are Often Driven by Launch Platform Compatibility.

238 Missile Design Guide

Fig. 6.173 Lightweight Missiles Provide Enhanced Firepower.

Brimstone Launcher

Brimstone / F-18 Carriage

Brimstone / Tornado Carriage

Note: Brimstone weight = 107 lb (48.5 kg), length = 71 in (1.8 m), and diameter = 7 in (0.178 m)

Fig. 6.174 Example of Lightweight, Small Size, and High Firepower Missile—Precision Strike Brimstone.
Note: See video supplement, Brimstone Firepower.

Note: Specifications applicable to EMC include MIL-STD-461G, 464C, 469A, 1512, 1857, 1795A, and MIL-A-17161D

Fig. 6.175 Missile Electromagnetic Compatibility (EMC) Is a Requirement for Launch Platform Integration.

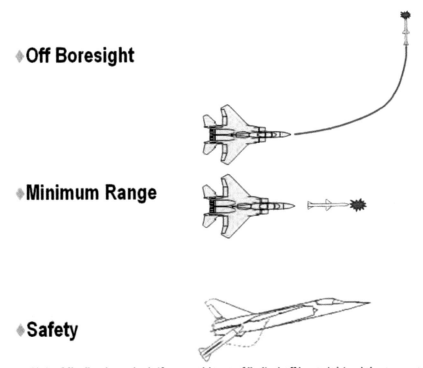

◆ **Off Boresight**

◆ **Minimum Range**

◆ **Safety**

Note: Missile - launch platform problems of limited off boresight, minimum range, and safety are applicable to all launch platforms (fixed wing aircraft, helicopter, ship, ground vehicles)

Fig. 6.176 Missile—Launch Platform Problems Include Limited Off Boresight, Minimum Range, and Safety.

Launch platform may obscure seeker field-of-regard ⇒ limited off-boresight for lock-on-before launch seekers

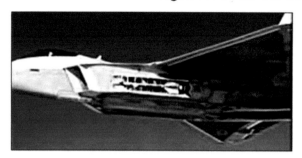

F-22 Side Bay with AIM-9X
Has Limited Off-Boresight

Apache Fire Control System and
Weapons Off-Boresight Not Obscured by
Launch Platform

Fig. 6.177a Examples of Missile—Launch Platform Integration Problems.
Note: See video supplement, F-16 Wing Tip Carriage of IRIS-T and Apache Fire Control System.

Lug / hanger clearance on rail ⇒ tip-off rotation @ launch ⇒ miss at minimum range

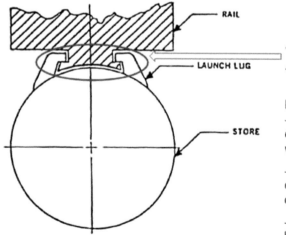

Missile Launch Lug / Rail Hangar Clearance

Typical Allowable Lug / Hanger Rail Clearance
- Pitch: +/- 0.04 (binding) to 0.08 deg (high tip-off)
- Yaw: +/- 0.02 (binding) to 0.05 deg (high tip-off)

Note:
- Most rail launchers have a snubber mounted to the back of the rail. Padded arms close on the missile to provide vibration damping and reduce shock.

- A Monte-Carlo probabilistic simulation is usually used to determine the effect of launch tip-off on missile miss distance / minimum range.

- A design tradeoff is two versus three lugs / hangers. Two lugs / hangers have less drag, but three lugs / hangers may be required to reduce tip-off error when the missile leaves the rail.

Source: MIL-STD-8591

Fig. 6.177b Examples of Missile—Launch Platform Integration Problems (cont).

Launcher aeroelasticity ⇒ tip-off @ launch ⇒ miss at minimum range

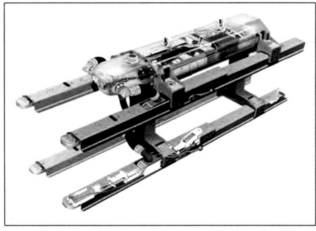

M299 Hellfire Lightweight (145 lb) Launcher

Fig. 6.177c Examples of Missile—Launch Platform Integration Problems (cont).
Note: See video supplement, Hellfire and Sidewinder Rail Launcher Elasticity.

Crosswind / downwash ⇒ tip-off and possible instability of missile leaving launcher ⇒ miss at min range and possible loss of control

ESSM Sea Sparrow Launch in Cross Wind

Helicopter Downwash

Fig. 6.177d Example of Missile—Launch Platform Integration Problems (cont).
Note: See video supplement, Crosswind/Downwash.

High crosswind / downwash ⇒ high angle of attack @ launch

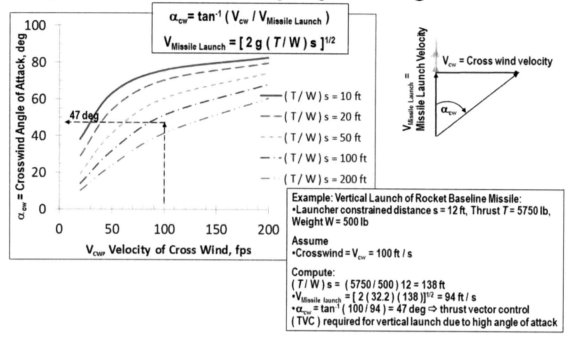

Fig. 6.177e Examples of Missile—Launch Platform Integration Problems (cont).

Launch dynamics ⇒ Delayed Global Positioning System (GPS) acquisition ⇒ Miss @ minimum range

Examples of Contributors to Delayed GPS Acquisition from Launch Dynamics

- Launcher Recoil

- Cross Wind at Launch

- Weapon Induced Dynamics During Launch
    - Surfaces deployment
    - Surfaces misalignment
    - Longitudinal / lateral acceleration
    - Roll rate
    - Thrust misalignment
    - Center-of-gravity misalignment

Fig. 6.177f Examples of Missile—Launch Platform Integration Problems (cont).

Launcher retention / release ⇒ potential inadvertent release / hang-fire

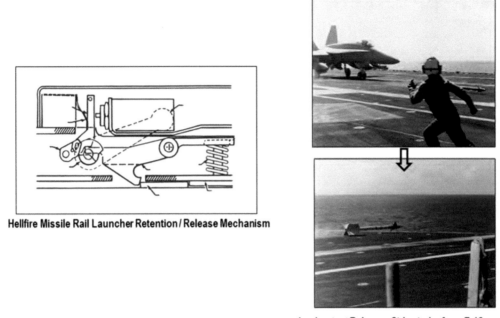

Fig. 6.177g Example of Missile—Launch Platform Integration Problems (cont).
Note: See video supplement, Release; Hang-fire.

Folded surfaces ⇒ potential unreliable deployment

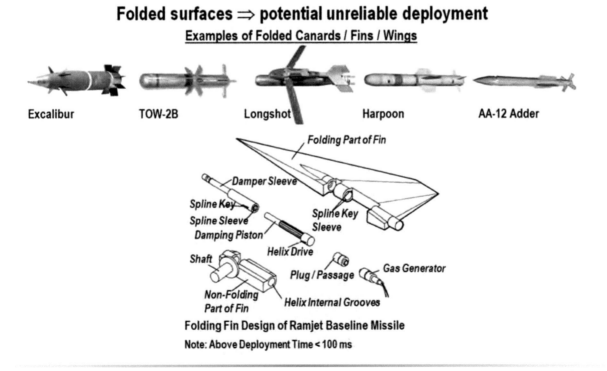

Fig. 6.177h Example of Missile—Launch Platform Integration Problems (cont).
Note: See video supplement, Fins/Wings Deployment: HOT, Longshot, Harpoon, Tomahawk and ALCM.

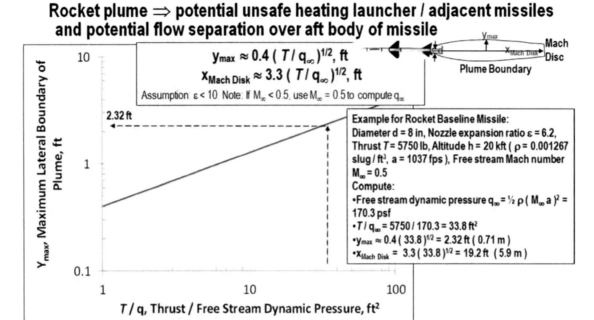

Fig. 6.177i Examples of Missile—Launch Platform Integration Problems (cont).

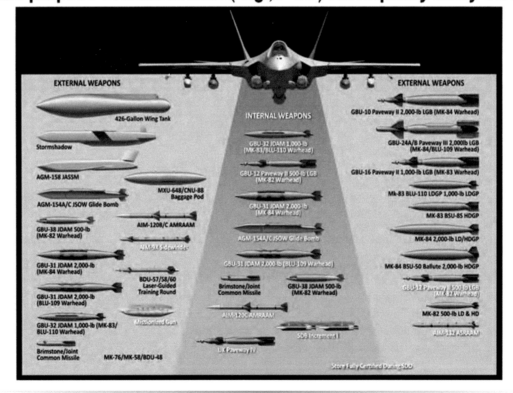

Fig. 6.177j Example of Missile—Launch Platform Integration Problems (cont).

Multi-purpose combat aircraft (e.g., F-18) must qualify many stores

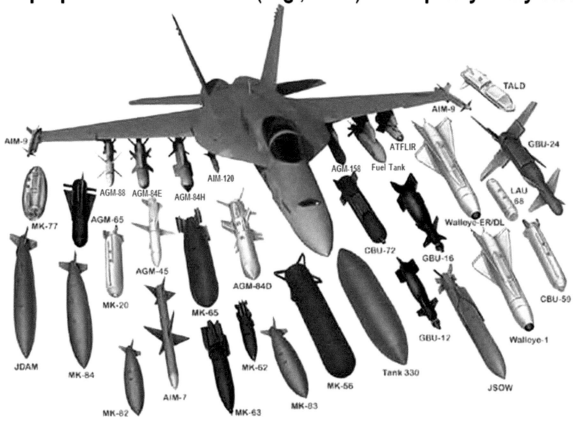

Fig. 6.177k Example Missile—Launch Platform Integration Problems (cont).

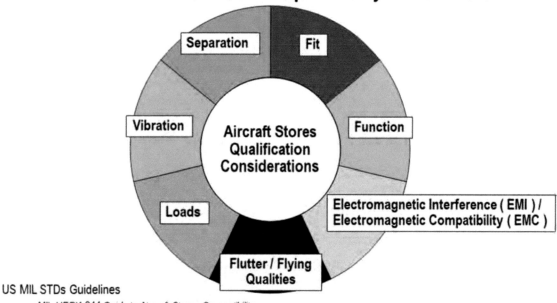

Fig. 6.177l Examples of Missile—Launch Platform Integration Problems (cont).

Aircraft Stores Qualification Is a Long Process

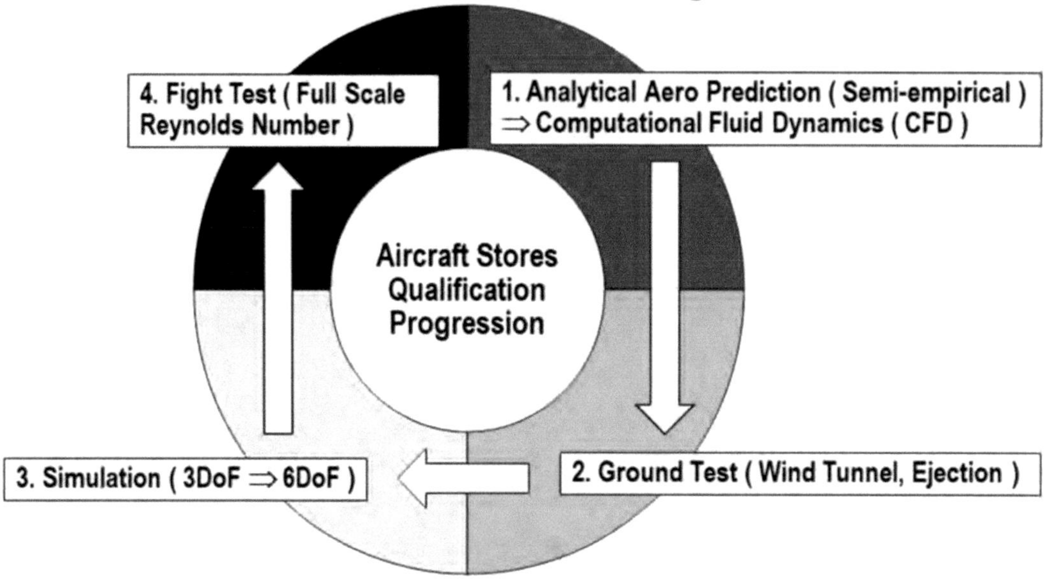

Aircraft Stores Qualification Progression

1. Analytical Aero Prediction (Semi-empirical) ⇒ Computational Fluid Dynamics (CFD)
2. Ground Test (Wind Tunnel, Ejection)
3. Simulation (3DoF ⇒ 6DoF)
4. Fight Test (Full Scale Reynolds Number)

Example of the Current Long Process for Aircraft Stores Qualification
- F-35 First Flight: Year 2006
- F-35 Begins First Store Qualification Flight Tests: Year 2016

Fig. 6.177m Examples of Missile—Launch Platform Integration Problems (cont).

Aircraft launch platform maneuvering / local flow field (angle of attack α, angle of sideslip β) and high dynamic pressure ⇒ potential unsafe separation

F-18 at High Angle of Attack   Unsafe Store Separation

Note:
- Aircraft store separation problems tend to occur for large stores, buoyant stores (e.g., empty fuel tanks), and close proximity / multiple adjacent stores
- Aircraft statically stable stores usually have fixed / locked flight control during separation while statically unstable stores require active flight control stabilization during launch (requires more reliable flight control design)

Fig. 6.177n Examples of Missile—Launch Platform Integration Problems (cont).
Note: See video supplement, Unsafe Store Separation.

Many Drivers to Qualify Aircraft Store for Safe Separation and Typically Have Large Uncertainty

- Safe Store Separation Typically Requires Store Normal Force Less Than Store Weight
    - $N \approx C_{N_\alpha} \alpha' q' S_{Ref} < W \cos \gamma$
- Safe Separation Typically Requires Store Static Stability and Pitch Down Moment
    - $M \approx C_{m_\alpha} \alpha' q' S_{Ref} < 0$
- Contributors to Store Uncertainties for Safe Separation
    - Local Angle of Attack (α')
    - Local Dynamic Pressure (q')
    - Static Stability ($C_{m_\alpha} < 0$)
- Major Contributors to Ejection Launcher Uncertainties for Safe Separation
    - Total Impulse
    - Pitch Down Rate
- Major Contributors to Launch Aircraft Uncertainties for Safe Separation
    - Mach Number
    - Angles of Attack, Sideslip, Roll
    - Flight Path Angle
    - Altitude / Reynolds Number
    - Load Factor
    - Rotation Rates
    - Stores Loading

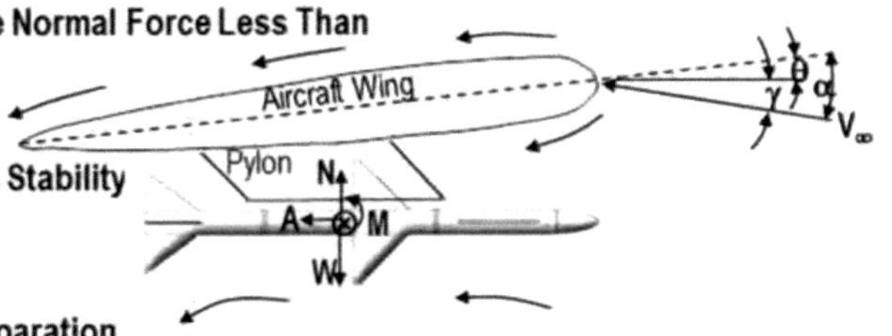

Symbols:
N = Normal Force, A = Axial Force, M = Pitching Moment, W = Weight, θ = Aircraft Pitch Attitude, γ = Aircraft Flight Path Angle, α = Aircraft Angle of Attack, V_∞ = Aircraft Velocity

Reference: MIL-HDBK-1763, Aircraft / Stores Compatibility

Fig. 6.177o Examples of Missile—Launch Platform Integration Problems (cont).

Aircraft launch platform has more aeroelasticity with heavy store ⇒ potential flutter / unsafe handling qualities / wing fatigue / store fatigue

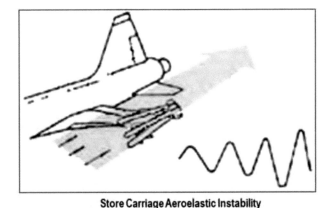

Store Carriage Aeroelastic Instability

Fig. 6.177p Examples of Missile—Launch Platform Integration Problems (cont).
Note: See video supplement, Store Wind Tunnel Flutter, Wind Tunnel Limit Cycle Oscillation, and Flight Test Limit Cycle Oscillation.

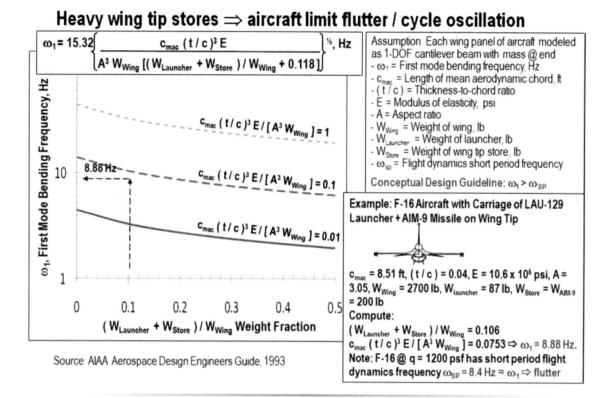

Fig. 6.177q Examples of Missile—Launch Platform Integration Problems (cont).

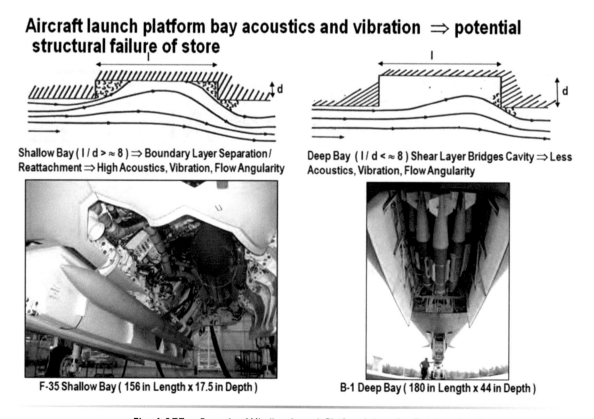

Fig. 6.177r Example of Missile—Launch Platform Integration Problems (cont).

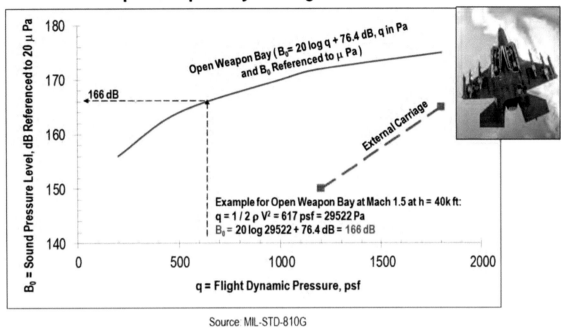

Fig. 6.177s Examples of Missile—Launch Platform Integration Problems (cont).

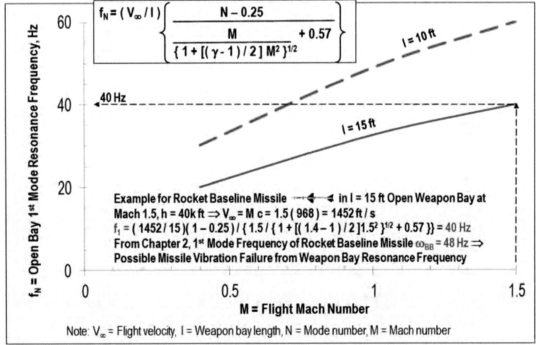

Fig. 6.177t Examples of Missile—Launch Platform Integration Problems (cont).

 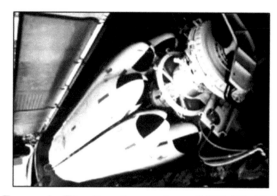

Example Rail Launcher: Hellfire / Brimstone **Example Ejection Launcher: AGM-86 ALCM**

Fig. 6.178 Examples of Aircraft Rail Launched and Ejection Launched Missiles.
Note: See video supplement, Hellfire/Brimstone Carriage/Launch; AGM-86 Carriage/Launch.

Store Weight / Parameter	30 Inch Suspension	14 Inch Suspension
• Weight Up to 100 lb	Not Applicable	Yes
• Lug height (in)	↓	0.75
• Min ejector area (in x in)		4.0 x 26.0
• Weight 101 to 1450 lb	Yes	Yes
• Lug height (in)	1.35	1.00
• Min ejector area (in x in)	4.0 x 36.0	4.0 x 26.0
• Weight 1451 to 3500 lb	Yes	Not Applicable
• Lug height (in)	1.35	↓
• Min ejector area (in x in)	4.0 x 36.0	
• Weight > 3500 lb		
• Sling suspended	Not Applicable	Not Applicable

Ejection Stroke

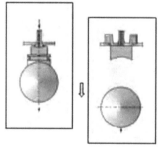

Fig. 6.179 MIL-STD-8591 Imposes Aircraft Store Suspension and Ejection Launcher Requirements.
Note: See video supplement, LAU-142 Ejection Launcher.

AMRAAM Rail Launch from F-16   **Rapid Bomb Drop from B-2**   **CALCM Ejection Launch from B-1**

Fig. 6.180 Examples of Safe Store Separation.
Note: See video supplement, Safe Store Separation.

Ejection Launcher Type	Optimum Pitch Rate for Launch (M, h, α)	Logistics	Ejection Velocity	Example	
Pyrotechnic Cartridges	No ☹	☹	☺	LAU-129 with AMRAAM	BRU-33/A with Mk-82
Pneumatic, Electromech, or Hydraulic	Yes ☺	☺	☺ ☹	LAU-142/A with AMRAAM	BRU-61/A with SDB

Note: Disadvantages of pyrotechnic cartridge launchers include fixed pitch rate and cartridges must be replaced after each launch
Launch conditions variation includes Mach number (M), altitude (h), and angle of attack (α)

Fig. 6.181 Smart Launchers May Replace Pyrotechnic Ejection Launcher for Aircraft with Light-to-Medium Stores.

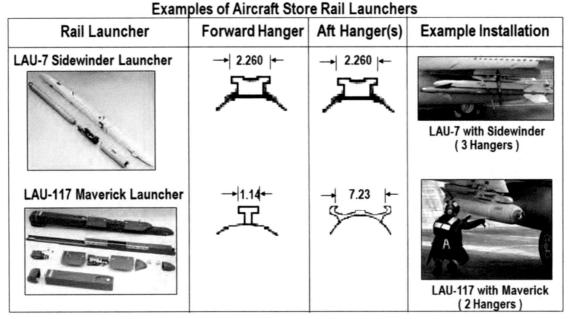

Fig. 6.182 MIL-STD-8591 Addresses Aircraft Store Rail Launchers.

F-18 Store Compatibility Test in AEDC 16T

AV-8 Store CTS Test in AEDC 4T

Types of Wind Tunnel Testing for Store Compatibility
1. Flow field mapping with pitot static pressure probe
2. Flow field mapping with instrumented store model
3. Store captive trajectory simulation (CTS)
4. Drop testing of store model
5. Store carriage loads measurement

Example Stores with Flow Field Interaction: Kh-41 + AA-10

Fig. 6.183 There Are Many Types of Store Separation Wind Tunnel Tests.

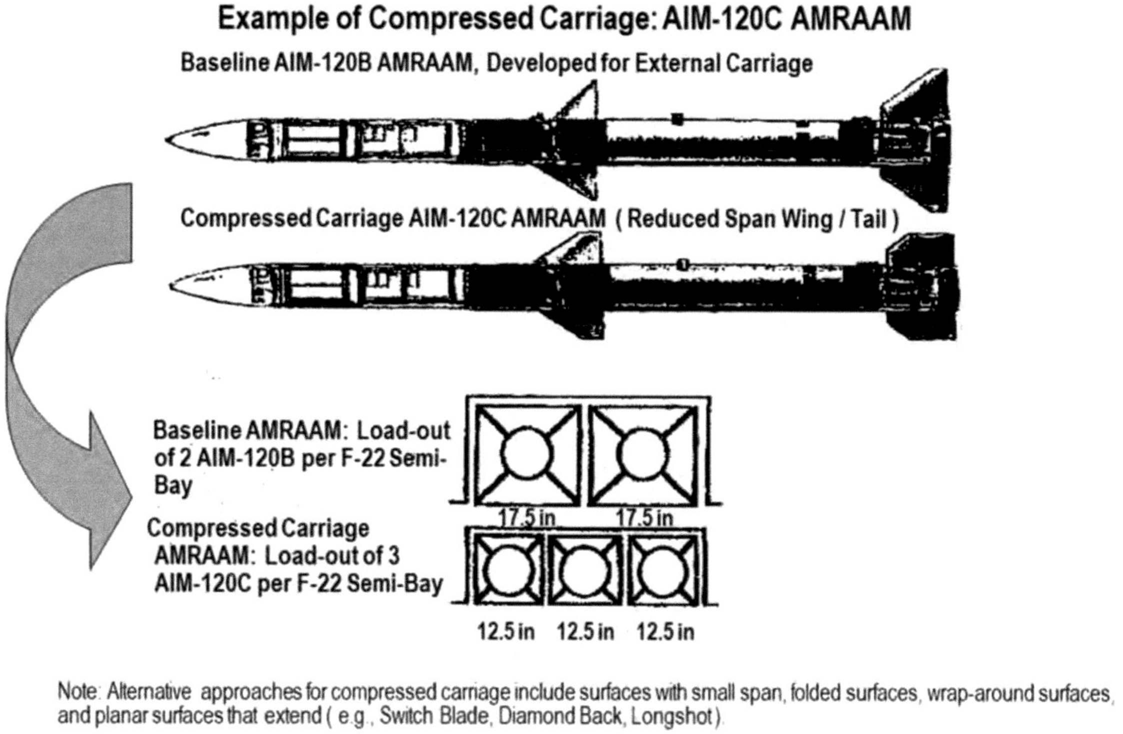

CHAPTER 6 Other Measures of Merit

Example of Compressed Carriage: AIM-120C AMRAAM

Baseline AIM-120B AMRAAM, Developed for External Carriage

Compressed Carriage AIM-120C AMRAAM (Reduced Span Wing / Tail)

Baseline AMRAAM: Load-out of 2 AIM-120B per F-22 Semi-Bay — 17.5 in, 17.5 in

Compressed Carriage AMRAAM: Load-out of 3 AIM-120C per F-22 Semi-Bay — 12.5 in, 12.5 in, 12.5 in

Note: Alternative approaches for compressed carriage include surfaces with small span, folded surfaces, wrap-around surfaces, and planar surfaces that extend (e.g., Switch Blade, Diamond Back, Longshot).

Fig. 6.184 Compressed Carriage Missiles Provide Higher Firepower.

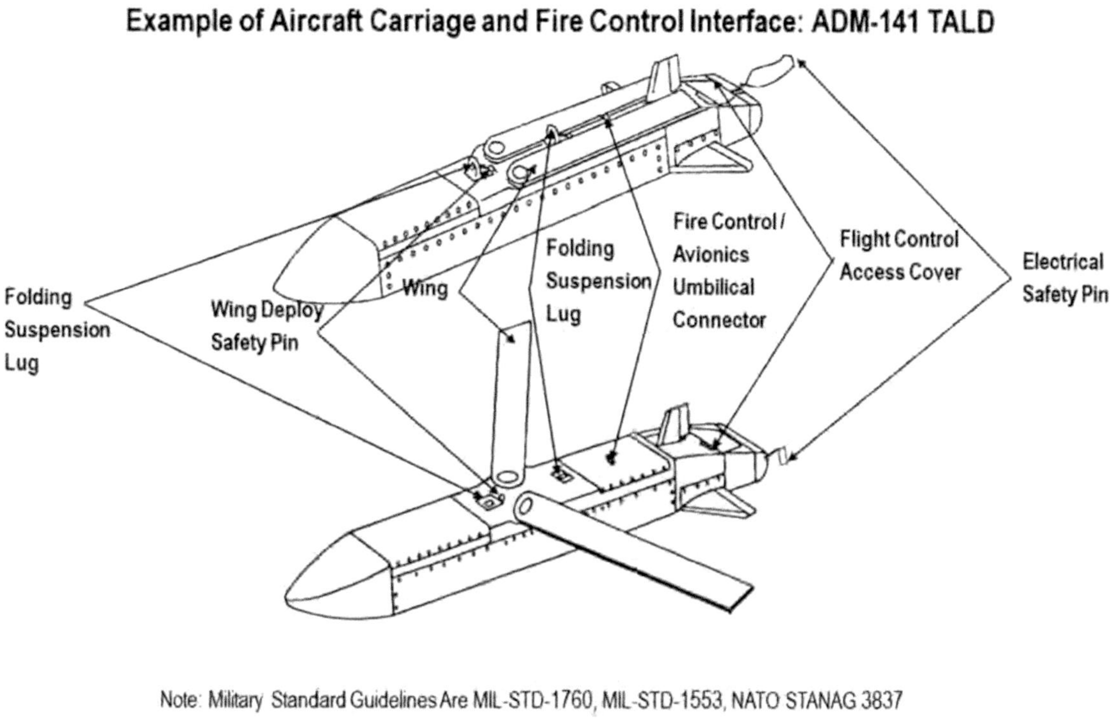

Example of Aircraft Carriage and Fire Control Interface: ADM-141 TALD

Labels: Folding Suspension Lug; Wing Deploy Safety Pin; Wing; Folding Suspension Lug; Fire Control / Avionics Umbilical Connector; Flight Control Access Cover; Electrical Safety Pin

Note: Military Standard Guidelines Are MIL-STD-1760, MIL-STD-1553, NATO STANAG 3837

Fig. 6.185a Examples of Missile Carriage and Fire Control Interface.

Example of Ground Carriage and Fire Control Interface: M270 Launcher for MLRS

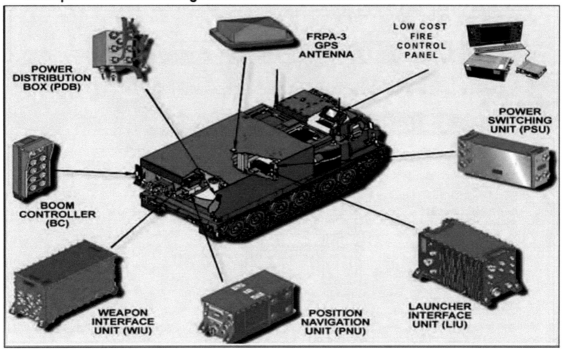

Fig. 6.185b Examples of Missile Carriage and Fire Control Interface (cont).

Example of Ground Carriage and Fire Control Interface: Patriot Air Defense System

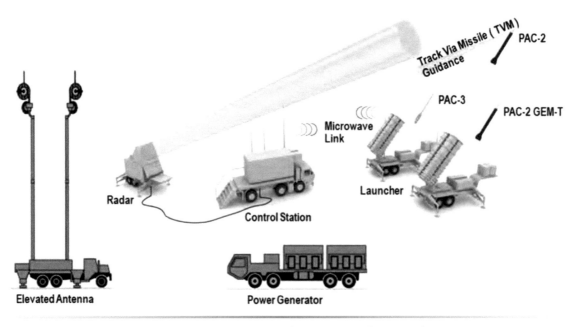

Fig. 6.185c Examples of Missile Carriage and Fire Control Interface (cont).

Examples of Network Centric Warfare Displays: Joint Tactical Interface Display System (JTIDS) / Link 16 / Multifunctional Information Distribution System (MIDS)

Fig. 6.186 Network Centric Warfare Display Provides Combat Situational Awareness.

THAAD

Patriot

MLRS

USS Vincennes (CG-49) Aegis

Fig. 6.187a Fire Control System Provides Weapon Selection, Release, and Firing.

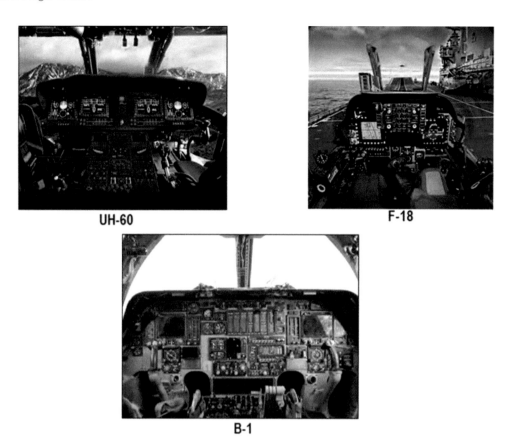

Fig. 6.187b Fire Control System Provides Weapon Selection, Release, Firing (cont).
Note: See video supplement, F-35.

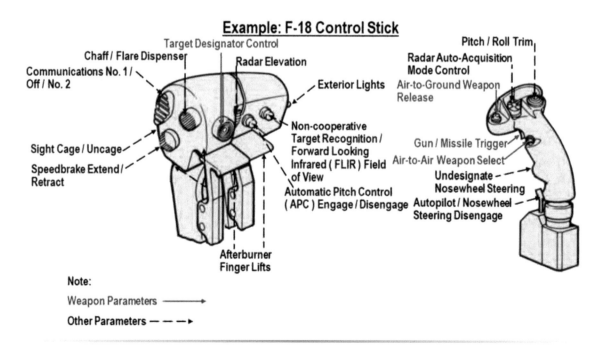

Fig. 6.188 An Aircraft Control Stick and Cockpit Display Provide Weapon Selection, Release, and Firing.
Note: See video supplement, F-18 Cockpit Display and Control Stick Function.

CHAPTER 6 Other Measures of Merit 257

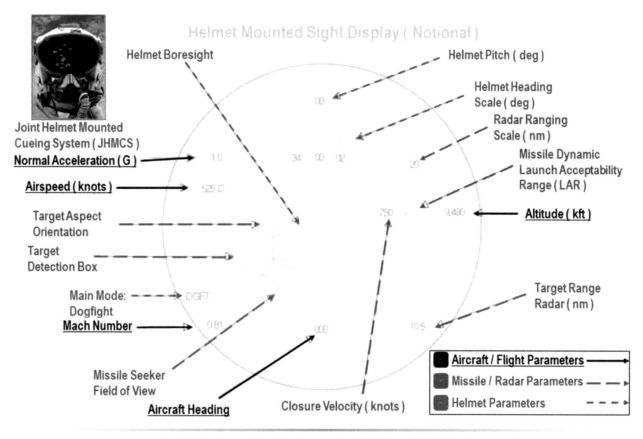

Fig. 6.189 A Helmet Mounted Sight Provides a Head-up Display of Missile Parameters.

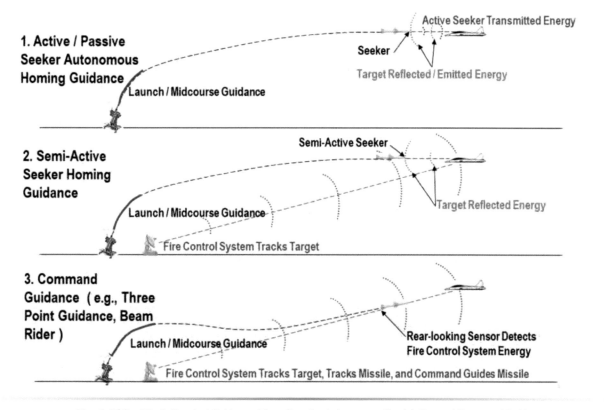

Fig. 6.190 Missile Terminal Guidance Alternatives Are Autonomous, Semi-Active, and Command Guidance.

Type of Terminal Guidance	Launch Platform Survivability	Missile Sensor Cost / Weight	Accuracy	Counter-Countermeasures (CCM)	Fire Control Correction / Fuzing
Active Homing: Radar Frequency (RF), Laser	◐	− ○	○ ●	◐	−
Passive Homing: Infrared (IR), RF, Anti-Radiation Homing (ARH)	●	○ ◐	◐ ●	− ●	−
Semi-Active Homing: RF, Laser	− ○	○ ◐	○ ◐	○	●
Command: RF, Laser, Wire	−	●	− ◐	●	●
Multi–Mode: Command Updates, Inertial, Active or Passive Homing	○	○	○ ●	○ ●	●

Note: ● Superior   ◐ Good   ○ Average   − Poor

Fig. 6.191 Missile Terminal Guidance Drivers Include Launch Platform Survivability, Missile Sensor Cost, and Accuracy.

Starstreak
Laser Command Guidance

RBS-70 / Bolide
Laser Command Guidance

Roland
Radar Frequency (RF) / FLIR Track – RF Command Guidance

TOW
Forward Looking Infrared (FLIR) Track - Wire Command Guidance

Fig. 6.192 A Command Guidance Fire Control System Tracks the Missile, Tracks Target, and Command Guides Missile.
Note: See video supplement, Starstreak, RBS-70 Roland, and TOW Command Guidance.

Example Semi-Active Radar Frequency (RF) Guidance: Sea Dart   Example Semi-Active RF Track-via-Missile Guidance: PAC-2

Example of Semi-Active Laser Guidance: SAL Maverick

Fig. 6.193   A Semi-Active Fire Control System Tracks the Target and Illuminates the Target for the Missile Seeker.
Note: See video supplement, Semi-Active Sea Dart, Patriot PAC-2, Maverick.

Example of INS, Data Link, Radar Seeker Guidance: Meteor   Example of INS, mmW Radar Seeker Guidance: Brimstone

Example of INS, Data Link, mmW Seeker Guidance: PAC-3

Fig. 6.194   Multi-mode Guidance Example—An Active Radar Seeker Augmented with an Inertial Navigation System and a Data Link.
Note: See video supplement, PAC-3 Multi-mode Guidance.

260 Missile Design Guide

F-35 AN / APG-81 AESA Radar

Fig. 6.195 A Phased Array Radar Fire Control System Has Higher Search Rate, Greater Coverage, and Higher Reliability.
Note: See video supplement, Raytheon Video: Air and Missile Defense Radar.

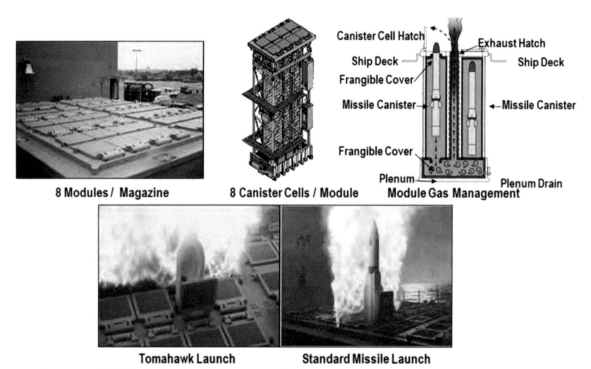

Advantages of Mk-41 VLS include high firing rate against all-aspect targets, low radar cross section, environment protection of weapons, multiple types of weapons. Disadvantages of Mk-41 VLS include below-deck safety and cannot reload at sea.

Fig. 6.196 An Example of Ship Carriage and Launcher Is the US Mk41 Vertical Launch System (VLS).
Note: See video supplement, Mk41 VLS.

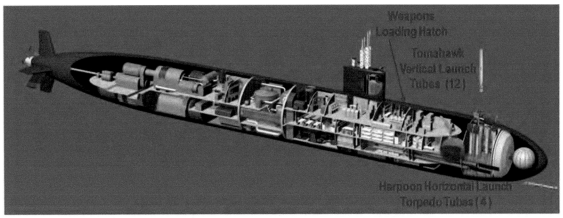

Example: US Los Angeles Class (688) Attack Submarine

Tomahawk Launch

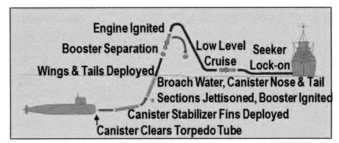

Harpoon Flight Profile

Fig. 6.197 Submarine Carriage and Launchers Have Vertical and Horizontal Launch Tubes.

Environment Param	Typical Requirement
♦ Surface Temperature	− 60 °F to 160 °F*
♦ Surface Humidity	5% to 100%
♦ Rain Rate	120 mm / h**
♦ Surface Wind	150 km / h***
♦ Salt Fog	3 g / mm² per year
♦ Dust / Sand / Dirt	2 g / m³, wind @ 18 m / s
♦ Vibration	10 g rms at 1000 Hz: MIL STD 810G, 648, 1670A
♦ Shock	~ 0.5 m concrete drop (half sine wave 100 g @ 10 ms)
♦ Acoustic	160 dB
♦ External Power Fluct	+/− 10%, MIL-HDBK-781
♦ Lightning Strike	MIL-STD 461, 464

ATACMS Launch

Note: MIL-HDBK-310 and earlier MIL-STD-210B suggest 1% world-wide climatic extreme typical requirement.

* Highest recorded ambient temperature 136 °F. Lowest recorded temperature = − 129 °F. ≈ 20% probability temperature lower than − 60 °F during worst month / location.

** Highest recorded rain rate = 436 mm / h. ≈ 0.5% probability greater than 120 mm / h during worst month / location.

*** Highest recorded wind = 407 km / h. ≈ 1% probability greater than 100 km / h during worst month / location.

Typical external carriage maximum hours for aircraft ~ 100 h. Typical external carriage max hours for helicopter ~ 1000 h.

Fig. 6.198 Missile and Launch Platform Climatic Environment Is Typically Based on the 1% Probability Extreme.
Note: See video supplement, Ground/Sea Environment.

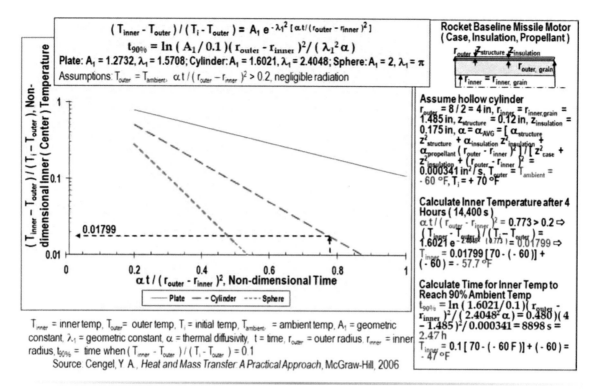

Fig. 6.199 Missile Environment Cooling Time Drivers Include Temperature, Dimensions, Geometry, and Diffusivity.

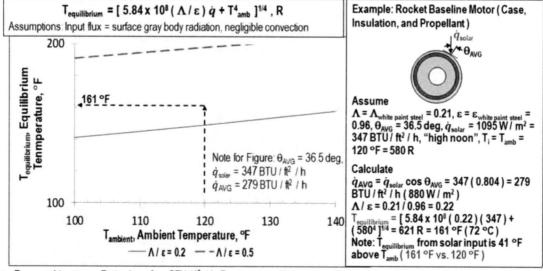

Fig. 6.200 Missile Exposure to Solar Input Requires Design for Higher Than Ambient Temperature.

Chapter 7 Sizing Examples and Sizing Tools

7.1 Introduction

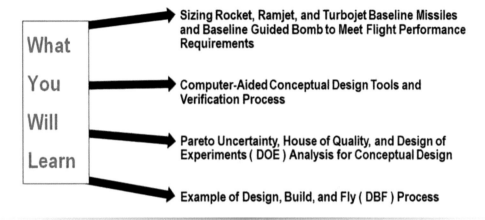

Fig. 7.1 Chapter 7 Missile Sizing Examples and Sizing Tools—What You Will Learn.

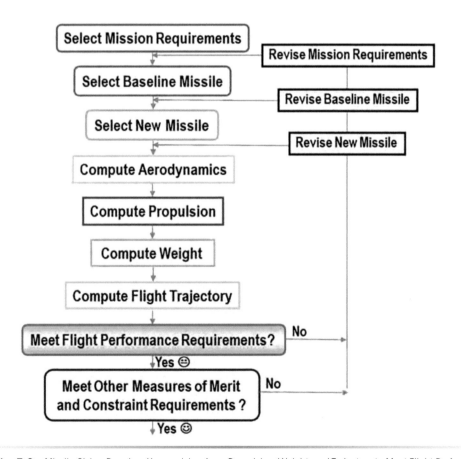

Fig. 7.2 Missile Sizing Requires Harmonizing Aero, Propulsion, Weight, and Trajectory to Meet Flight Performance.

7.2 Rocket Baseline Missile

Examples of Sparrow Missile Launch Platforms

Fig. 7.3 Rocket Baseline Missile Is AIM-7 Sparrow.
Note: See video supplement, Historical Video of Sparrow Missile - Circa 1956.

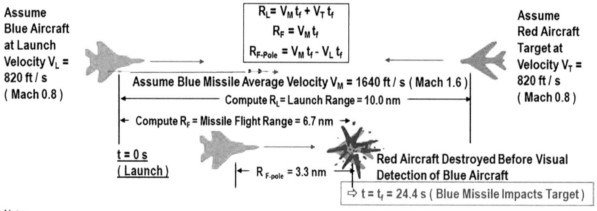

Note:
- $h = 20k\,ft$ (Assumed altitude of engagement)
- $R_{F\text{-pole}}$ = Range between blue aircraft and red aircraft when blue missile impacts red aircraft
- $R_{F\text{-pole}} = 3.3\,nm = 6.1\,km$ (Assumed visual detection range of blue aircraft by red aircraft)
- $t_f = 24.4\,s$ (Time required for blue missile to fly 6.7 nm standoff range, which drives blue missile flight performance design)

Fig. 7.4 Rocket Baseline Missile Provides Beyond-Visual-Range (BVR) Intercept.
Note: See video supplement, Illustrating R_L, R_f, and $R_{F\,pole}$.

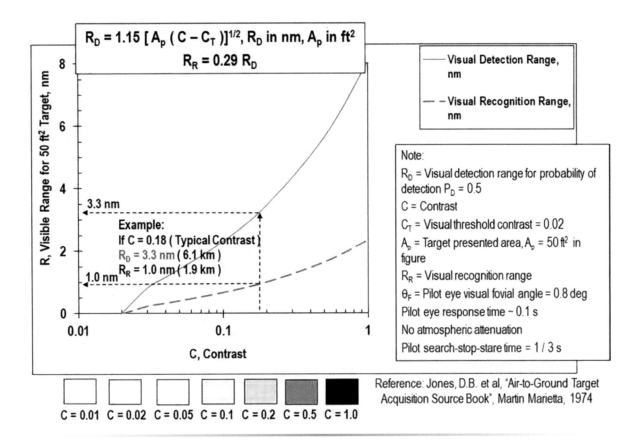

Fig. 7.5 Target Contrast and Size Drive Visual Detection and Recognition Range.

Example of F-16 "Camouflage" (Note Store Contrast and Low Texture. Store Color is Based on MIL-STD-709D)

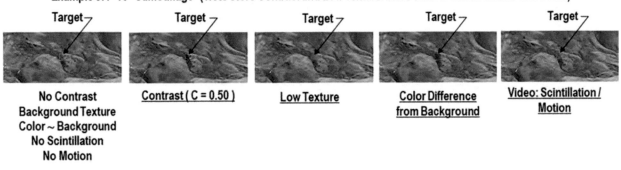

Fig. 7.6 Target Contrast, Low Texture, Color, Scintillation, and Motion Increase Visual Detection in Clutter.
Note: See video supplement, Scintillation/Motion.

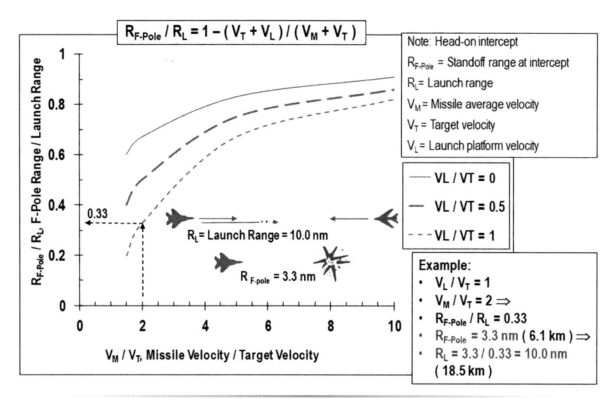

Fig. 7.7 High Missile Velocity Improves Air Intercept Standoff Range.

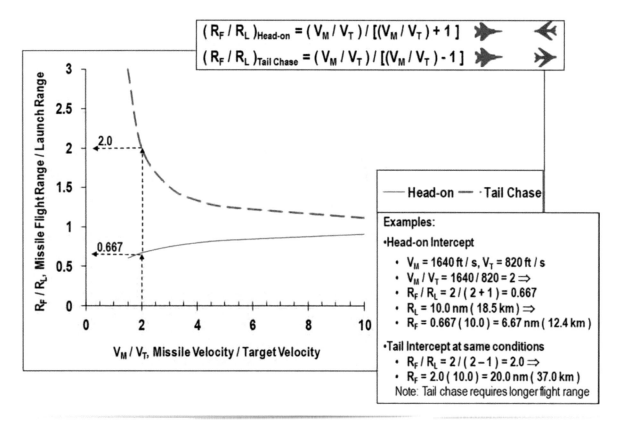

Fig. 7.8 The Requirement for Missile Flight Range Is Greater for a Tail Chase Intercept.

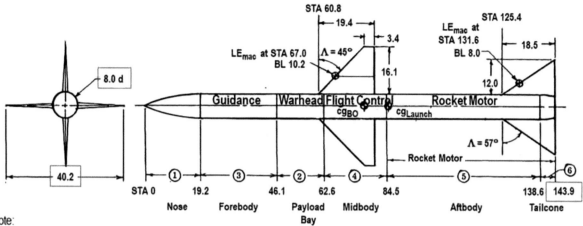

Fig. 7.9 Drawing of Rocket Baseline Missile Configuration.

Note:
- Dimensions in Inches
- d = Diameter
- cg_{BO} = Center-of-Gravity at Burnout
- cg_{Launch} = Center-of-Gravity at Launch
- LE_{mac} = Leading Edge of Mean Aerodynamic Chord
- Λ = Leading Edge Sweep Angle
- STA = Station Longitudinal Location from Leading Edge of Nose
- BL = Body Lateral Location from Centerline of Missile

Source: Bithell, R.A. and Stoner, R.C., "Rapid Approach for Missile Synthesis, Vol. 1, Rocket Synthesis Handbook," AFWAL-TR-81-3022, Vol. 1, March 1982.

Component	Weight, lb	Center of Gravity (cg) Station, in
① Nose (Radome)	4.1	12.0
③ Forebody Structure	12.4	30.5
Guidance	46.6	32.6
② Payload Bay Structure	7.6	54.3
Warhead	77.7	54.3
④ Midbody Structure	10.2	73.5
Flight Control Actuation System	61.0	75.5
⑤ Aftbody Structure	0.0	—
Rocket Motor Case	47.3	107.5
Insulation (EPDM – Silica)	23.0	117.2
⑥ Tailcone Structure	6.5	141.2
Nozzle	5.8	141.2
Fixed Surfaces	26.2	137.8
Movable Surfaces	38.6	75.5
Burnout Total	367.0	76.2
Propellant	133.0	107.8
Launch Total	500.0	84.5

Note: Configuration locations ① ② ③ ④ ⑤ ⑥ shown on previous figure
Source: Bithell and Stoner

Fig. 7.10 Mass Properties of Rocket Baseline Missile.

Body
- Dome material — Pyroceram
- Airframe structure material — Aluminum (2219-T81)
- Length, in — 143.9
- Diameter, in — 8.0
- Airframe structure thickness, in — 0.16
- Fineness ratio — 17.99
- Volume, ft^3 — 3.82
- Wetted area, ft^2 — 24.06
- Nozzle exit area, ft^2 — 0.078
- Boattail fineness ratio — 0.38
- Nose fineness ratio — 2.40
- Nose bluntness — 0.0
- Boattail angle, deg — 7.5

Movable surfaces (forward)
- Material — Aluminum 2219-T81
- Planform area, ft^2 (2 panels exposed) — 2.55
- Wetted area, ft^2 (4 panels) — 10.20
- Aspect ratio (2 panels exposed) — 2.82
- Taper ratio — 0.175
- Root chord, in — 19.4
- Tip chord, in — 3.4
- Span, in (2 panels exposed) — 32.2
- Leading edge sweep, deg — 45.0

Source: Bithell and Stoner

Fig. 7.11a Rocket Baseline Missile Definition.

Movable surfaces (continued)
- Mean aerodynamic chord, in — 13.3
- Thickness ratio — 0.044
- Section type — Modified double wedge
- Section leading edge total angle, deg — 10.01
- Longitudinal location of leading edge of mean aerodynamic chord x_{mac}, in — 67.0
- Lateral location of mean aerodynamic chord y_{mac}, in (from root chord) — 6.2
- Actuator rate limit, deg/s — 360.0

Fixed surfaces (aft)
- Material — Aluminum 2219-T81
- Modulus of elasticity, 10^6 psi — 10.5
- Planform area, ft^2 (2 panels exposed) — 1.54
- Wetted area, ft^2 (4 panels) — 6.17
- Aspect ratio (2 panels exposed) — 2.59
- Taper ratio — 0.0
- Root chord, in — 18.5
- Tip chord, in — 0.0
- Span, in (2 panels exposed) — 24.0
- Leading edge sweep, deg — 57.0
- Mean aerodynamic chord, in — 12.3
- Thickness ratio — 0.027
- Section type — Modified double wedge
- Section leading edge total angle, deg — 6.17
- Longitudinal location of leading edge of mean aerodynamic chord x_{mac}, in — 131.6
- Lateral location of mean aerodynamic chord y_{mac}, in (from root chord) — 4.0

Source: Bithell and Stoner

Fig. 7.11b Rocket Baseline Missile Definition (cont).

References Values
- Reference area, ft^2 — 0.349
- Reference length, ft — 0.667
- Pitch / Yaw Moment of inertia at launch, slug-ft^2 — 117.0
- Pitch / Yaw Moment of inertia at burnout, slug-ft^2 — 94.0

Rocket Motor Performance (altitude = 20k ft, temperature = 70° F)
- Burning time, sec (boost / sustain) — 3.69 / 10.86
- Maximum pressure, psi — 2042
- Average pressure, psi (boost / sustain) — 1769 / 301
- Average thrust, lbf (boost / sustain) — 5750 / 1018
- Total impulse, lbf-s (boost / sustain) — 21217 / 11055
- Specific impulse, lbf-s / lbm (boost / sustain) — 250 / 230.4

Propellant
- Weight, lbm (boost / sustain) — 84.8 / 48.2
- Flame temperature @ 1,000 psi, °F (boost / sustain) — 5282 / 5228
- Propellant density, lbm / in^3 — 0.065
- Characteristic velocity, ft / s — 5200
- Burn rate @ 1000 psi, in / s — 0.5
- Burn rate pressure exponent — 0.3

Source: Bithell and Stoner

Fig. 7.11c Rocket Baseline Missile Definition (cont).

Propellant (continued)
- Burn rate sensitivity with temperature, % / °F — 0.10
- Pressure sensitivity with temperature, % / °F — 0.14

Rocket Motor Case
- Yield / ultimate tensile strength, psi — 170,000 / 190,000
- Material — 4130 Steel
- Modulus of elasticity, psi — 29.5×10^6 psi
- Length, in — 59.4
- Outside diameter, in — 8.00
- Thickness, in (minimum) — 0.074
- Burst pressure, psi — 3140
- Volumetric efficiency — 0.76
- Grain configuration — Three slots + web
- Dome ellipse ratio — 2.0

Nozzle
- Housing material — 4130 Steel
- Exit geometry — Contoured (equiv. 15°)
- Throat area, in^2 — 1.81
- Expansion ratio — 6.2
- Length, in — 4.9
- Exit diameter, in — 3.78

Source: Bithell and Stoner

Fig. 7.11d Rocket Baseline Missile Definition (cont).

270 Missile Design Guide

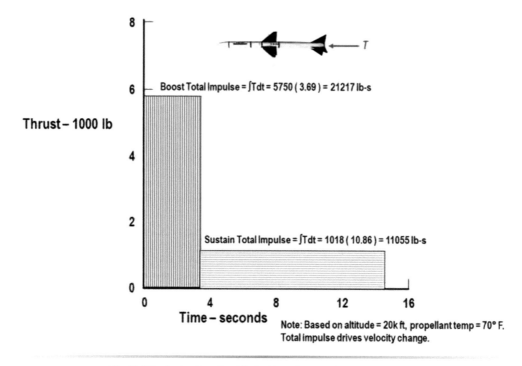

Fig. 7.12 Rocket Baseline Missile Has a Boost—Sustain Thrust Time History.

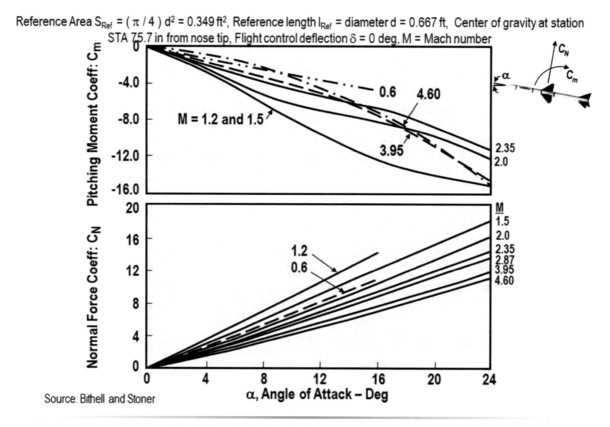

Fig. 7.13a Rocket Baseline Missile Aerodynamic Characteristics.

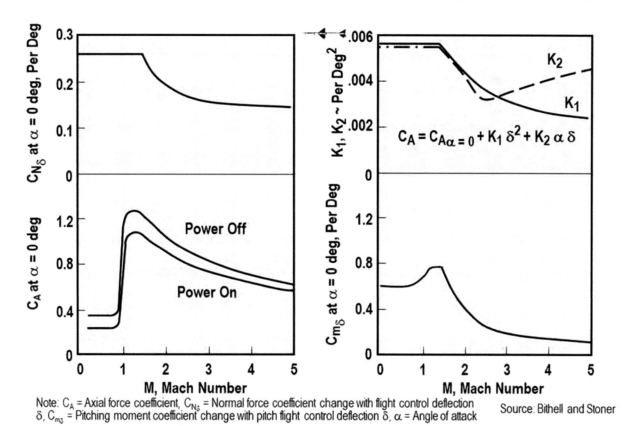

Fig. 7.13b Rocket Baseline Missile Aerodynamic Characteristics (cont).

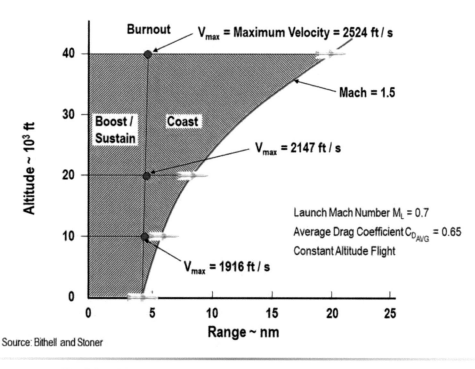

Fig. 7.14 High Altitude Launch Enhances the Rocket Baseline Missile Flight Range.

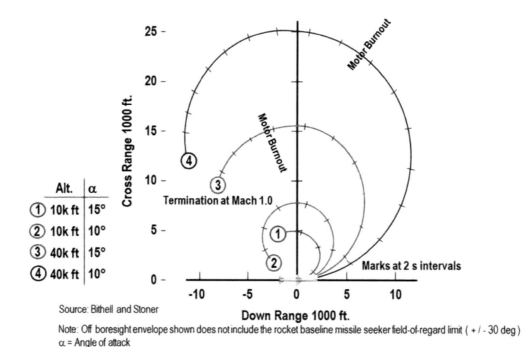

Fig. 7.15 Low Altitude Maneuvers Have Smaller Turn Radius for Rocket Baseline Missile.

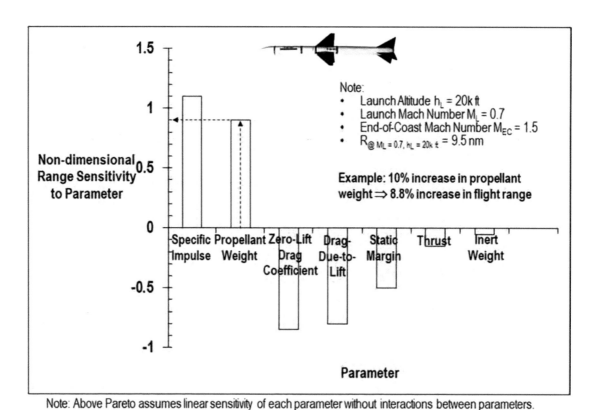

Fig. 7.16 Rocket Baseline Missile Range Is Driven by Specific Impulse, Propellant Weight, Drag, and Static Margin.

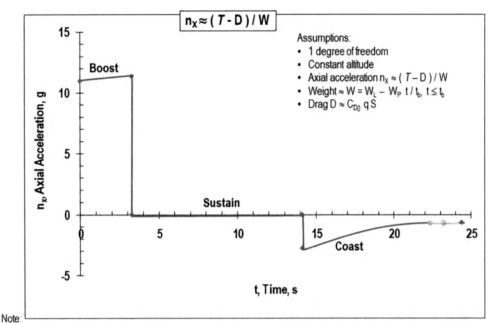

Fig. 7.17 Rocket Baseline Missile Flight Has Boost, Sustain, and Coast Axial Acceleration.

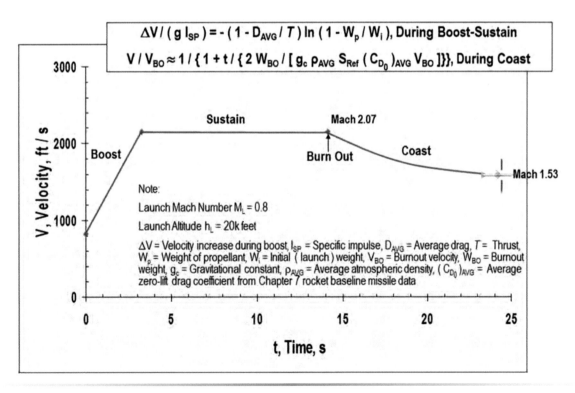

Fig. 7.18 The Rocket Baseline Missile Has a Boost—Sustain—Coast Velocity Profile.

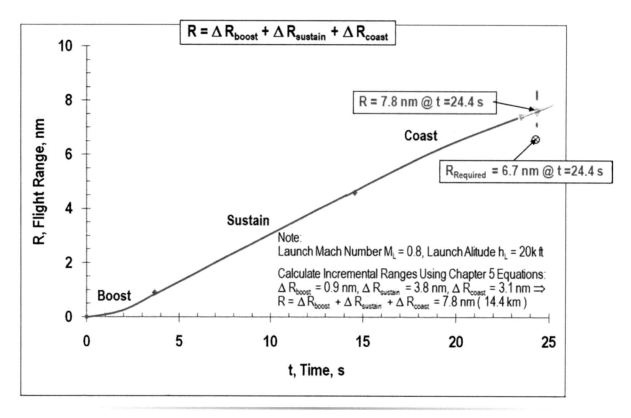

Fig. 7.19 Flight Range and Time-to-Target of the Rocket Baseline Missile Satisfy Assumed Requirements.

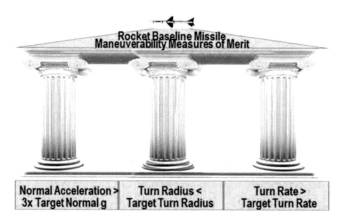

Note: Roland leads the target and has higher velocity, higher turn rate, and smaller turn radius than the target

Fig. 7.20 Maneuverability Measures of Merit for Rocket Baseline Missile Against a Maneuvering Target.
Note: See video supplement, Roland Maneuvering to Intercept a Maneuvering Aircraft.

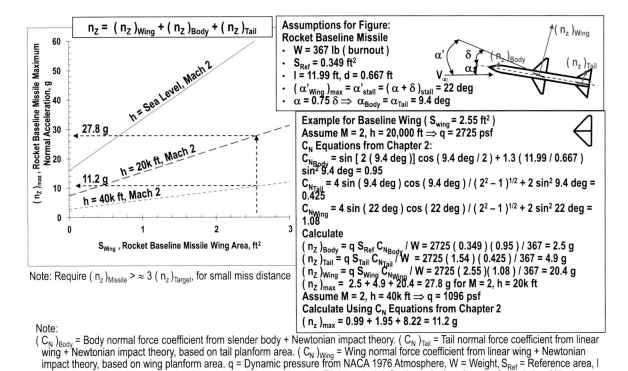

Fig. 7.21 Rocket Baseline Missile Wing Area Is a Driver for Normal Acceleration Maneuverability.

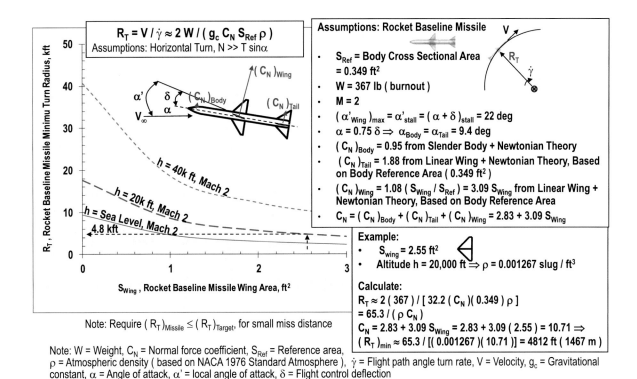

Fig. 7.22 Rocket Baseline Missile Wing Area Is a Driver for Turn Radius Maneuverability.

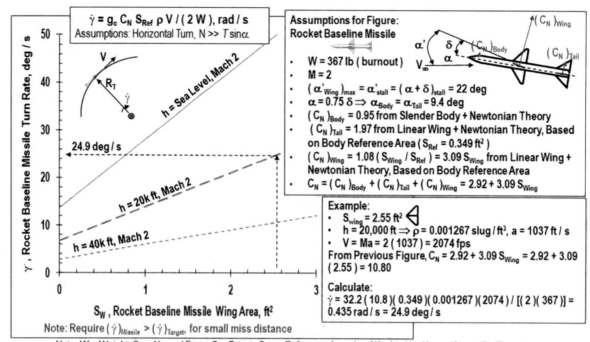

Fig. 7.23 Rocket Baseline Missile Wing Area Is a Driver for Turn Rate Maneuverability.

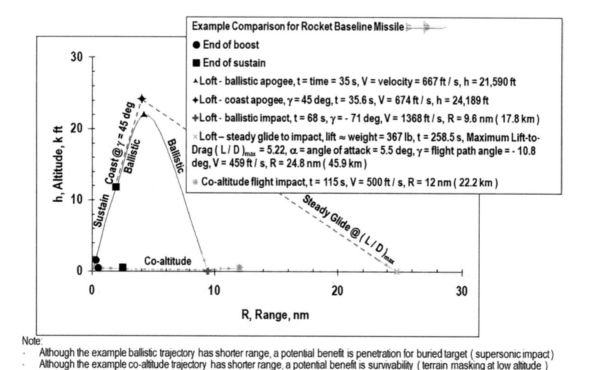

Fig. 7.24 A Lofted Boost—Glide May Provide Extended Range for the Rocket Baseline Missile.

7.3 Ramjet Baseline Missile

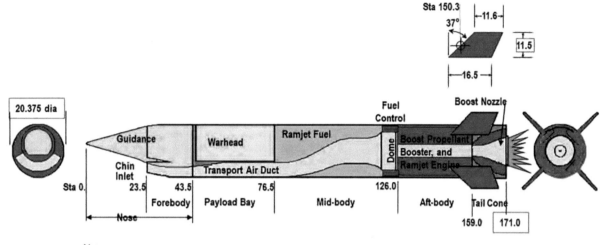

Source: Bithell, R.A. and Stoner, R.C. "Rapid Approach for Missile Synthesis", Vol. II, Air-breathing Synthesis Handbook, AFWAL TR 81-3022, Vol. II, March 1982.

Fig. 7.25 The Ramjet Baseline Missile Is a Chin Inlet Integral Rocket Ramjet (IRR).

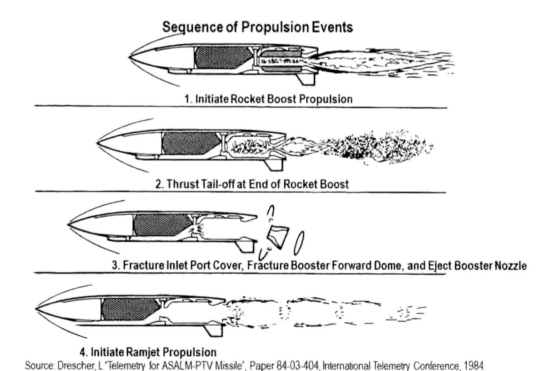

Source: Drescher, L "Telemetry for ASALM-PTV Missile", Paper 84-03-404, International Telemetry Conference, 1984.

Fig. 7.26 Ramjet Baseline Missile Has an Integral Rocket Booster—Ramjet Combustor.

Component	Weight, lb	Center of Gravity (cg) Station, in
Nose	15.9	15.7
Forebody Structure	42.4	33.5
Guidance	129.0	33.5
Payload Bay Structure	64.5	60.0
Warhead	510.0	60.0
Midbody Structure	95.2	101.2
Inlet	103.0	80.0
Electrical	30.0	112.0
Hydraulic System for Control Actuation	20.0	121.0
Fuel Distribution	5.0	121.0
Aftbody Struct (Combustor / Booster Case)	44.5	142.5
Engine (Injectors, Igniter, Flameholder)	33.5	142.5
Tailcone Structure	31.6	165.0
Ramjet Nozzle	31.0	165.0
Flight Control Actuators	37.0	164.0
Fins (4)	70.0	157.2
End of Cruise	1262.6	81.8
Ramjet Fuel (RJ-5 Hydrocarbon, 11900 in^3)	476.0	87.0
Start of Cruise	1738.6	83.2
Boost Nozzle (Ejected)	31.0	164.0
Frangible Port	11.5	126.0
End of Boost	1781.1	84.9
Boost Propellant	449.0	142.5
Booster Ignition	2230.1	96.5

Source: Bithell and Stoner

Fig. 7.27 Mass Properties of Ramjet Baseline Missile.

Inlet
   Type   Mixed compression
   Material   Ti-6Al-2Sn-4Zr-2Mo-Si
   Yield internal pressure load @ 1300 °F, psi   200
   Backside insulation   3 / 16 in Min-K
   Conical forebody half angle, deg   17.7
   Ramp wedge angle, deg   8.36
   Cowl angle, deg   8.24
   Internal contraction ratio   12.2 Percent
   Capture area, ft^2 0.79
   Throat area, ft^2 0.29
   Exit area, ft^2 0.54
   Length, in   102.5
Body
   Seeker dome material   Silicon nitride
   Forebody airframe material   Titanium Ti-6424S
   Airframe insulation   Min-K
   Combustor material   Inconel 718
   Combustor Insulation   DC 93-104
   Nozzle material   Silica Phenolic
   Length, in   171.0
   Diameter, in   20.375
   Fineness ratio   8.39
   Volume, ft^3 28.33
   Wetted area, ft^2 68.81
   Base area, ft^2 (cruise)   0.58
   Nose half angle, deg   17.7
   Nose length, in   23.5

Source: Bithell and Stoner

Fig. 7.28a Ramjet Baseline Missile Definition.

Tail (Exposed)	
Material	Titanium Ti-6424S
Planform area (2 panels), ft^2	2.24
Wetted area (4 panels), ft^2	8.96
Aspect ratio (2 panels exposed)	1.64
Taper ratio	0.70
Root chord, in	16.5
Span (2 panels exposed), in	23.0
Leading edge sweep, deg	37.0
Mean aerodynamic chord, in	14.2
Thickness ratio	0.04
Section type	Modified double wedge
Section leading edge total angle, deg	9.1
Longitudinal location of mean aerodynamic chord (x_{mac}), in	150.3
Lateral location of mean aerodynamic chord (y_{mac}), in (from root chord)	5.4
Reference values	
Reference area, ft^2	2.264
Reference length, ft	1.698

Source: Bithell, and Stoner

Fig. 7.28b Ramjet Baseline Missile Definition (cont).

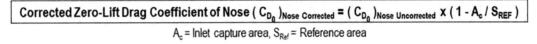

Corrected Zero-Lift Drag Coefficient of Nose $(C_{D_0})_{\text{Nose Corrected}} = (C_{D_0})_{\text{Nose Uncorrected}} \times (1 - A_c / S_{REF})$

A_c = Inlet capture area, S_{Ref} = Reference area

Ramjet Baseline Missile Engine Station Identification and Flow Path

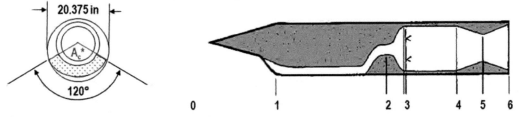

Subscripts
0 Free stream flow into inlet (Example, Ramjet Baseline at Mach 4, Angle of attack α = 0 deg $\Rightarrow A_0$ = 104 in^2. Note: A_c = 114 in^2)
1 Inlet throat (Ramjet Baseline $A_1 = A_{IT}$ = 41.9 in^2)
2 Diffuser exit (Ramjet Baseline A_2 = 77.3 in^2)
3 Flame holder plane (Ramjet Baseline A_3 = 287.1 in^2)
4 Combustor exit (Ramjet Baseline A_4 = 287.1 in^2)
5 Nozzle throat (Ramjet Baseline A_5 = 103.1 in^2)
6. Nozzle exit (Ramjet Baseline A_6 = 233.6 in^2)

Note:
*Ramjet Baseline Missile Inlet Capture Area A_c = 114 in^2 ($\approx$ Bottom 1/3 of Body Nose)
Ramjet Baseline Missile Reference Area (Body Cross-sectional Area) S_{Ref} = 326 in^2
Ramjet Baseline Missile Reference Length (Body Diameter) d = 20.375 in

Source: Bithell, and Stoner

Fig. 7.29 Ramjet Baseline Missile Engine Nomenclature, Flow Path Geometry and Nose Drag Correction.

280 Missile Design Guide

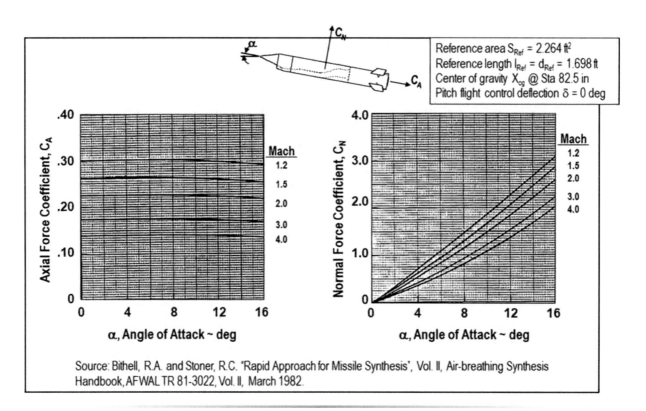

Fig. 7.30a Aerodynamic Characteristics of Ramjet Baseline Missile.

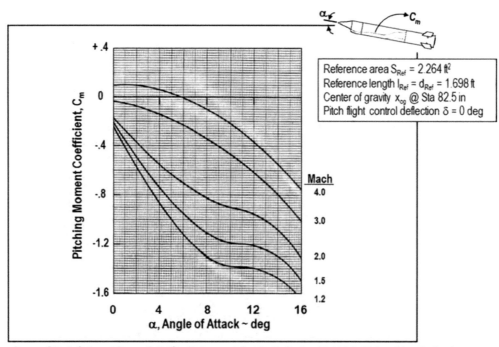

Fig. 7.30b Aerodynamic Characteristics of Ramjet Baseline Missile (cont).

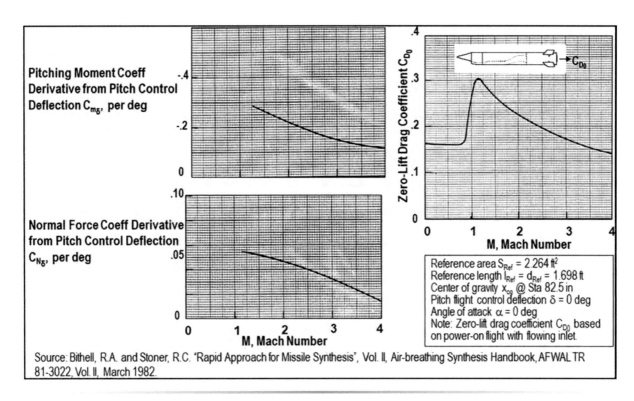

Fig. 7.30c Aerodynamic Characteristics of Ramjet Baseline Missile (cont).

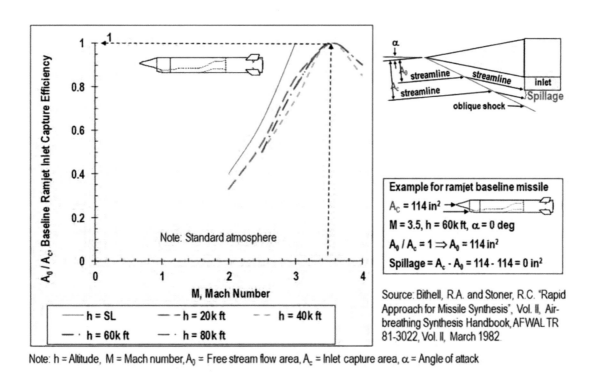

Fig. 7.31 Maximum Efficiency of the Ramjet Baseline Missile Inlet Occurs at About Mach 3.5.

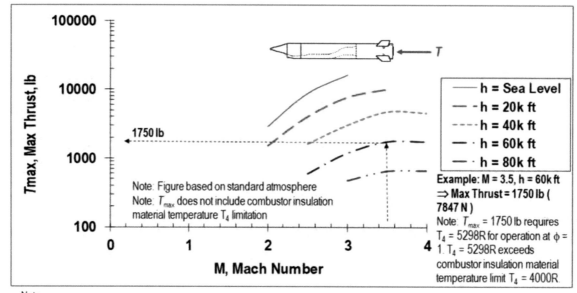

Note:
$T \approx [\varphi / (A_0 / A_C)] T_{max}$ if $\varphi \leq A_0 / A_C$
$T \approx T_{max}$ if $\varphi \geq A_0 / A_C$
T = Thrust, T_{max} = Maximum thrust, φ = Equivalence ratio (actual fuel-to-air ratio compared to stochiometric fuel-to-air ratio), A_0 = Free stream air flow area, A_C = Capture area of inlet, T_4 = Combustion temperature, h = Altitude

Figure Based on Bithell and Stoner

Fig. 7.32 Maximum Thrust of the Ramjet Baseline Missile Occurs at About Mach 3.5.

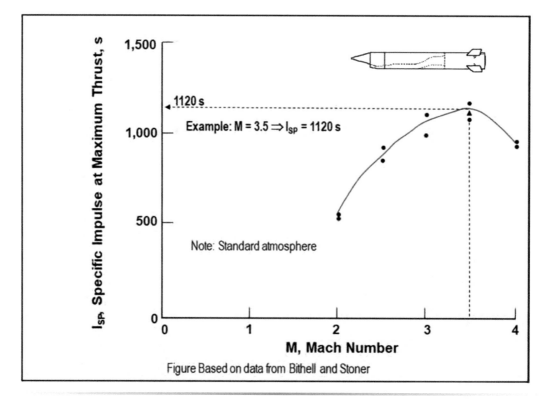

Fig. 7.33 Maximum Specific Impulse of the Ramjet Baseline Missile Occurs at About Mach 3.5.

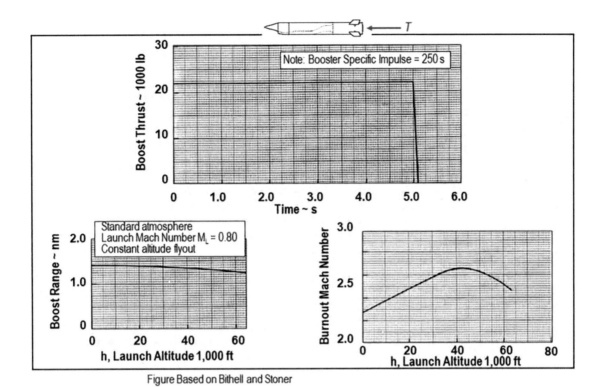

Fig. 7.34 Rocket Booster of Ramjet Baseline Missile Provides a Boost Greater Than Mach 2.2.

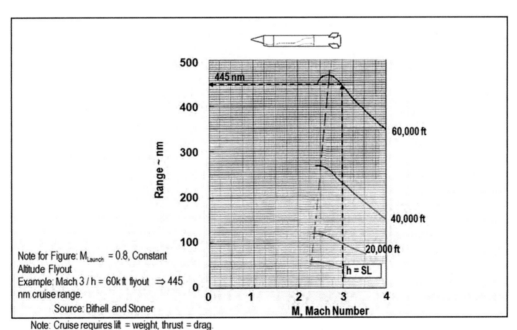

Note: Cruise requires lift = weight, thrust = drag.
Breguet range prediction for Mach 3 / h = 60k ft flyout is R = V I$_{SP}$ (L / D) ln [W$_{BC}$/ (W$_{BC}$ - W$_f$)] = 2901 (1040) (3.15) ln (1739 / (1739 - 476)) = 3,039,469 ft = 500 nm. Breguet range prediction is typically ≈ 10% greater than actual cruise range.

Fig. 7.35 Ramjet Baseline Missile Has Best Performance at High Altitude Cruise.

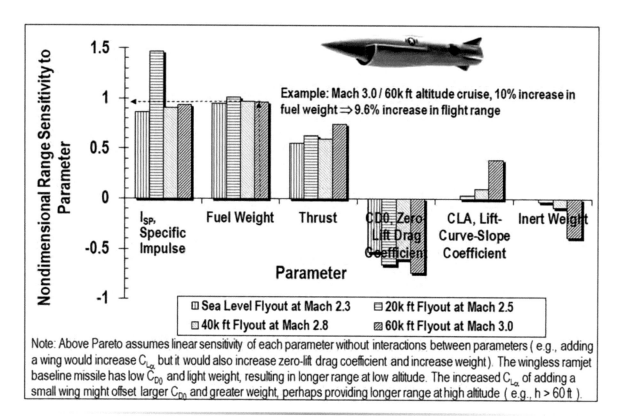

Fig. 7.36 Pareto Analysis Shows Ramjet Baseline Missile Range Is Driven by Specific Impulse, Fuel Weight, Thrust, and Drag.

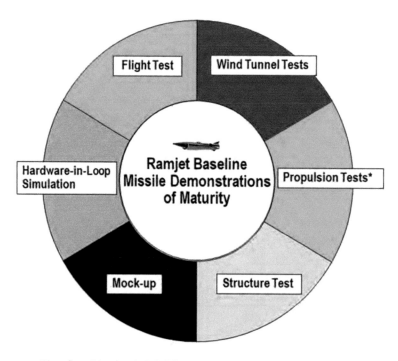

*Note: Propulsion tests included direct connect, freejet, and booster firing

Fig. 7.37 Ramjet Baseline Missile Has Ground, System Integration, System Engineering, and Flight Test Demos.

Parameter	Baseline Value at Mach 3.0 / 60k ft	Uncertainty in Parameter	$\Delta R / R$ from Uncertainty
1. Specific Impulse I_{SP}	1040 s	+/- 5%, 1σ	+/- 5%, 1σ
2. Ramjet Fuel Weight W_f	476 lb	+/- 1%, 1σ	+/- 1%, 1σ
3. Cruise Thrust T_{cruise} (Equivalence Ratio $\phi = 0.39$)	458 lb	+/- 5%, 1σ	+/- 2%, 1σ
4. Zero-Lift Drag Coefficient C_{D0}	0.17	+/- 5%, 1σ	+/- 4%, 1σ
5. Lift Curve Slope Coefficient $C_{L\alpha}$	0.13 / deg	+/- 3%, 1σ	+/- 1%, 1σ
6. Inert Weight W_{inert}	1205 lb	+/- 2%, 1σ	+/- 0.8%, 1σ

• **Total Flight Range Uncertainty of Ramjet Baseline Missile at Mach 3.0 / 60k ft Flyout is**
  - $\Delta R / R = [(\Delta R/R)_1^2 + (\Delta R/R)_2^2 + (\Delta R/R)_3^2 + (\Delta R/R)_4^2 + (\Delta R/R)_5^2 + (\Delta R/R)_6^2]^{1/2} = +/- 6.9\%, 1\sigma$
  - R = 445 nm +/- 31 nm, 1σ $\Rightarrow$ 352 nm < $R_{3\sigma}$ < 538 nm
• **From Above, Drivers of Uncertainty in Ramjet Baseline Missile Flight Range Are Uncertainties in I_{SP} and C_{D0}**
  - +/- 5%, 1σ Uncertainty in I_{SP} $\Rightarrow$ +/- 5%, 1σ Uncertainty in Flight Range
  - +/- 5%, 1σ Uncertainty in C_{D0} $\Rightarrow$ +/- 4%, 1σ Uncertainty in Flight Range
• **To Meet a Customer Flight Range Requirement of $R_{3\sigma}$ > 400 nm, the Development Program Could Either**
  1) Provide More Program Development Funding to Reduce Uncertainties in the Primary Parameters (e.g., I_{sp}, C_{D0})
  2) Overdesign the Missile for a Longer Nominal Range $\Rightarrow$ Higher Production Cost and Logistics Cost

*Note: $\Delta R / R$ Range uncertainty due to uncertainty in parameter was computed from previous Pareto for ramjet baseline missile (e.g., 10% increase in fuel weight $\Rightarrow$ 9.6% increase in flight range) $\Rightarrow$ $\Delta R / R$ from +/- 1%, 1σ uncertainty in fuel weight (e.g., ullage) $\Rightarrow$ 0.96 (+/- 1%, 1σ) = +/- 1%, 1σ uncertainty in flight range

Fig. 7.38 Ramjet Baseline Missile Flight Range Uncertainty Is Approximately 7%, 1 Sigma.

$$R_{cruise} = V I_{SP} (L/D) \ln [W_{BC} / (W_{BC} - W_f)]$$

Propulsion / Configuration	Fuel Type / Volumetric Performance (BTU / in3) / Density (lbm / in3)	Fuel Volume (in3) / Fuel Weight (lb)	ISP (s) / Cruise Range at Mach 3.5, 60k ft (nm)
Liquid Fuel Ramjet	RJ-5 / 693 / 0.040	11900 / 476	1120 / 390
Ducted Rocket (Low Smoke)	Solid Hydrocarbon / 1132 / 0.075	7922 / 594	677 / 294
Ducted Rocket (High Performance)	Boron / 2040 / 0.082	7922 / 649	769 / 366
Solid Fuel Ramjet	Boron / 2040 / 0.082	7056 / 579	1170 / 496
Slurry Fuel Ramjet	40% JP-10, 60% Boron Carbide / 1191 / 0.050	11900 / 595	1835 / 770

Note: ■ Flow Path ▫ Available Fuel
Note: R_{cruise} Computed from above Breguet range equation. Assumptions are lift = weight and thrust = drag
R_{cruise} = Cruise range, V = Velocity, I_{SP} = specific impulse, L = Lift, D = Drag, W_{BC} = Weight at beginning of cruise, W_f = Weight of fuel

Fig. 7.39 High Density Fuel and Efficient Packaging Provide Extended Range for Ramjet Baseline Missile.

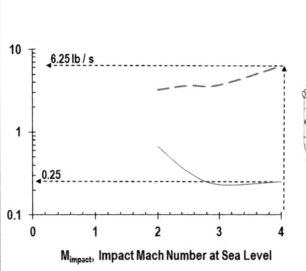

Fig. 7.40 Ramjet Baseline Missile Achieves Velocity Control Through Fuel Flow Rate Control.

7.4 Turbojet Baseline Missile

Fig. 7.41 Turbojet Baseline Missile Has Ship, Submarine, Aircraft, and Ground Vehicle Launch Platforms.
Note: See video supplement, Harpoon Anti-Ship Missile.

Missile	← 9 m (29.5 ft) →	Country	Propulsion	Guidance
Harpoon		USA	Turbojet	RF, INS, GPS
LRASM		USA	Turbofan	RF, IIR, INS, GPS, Data Link
SS-N-25 / kH-35		Russia	Turbojet / Turbofan	RF, INS, GLONASS
Exocet MM-40		France	Turbojet	RF, INS
Naval Strike Missile		Norway	Turbojet	IIR, INS, GPS, Data Link
Taurus		Germany	Turbojet	IIR, INS, GPS, Data Link
RBS-15		Sweden	Turbojet	RF, INS, GPS, Data Link
SS-N-19 / 3M45		Russia	Turbojet	RF, INS, GLONASS?
Brahmos		India	Ramjet	RF, INS, GLONASS
Hsiung Feng 3		Taiwan	Ramjet	RF, IIR, INS, GPS
YJ-12		China	Ramjet	RF, INS, BeiDou
SS-N-22 / 3M-80		Russia	Ramjet	RF, INS, GLONASS?
AS-17 / kH-31		Russia	Ramjet	RF, INS, GLONASS?
SS-N-26 / 3M55		Russia	Ramjet	RF, INS, GLONASS?

Definitions: RF: Radar frequency, INS: Inertial navigation system, GPS: Global Positioning System, IIR: Imaging infrared, GLONASS: Global Navigation System

Fig. 7.42 Most Anti-Ship Missiles Have Air-Breathing Propulsion, Radar Seeker, Inertial Navigation System, and Satellite Navigation.

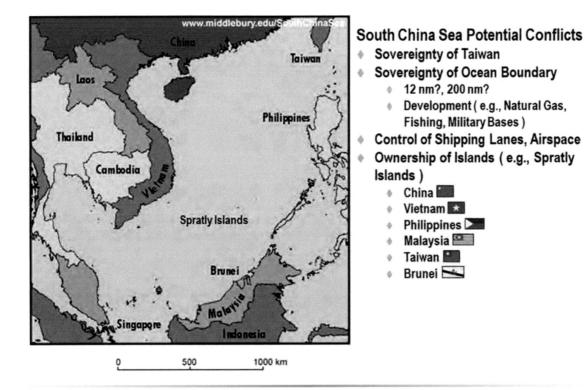

South China Sea Potential Conflicts
- Sovereignty of Taiwan
- Sovereignty of Ocean Boundary
  - 12 nm?, 200 nm?
  - Development (e.g., Natural Gas, Fishing, Military Bases)
- Control of Shipping Lanes, Airspace
- Ownership of Islands (e.g., Spratly Islands)
  - China
  - Vietnam
  - Philippines
  - Malaysia
  - Taiwan
  - Brunei

Fig. 7.43 Example of a Potential Conflict Area that Drives Anti-Ship Missile Requirements is the South China Sea.

288 Missile Design Guide

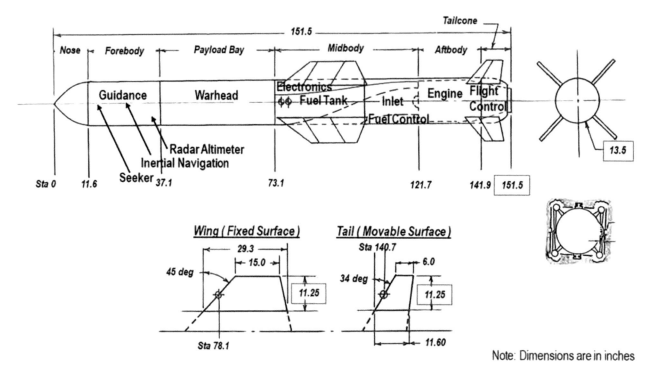

Fig. 7.44 Drawing of Turbojet Baseline Missile Configuration and Subsystems.

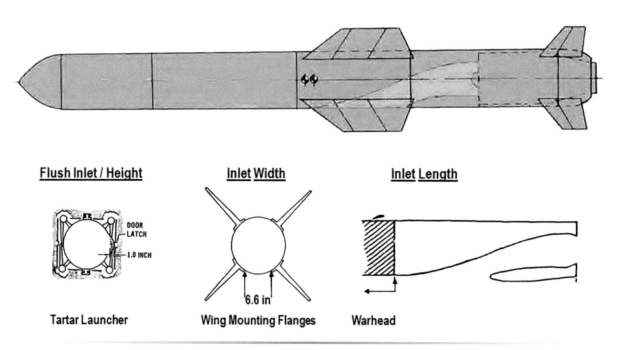

Fig. 7.45 Turbojet Baseline Missile Inlet Has Geometry Constraints.

Component	Weight, lb	Center-of-Gravity (cg) Station, in
Nose	4.6	7.7
Forebody Structure	21.5	24.0
Guidance	89.0	22.0
Instrumentation	5.5	34.0
Payload Bay Structure	0.0	-
Warhead	510.0	55.1
Midbody Structure	124.0	97.4
Electrical	49.9	94.2
Fuel Distribution	10.0	104.0
Inlet	15.0	107.0
Aftbody Structure	19.0	131.8
Engine	101.5	136.0
Tailcone Structure	15.0	146.7
Flight Control Actuators	33.8	144.2
Wings (4 Fixed Surfaces)	41.4	89.6
Tails (4 Movable Surfaces)	16.3	145.3
End of Cruise	1056.5	75.4
Fuel	108.5	97.4
Start of Cruise	1165.0	77.4

Source: Bithell, R.A. and Stoner, R.C. "Rapid Approach for Missile Synthesis", Vol. II, Air-breathing Synthesis Handbook, AFWAL TR 81-3022, Vol. II, March 1982.

Fig. 7.46 Mass Properties of Turbojet Baseline Missile.

Body	
Length, in	151.5
Diameter, in	13.5
Fineness ratio	11.2
Volume, ft^3	12.02
Wetted area, ft^2	43.45
Base area, ft^2	0.567
Boattail fineness ratio	0.44
Nose fineness ratio	0.86
Nose bluntness	0.0
Wing (Fixed Surface)	
Planform area (2 panels exposed), ft^2	3.46
Wetted area (4 panels), ft^2	13.84
Aspect ratio (2 panels exposed)	1.02
Taper ratio	0.512
Root chord, in	29.30
Tip chord, in	15.00
Span (2 panels exposed), in	22.50
Leading edge sweep, deg	45.0
Mean aerodynamic chord, in	22.92
Thickness ratio	0.06
Horizontal distance from nose tip to beginning of wing mean aerodynamic chord x_{mac}, in	78.1
Vertical distance to wing mean aerodynamic chord y_{mac} (from root chord), in	5.02

Source: Bithell and Stoner

Fig. 7.47a Turbojet Baseline Missile Geometry Definition.

Tail (Movable Surfaces)	
Planform area (2 panels exposed), ft^2	1.375
Wetted area (4 panels), ft^2	5.50
Aspect ratio (2 panels exposed)	2.56
Taper ratio	0.52
Root chord, in	11.60
Tip chord, in	6.00
Span (2 panels exposed), in	22.50
Leading edge sweep, deg	34.0
Mean aerodynamic chord, in	9.11
Thickness ratio	0.09
Horizontal distance from nose tip to beginning of tail mean aerodynamic chord x_{mac}, in	140.7
Vertical distance to tail mean aerodynamic chord y_{mac} (from root chord), in	5.03
Reference Values	
Reference area, ft^2	0.994
Reference length, ft	1.125

Source: Bithell, R.A. and Stoner, R.C. "Rapid Approach for Missile Synthesis", Vol. II, Air-breathing Synthesis Handbook, AFWAL TR 81-3022, Vol. II, March 1982.

Fig. 7.47b Turbojet Baseline Missile Geometry Definition (cont).

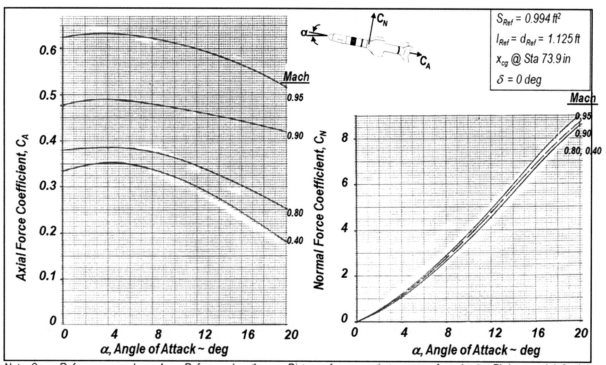

Note: S_{Ref} = Reference area, l_{Ref} = d_{Ref} = Reference length, x_{cg} = Distance from nose tip to center of gravity, δ = Pitch control deflection
Source: Bithell, R.A. and Stoner, R.C. "Rapid Approach for Missile Synthesis", Vol. II, Air-breathing Synthesis Handbook, AFWAL TR 81-3022, Vol. II, March 1982

Fig. 7.48a Aerodynamic Characteristics of Turbojet Baseline Missile.

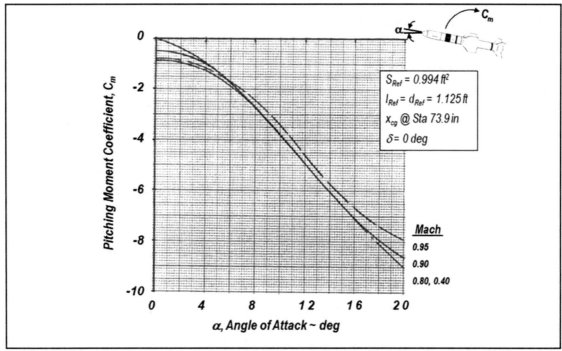

Note: S_{Ref} = Reference area, l_{Ref} = d_{Ref} = Reference length, x_{cg} = Distance from nose tip to center of gravity, δ = Pitch control deflection
Source: Bithell, R.A. and Stoner, R.C. "Rapid Approach for Missile Synthesis", Vol. II, Air-breathing Synthesis Handbook, AFWAL TR 81-3022, Vol. II, March 1982

Fig. 7.48b Aerodynamic Characteristics of Turbojet Baseline Missile (cont).

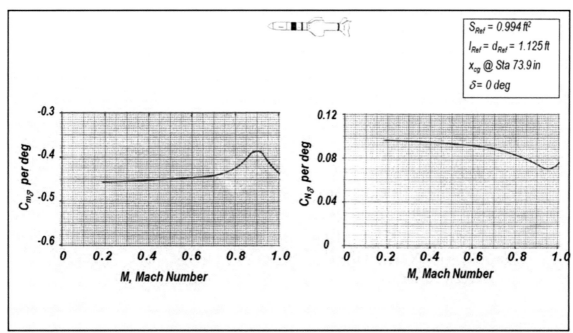

Note: S_{Ref} = Reference area, l_{Ref} = Reference length, x_{cg} = Distance from nose tip to center of gravity, δ = Flight control deflection, C_{m_δ} = Pitching moment coefficient derivative from pitch control deflection, C_{N_δ} = Normal force coefficient derivative from pitch control deflection
Source: Bithell, R.A. and Stoner, R.C. "Rapid Approach for Missile Synthesis", Vol. II, Air-breathing Synthesis Handbook, AFWAL TR 81-3022, Vol. II, March 1982

Fig. 7.48c Aerodynamic Characteristics of Turbojet Baseline Missile (cont).

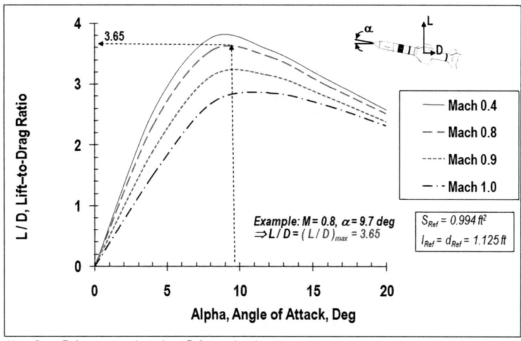

Note: S_{Ref} = Reference area, $I_{Ref} = d_{Ref}$ = Reference length

Source: Bithell, R.A. and Stoner, R.C. "Rapid Approach for Missile Synthesis", Vol. II, Air-breathing Synthesis Handbook, AFWAL TR 81-3022, Vol. II, March 1982

Fig. 7.48d Aerodynamic Characteristics of Turbojet Baseline Missile (cont).

- **Applications**
  - AGM-84A and RGM-84 Harpoon
  - AGM-84E SLAM and AGM-84H SLAM-ER
- **Components**
  - 2-Stage, axial – centrifugal compressor (entrance diameter $d_{entrance}$ = 8.4 in, pressure ratio p_3 / p_2 = 5.4)
  - Annular combustor (temperature T_4 = 1800 R max combustion temperature)
  - Single stage axial flow turbine
  - Convergent nozzle (exit temperature T_{exit} = 1090 R, exit diameter d_{exit} = 8.4 in)
- **Other**
  - Dry weight: 101.5 lb (46.0 kg)
  - Low parts count (castings)
  - Shelf life: 20 years
  - Reliability: 99%
  - Cruise rotational speed: 37,750 rpm
  - Fuel: JP-4, JP-5, JP-8, JP-10

Width: 12.52 in

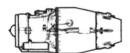
Length: 29.44 in

Source: Teledyne Turbine Engines Brochure, "Model 370, J402-CA-400 Performance," 2006

Fig. 7.49 Turbojet Baseline Missile Engine Is the Teledyne J402-CA-400.

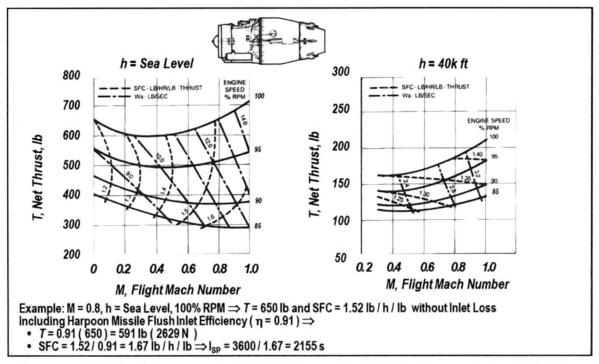

Fig. 7.50 Turbojet Baseline Missile Engine Performance—Teledyne J402-CA-400.

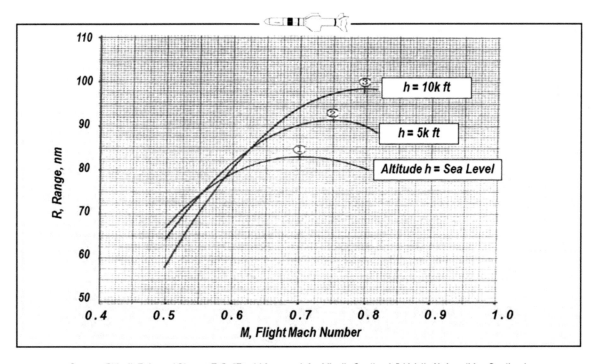

Fig. 7.51 Turbojet Baseline Missile Cruise Performance.

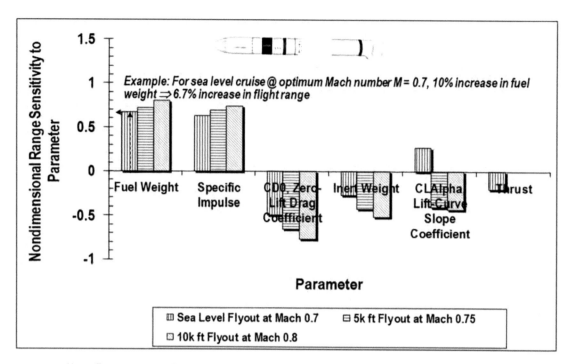

Above Pareto assumes linear sensitivity of each parameter without interactions between parameters

Fig. 7.52 Turbojet Baseline Missile Range Drivers Are Specific Impulse, Fuel Weight, and Zero-Lift Drag Coefficient.

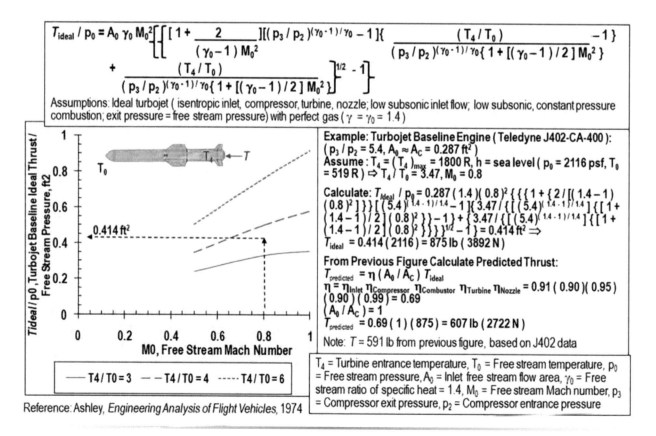

Fig. 7.53 Turbojet Baseline Missile Thrust Is a Function of Mach Number, Altitude, and Combustion Temperature.

$$[g c_p T_0 / (a_0 H_f)] (I_{SP})_{Ideal} = [T / (p_0 A_0)] / \{\gamma_0 M_0 [(T_4/T_0) - (p_3/p_2)^{(\gamma_0-1)/\gamma_0}]\}$$

Ideal turbojet (isentropic inlet, compressor, turbine, nozzle flow; low subsonic, constant pressure combustion; exit pressure = free stream pressure) with perfect gas ($\gamma = \gamma_0 = 1.4$)

Example for Turbojet Baseline Engine (Teledyne J402-CA-400): ($p_3/p_2 = 5.4$, $A_0 \approx A_C = 0.287$ ft²)
Assume: $H_f = 18{,}100$ BTU/lbm (JP-10), $c_p = 0.276$ BTU/lbm/R, $T_4 = (T_4)_{max} = 1800$ R, h = sea level ($p_0 = 2116$ psf, $a_0 = 1116$ fps/s, $T_0 = 519$ R) ⇒ $T_4/T_0 = 3.47$, $M_0 = 0.8$
From prior example: $T_{ideal} / p_0 = 0.414$ ft²

Calculate: $[g c_p T_0 / (a_0 H_f)] (I_{SP})_{ideal} = 0.414 / \{0.287 (1.4)(0.8)[3.47 - (5.4)^{(1.4-1)/1.4}]\} = 0.696$
$(I_{SP})_{ideal} = 0.696 / \{32.2 (0.276)(519) / [1116 (18100)]\} = 0.696 / 0.000221 = 3149$ s

From Previous Figure Calculate Predicted Specific Impulse
$(I_{SP})_{predicted} = \eta (I_{SP})_{ideal}$
$\eta = \eta_{Inlet} \eta_{Compressor} \eta_{Combustor} \eta_{Turbine} \eta_{Nozzle} = 0.91 (0.90)(0.95)(0.90)(0.99) = 0.69$
$(I_{SP})_{predicted} = 0.69 (3149) = 2173$ s

Note: $I_{SP} = 2155$ s from previous figure, based on J402 data

Nomenclature: g = Acceleration of gravity = 32.2 ft/s, c_p = Specific heat at constant pressure, a_0 = Free stream speed of sound, H_f = Heating value of fuel, $(I_{SP})_{Ideal}$ = Ideal specific Impulse, T = Thrust, T_4 = Turbine entrance temperature, T_0 = Free stream temperature, p_0 = Free stream pressure, A_0 = Inlet free stream flow area, γ_0 = Free stream ratio of specific heat = 1.4, M_0 = Free stream Mach number, p_3 = Compressor exit pressure, p_2 = Compressor entrance pressure

Reference: Ashley, H., *Engineering Analysis of Flight Vehicles*, Dover Publications, Inc., 1974

Fig. 7.54 Turbojet Baseline Missile Specific Impulse Is a Function of Mach Number, Altitude, and Combustion Temperature.

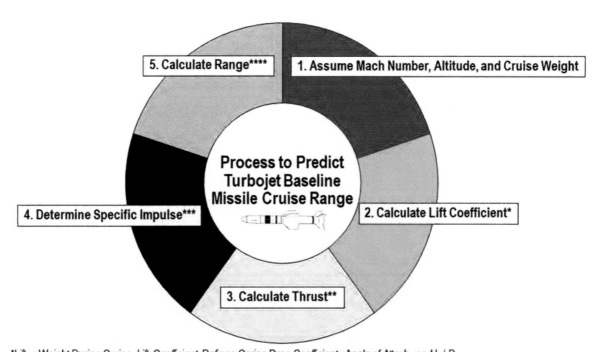

*Lift = Weight During Cruise. Lift Coefficient Defines Cruise Drag Coefficient, Angle of Attack, and L/D.
**Thrust = Drag During Cruise.
***Specific Impulse May Be Calculated from Previous Figure. It Is Also Available from J402 Engine Data in previous figure.
****Flight Range Calculated by Breguet Range Equation.

Fig. 7.55a Process for Turbojet Baseline Missile Cruise Range Prediction.

296 Missile Design Guide

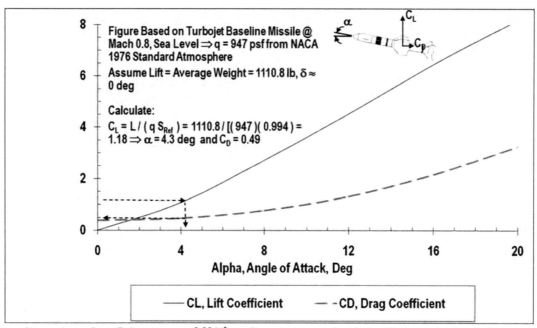

Fig. 7.55b   Process for Turbojet Baseline Missile Cruise Range Prediction (cont).

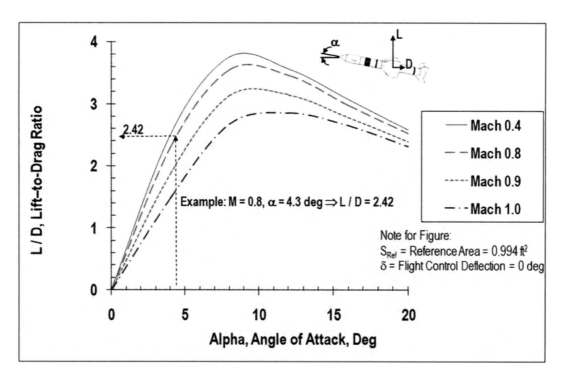

Fig. 7.55c   Process for Turbojet Baseline Missile Cruise Range Prediction (cont).

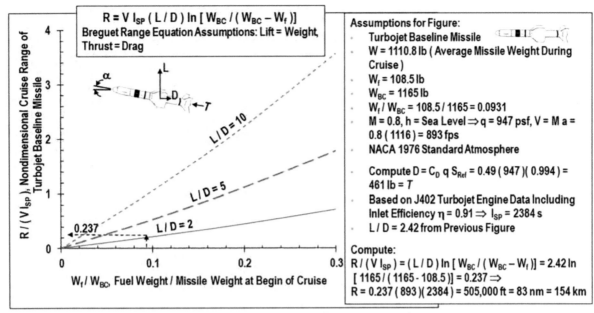

Fig. 7.55d   Process for Turbojet Baseline Missile Cruise Range Prediction (cont).

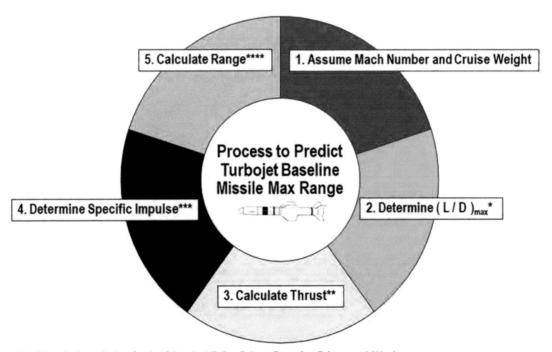

*$(L/D)_{max}$ Defines Cruise Angle of Attack, Lift Coefficient, Drag Coefficient, and Altitude
**Thrust = Drag During Cruise
***Specific Impulse Available from J402 Engine Data
****Flight Range Calculated by Breguet Range Equation

Fig. 7.56a   Process for Turbojet Baseline Missile Maximum Range Prediction.

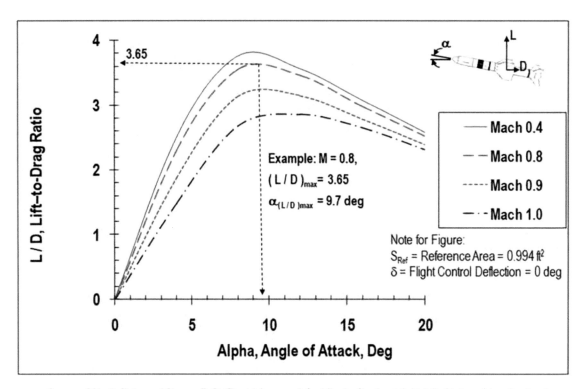

Fig. 7.56b Process for Turbojet Baseline Missile Maximum Range Prediction (cont).

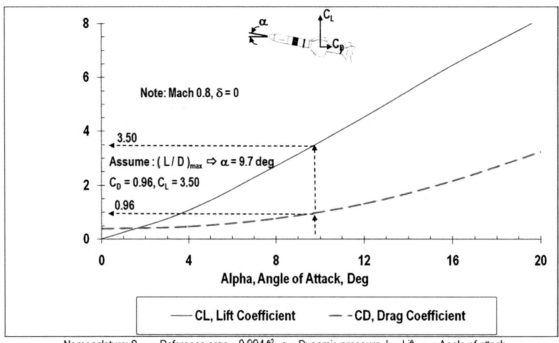

Fig. 7.56c Process for Turbojet Baseline Missile Maximum Range Prediction (cont).

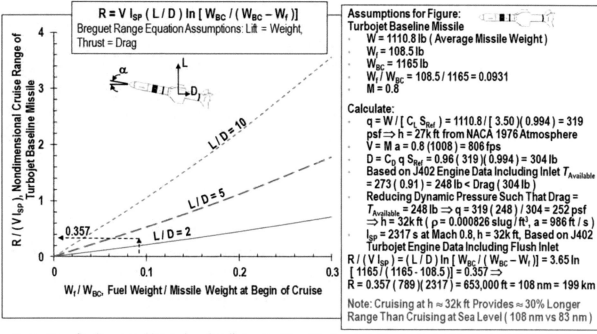

Fig. 7.56d Process for Turbojet Baseline Missile Maximum Range Prediction (cont).

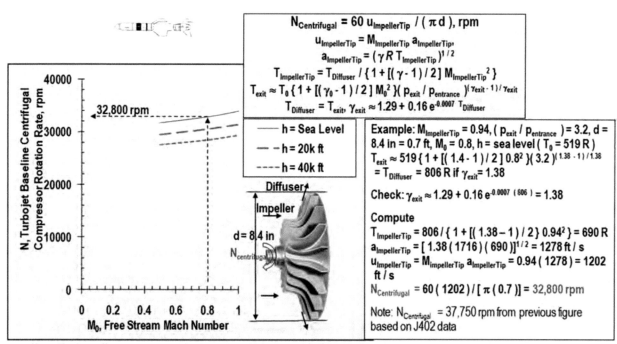

Fig. 7.57 Turbojet Baseline Missile Centrifugal Compressor Has Nearly Constant Rotational Speed.

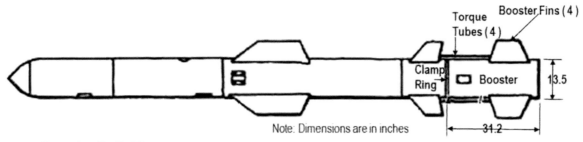

Booster Definition:
- Booster Weight = 360 lb
- Propellant Weight = 143 lb
- Total Impulse = 36,000 lb-s
- Maximum Thrust = 14,470 lb
- Specific Impulse = 251.7 s

Fig. 7.58 For Surface Launch, Turbojet Baseline Missile Has a Drop-Off Booster.
Note: See video supplement, Littoral Combat Ship (LCS) Launch of Harpoon Missile, Courtesy of Journey2Defence.

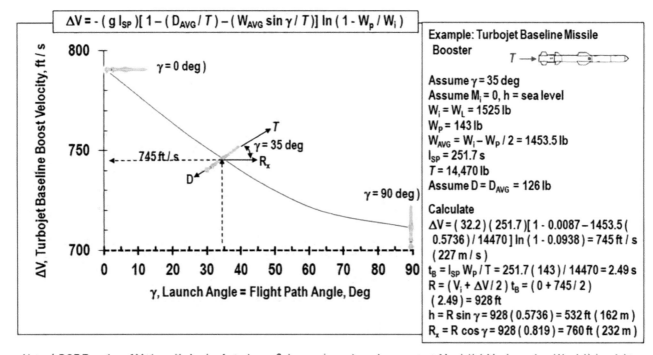

Note: 1-DOF Equation of Motion with Angle of attack $\alpha \approx 0$ deg, γ = Launch angle = constant, M_i = Initial Mach number, W_i = Initial weight, W_{AVG} = Average weight, W_P = Propellant weight, I_{SP} = Specific impulse, T = Thrust, h = Altitude, D_{AVG} = Average drag, ΔV = Incremental velocity, g = Acceleration of gravity = 32.2 ft/s^2, t_B = Burn time, R = Slant range, V_i = Initial velocity, h = Altitude, R_x = Horizontal range

Fig. 7.59 Booster for Turbojet Baseline Missile Provides Approximately 745 fps Boost Velocity.

7.5 Baseline Guided Bomb

Fig. 7.60 Baseline Guided Bomb (GBU-31 JDAM) Has Many Fighter and Bomber Launch Platforms.
Note: See video supplement, Joint Direct Attack (JDAM) Production.

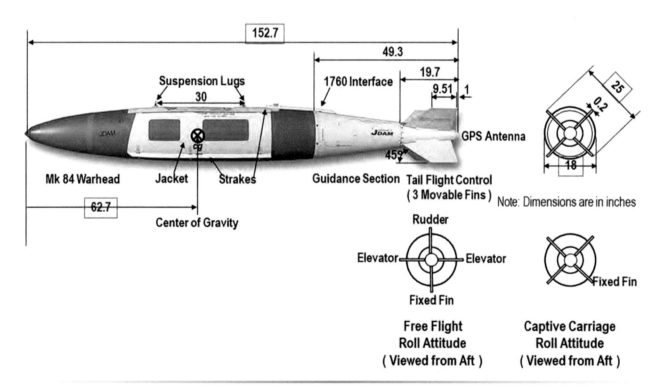

Fig. 7.61 Drawing of Baseline Guided Bomb Configuration.

Component	Weight, lb	C.G. STA, in
Mk 84 Warhead Case	976	
Mk 84 Explosive	993	
Tail Cone Structure and Fins	24	
Guidance (GPS / INS)	66	
Total	2059	62.7

Nomenclature: C.G. = Center of gravity, GPS = Global Positioning System, INS = Inertial navigation system

Fig. 7.62 Mass Properties of Baseline Guided Bomb.

Body
- Case material: Steel
- Explosive material: Tritonal
- Length, in: 152.7
- Diameter, in: 18
- Fineness ratio: 8.48
- Nose fineness ratio: 3.00
- Boattail angle, deg: 6.5

Tails
- Planform area, ft^2: 2.54
- Aspect ratio (2 panels exposed): 1.71
- Root chord, in: 19.73
- Tip chord, in: 9.53
- Span, in (2 panels exposed): 25.0
- Leading edge sweep, deg: 45.0

References values
- Reference area, ft^2: 1.767
- Reference length, ft: 1.5
- Center-of-gravity, in: 62.7
- Pitch / yaw moment of inertia, slug-ft^2: 406.6

Fig. 7.63 Baseline Guided Bomb Definition.

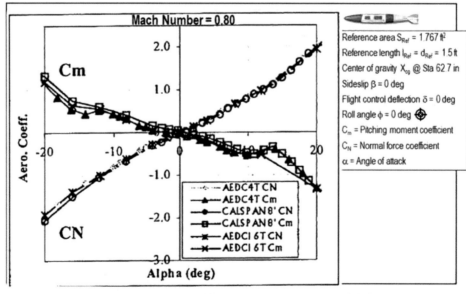

Source: Ray, E., "Comparison of AEDC 4T and Calspan 8-ft Wind Tunnels for FA-18C / JDAM", AIAA-2000-0793, January 2000.

Fig. 7.64a Aerodynamic Characteristics of Baseline Guided Bomb.

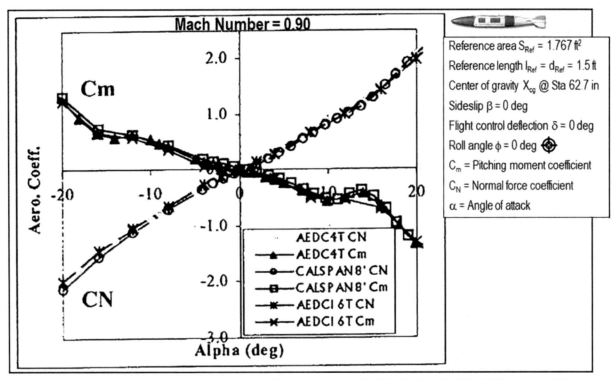

Fig. 7.64b Aerodynamic Characteristics of Baseline Guided Bomb (cont).

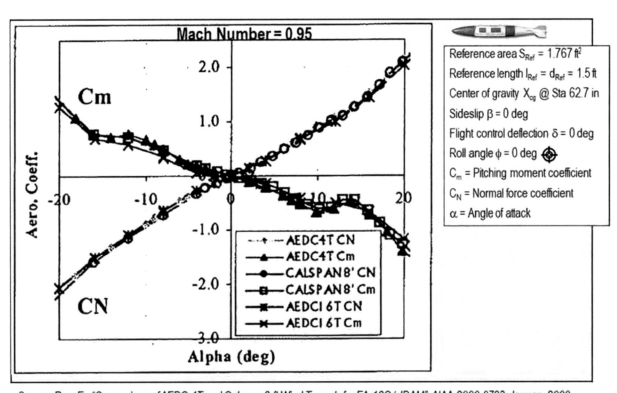

Fig. 7.64c Aerodynamic Characteristics of Baseline Guided Bomb (cont).

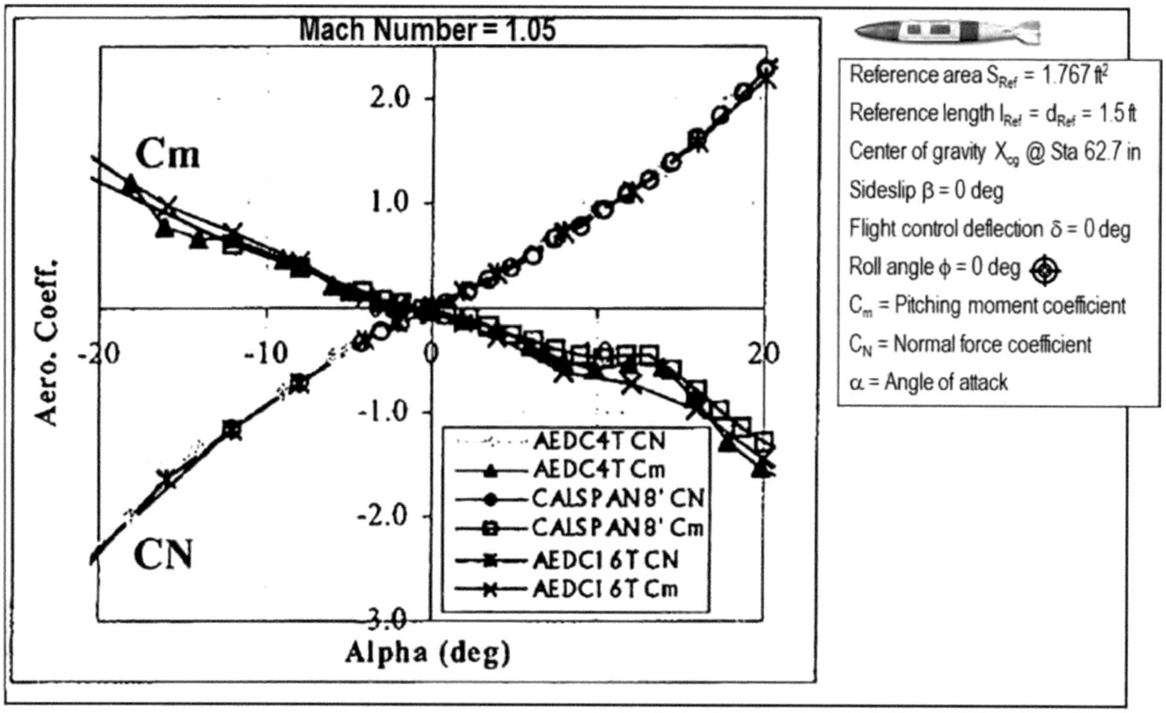

304 Missile Design Guide

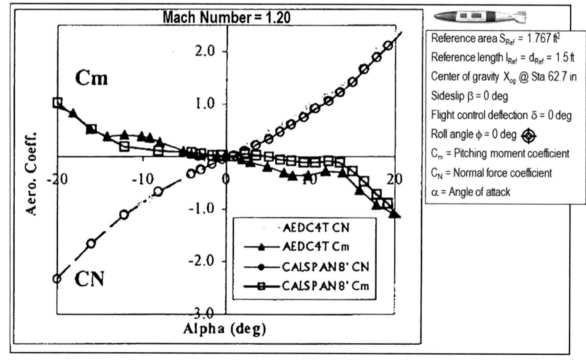

Fig. 7.64d Aerodynamic Characteristics of Baseline Guided Bomb (cont).

Fig. 7.64e Aerodynamic Characteristics of Baseline Guided Bomb (cont).

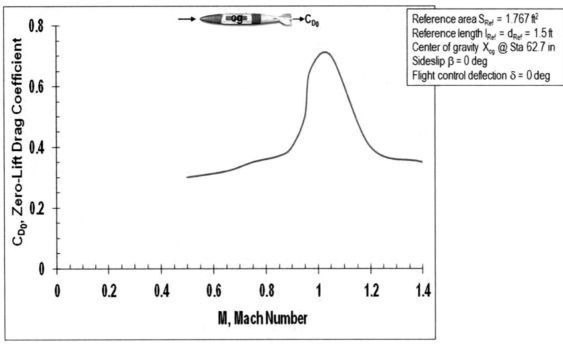

Source: Donegan, T. L., et al, "Computational Study of Afterbody Effects for the Joint Direct Attack Munition (JDAM)", AIAA 98-2799, July 1998.

Fig. 7.64f Aerodynamic Characteristics of Baseline Guided Bomb (cont).

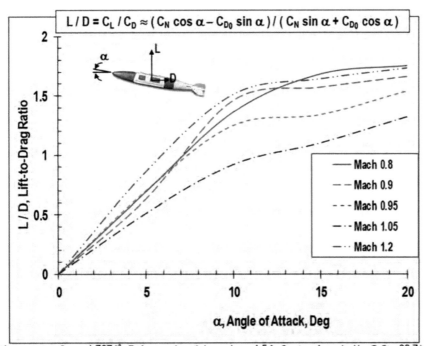

Note: Reference area S_{Ref} = 1.767 ft², Reference length l_{Ref} = d_{Ref} = 1.5 ft, Center of gravity X_{cg} @ Sta 62.7 in, Sideslip β = 0 deg, Flight control deflection δ = 0 deg, Roll angle ϕ = 0 deg, α = Angle of attack, C_L = Lift coefficient, C_D = Drag coefficient, C_N = Normal force coefficient, C_{D0} = Zero-lift Drag coefficient

Fig. 7.64g Aerodynamic Characteristics of Baseline Guided Bomb (cont).

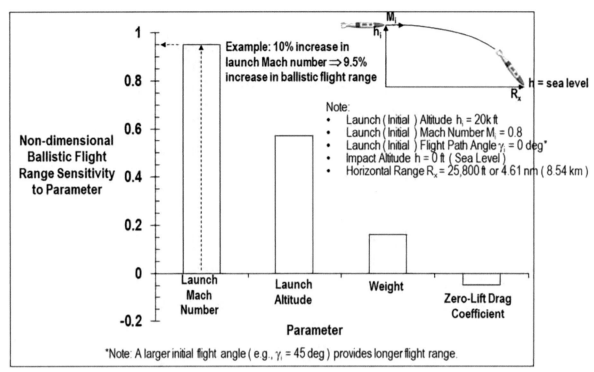

Fig. 7.65 Baseline Guided Bomb Ballistic Flight Range Is Driven by Launch (Initial) Velocity and Altitude.

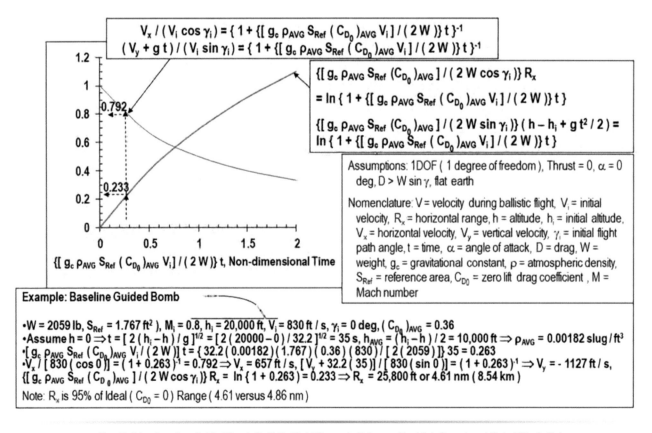

Fig. 7.66 Baseline Guided Bomb Ballistic Flight Range Is Enhanced by High-Speed and High-Altitude Release.

CHAPTER 7 Sizing Examples and Sizing Tools

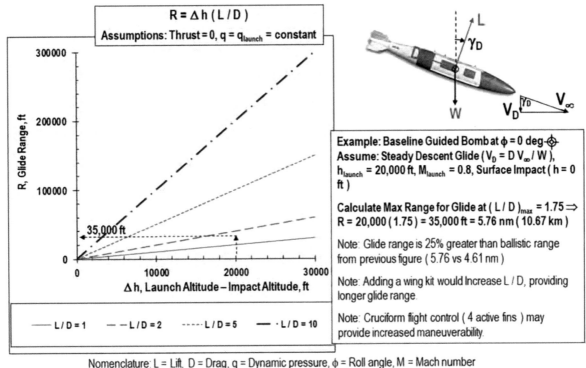

Fig. 7.67 Baseline Guided Bomb Glide Range Is Driven by Lift-to-Drag Ratio and Initial Glide Altitude.

7.6 Computer Aided Conceptual Design Sizing Tools

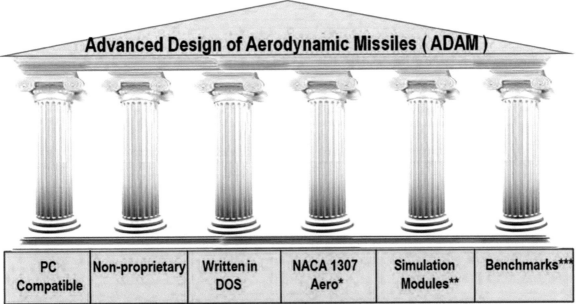

Fig. 7.68 ADAM Is a Missile Conceptual Design Sizing Code.

308 Missile Design Guide

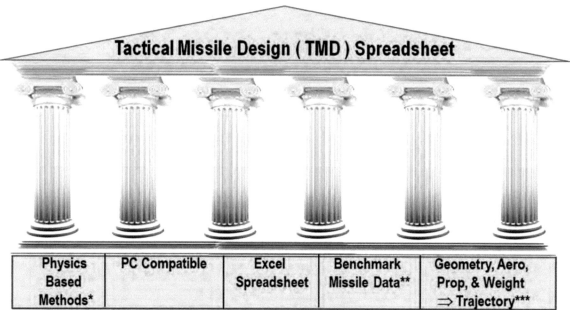

Fig. 7.69a Tactical Missile Design (TMD) Spreadsheet Is a Missile Conceptual Design Sizing Code.

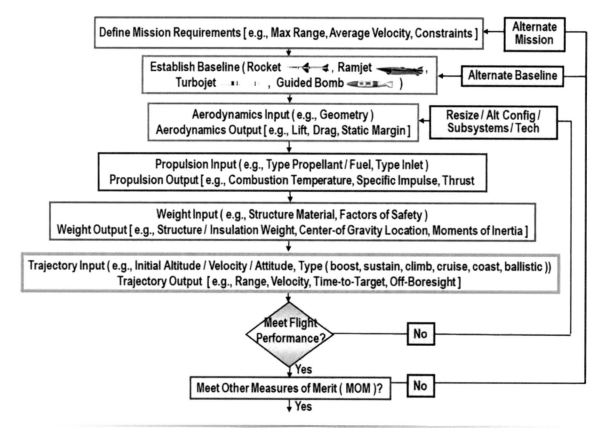

Fig. 7.69b Tactical Missile Design (TMD) Spreadsheet Is a Missile Conceptual Design Sizing Code (cont).

CHAPTER 7 Sizing Examples and Sizing Tools

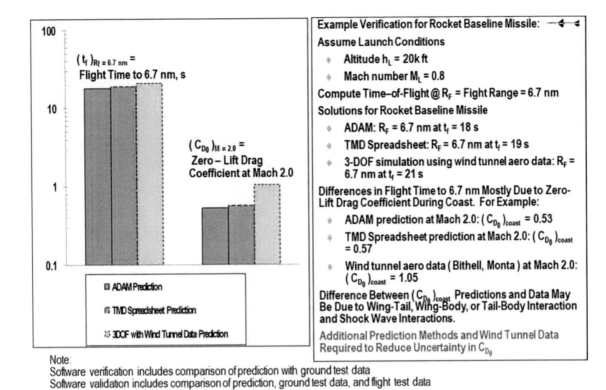

Fig. 7.70 TMD Spreadsheet Verification Is Based on Comparing with Alternative Prediction and Test Data.

7.7 Soda Straw Rocket (DBF, Pareto, Uncertainty Analysis, HOQ, DOE)

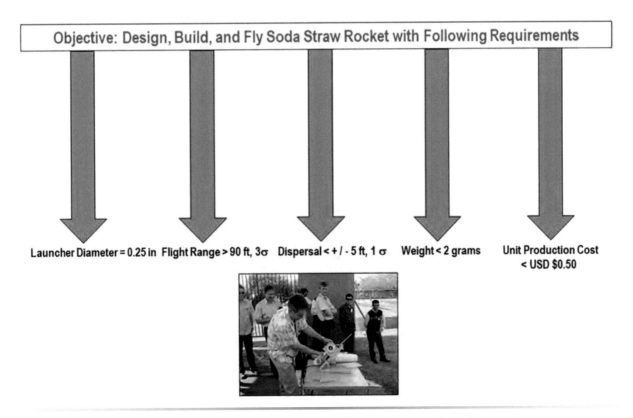

Fig. 7.71a Soda Straw Rocket Customer Requirements for Design, Build, and Fly.

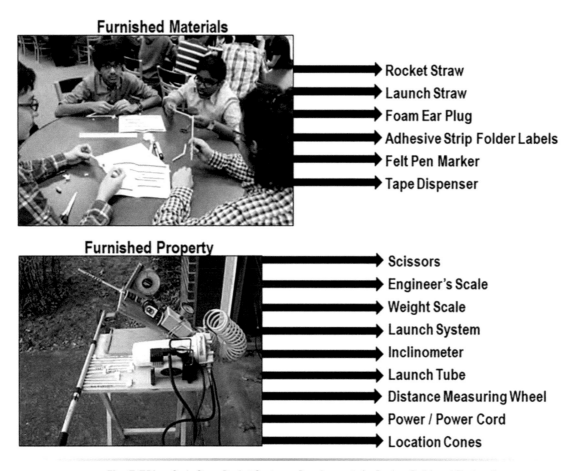

Fig. 7.71b Soda Straw Rocket Customer Requirements for Design, Build, and Fly (cont).

- **Specified Launch Constraints / Conditions**
  - Launch Pressure: 30 psi
  - Launch Elevation Angle: 30 deg
- **Compare Predictions with Test Data**

Fig. 7.71c Soda Straw Rocket Customer Requirements for Design, Build, and Fly (cont).
Note: See video supplement, Examples of Soda Straw Rocket Design, Build, and Fly Competition.

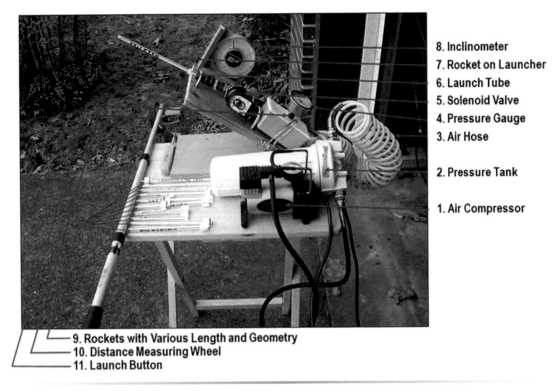

8. Inclinometer
7. Rocket on Launcher
6. Launch Tube
5. Solenoid Valve
4. Pressure Gauge
3. Air Hose

2. Pressure Tank

1. Air Compressor

9. Rockets with Various Length and Geometry
10. Distance Measuring Wheel
11. Launch Button

Fig. 7.72 Soda Straw Rocket Has an Air Pressure Launcher.

Assume "Giant" Soda Straw Rocket for Initial Baseline

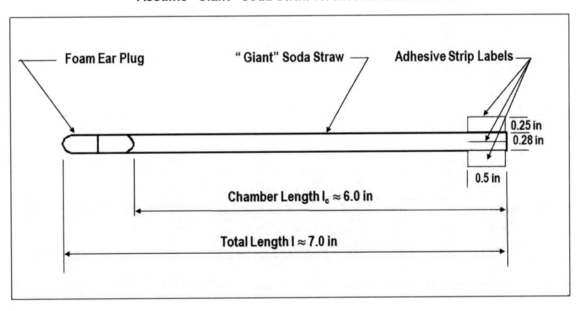

Foam Ear Plug · "Giant" Soda Straw · Adhesive Strip Labels

0.25 in
0.28 in
0.5 in

Chamber Length $l_c \approx 6.0$ in

Total Length $l \approx 7.0$ in

Note:
- 0.28 in diameter soda straw rocket is compatible with 0.25 in diameter launcher
- Chamber length may be less than 6.0 in and total length may be less than 7.0 in, depending upon shaping and insertion of ear plug

Fig. 7.73 Soda Straw Rocket Initial Baseline Configuration.

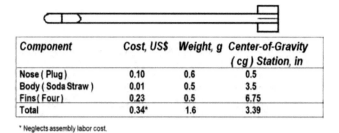

Component	Cost, US$	Weight, g	Center-of-Gravity (cg) Station, in
Nose (Plug)	0.10	0.6	0.5
Body (Soda Straw)	0.01	0.5	3.5
Fins (Four)	0.23	0.5	6.75
Total	0.34*	1.6	3.39

* Neglects assembly labor cost.

Fig. 7.74 Soda Straw Rocket Initial Baseline Cost, Weight, and Balance.

Body
Material type	HDPE Plastic
Material density, lbm / in³	0.043
Material strength (σ_{max}), psi	4600
Thickness (t), in	0.004
Length, in	≈ 7.0
Diameter, in	0.28
Fineness ratio (Length / Diameter)	≈ 25.0
Nose fineness ratio	0.5
Burst pressure ($p_{burst} \approx \sigma_{max}\, t / r$) = 4600 (0.004) / (0.14), psi	131

Fins
Material	Plastic
Planform area (2 panels exposed), in²	0.25
Wetted area (4 panels), in²	1.00
Aspect ratio (2 panels exposed)	1.00
Taper ratio	1.0
Chord, in	0.5
Span (exposed), in	0.5
Span (total including body), in	0.78
Leading edge sweep, deg	0
Longitudinal location of leading edge of mean aerodynamic chord x_{mac}, in	≈ 6.625

Fig. 7.75a Soda Straw Rocket Initial Baseline Definition.

Nose
Material type	Foam
Material density, lbm / in³	0.012
Average diameter (before shaping)	0.39 in
Length	≈ 0.90 in

Reference Values
Reference area (body cross section), in²	0.0616
Reference dimension (Body diameter), in	0.28

Thrust Performance
Inside cavity length, in	≈ 6.0
Typical pressure, psi	30
Maximum thrust @ 30 psi pressure, lb	1.47
Launcher solenoid time constant (standard temperature), s	0.025

Fig. 7.75b Soda Straw Rocket Initial Baseline Definition (cont.).

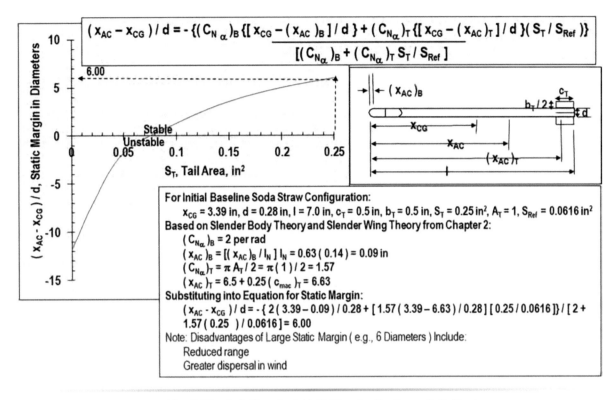

Fig. 7.76 Soda Straw Rocket Initial Baseline Has Excess Static Margin.

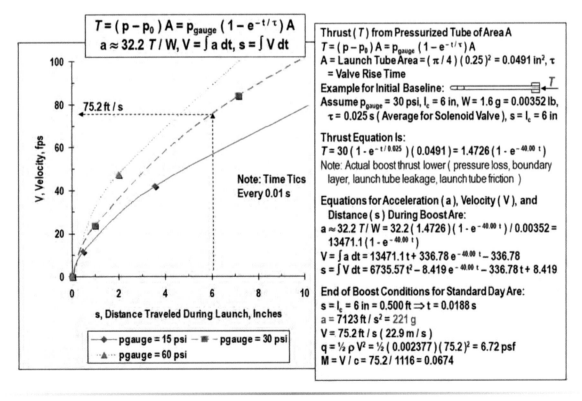

Fig. 7.77 Soda Straw Rocket Has High Acceleration Boost Performance.

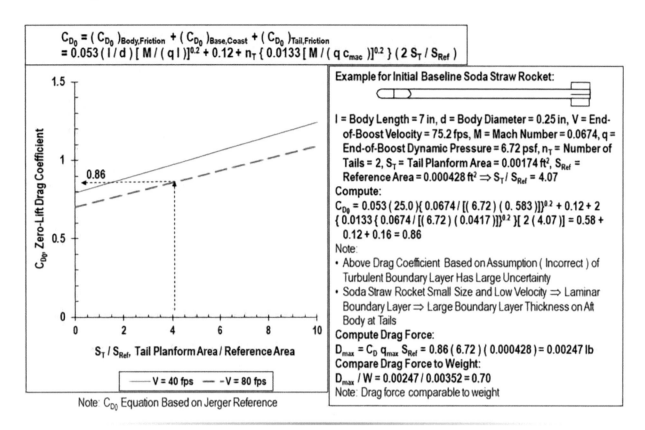

Fig. 7.78 Most of the Soda Straw Rocket Drag Coefficient Is from Skin Friction.

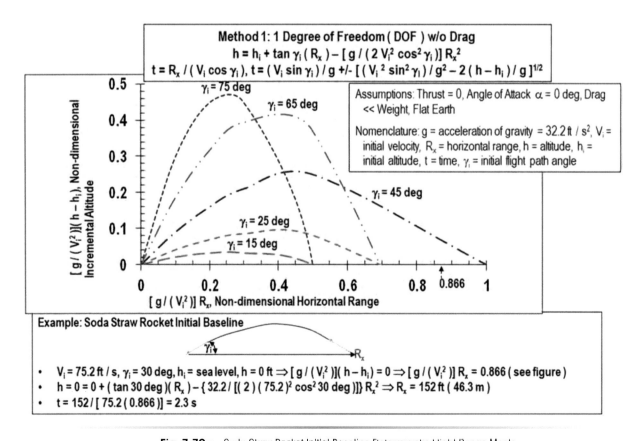

Fig. 7.79a Soda Straw Rocket Initial Baseline Deterministic Flight Range Meets 90 Feet Range Requirement.

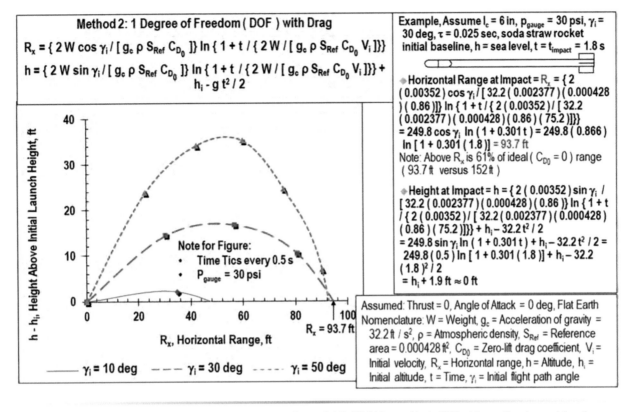

Fig. 7.79b Soda Straw Rocket Initial Baseline Deterministic Flight Range Meets 90 Feet Range Requirement (cont).

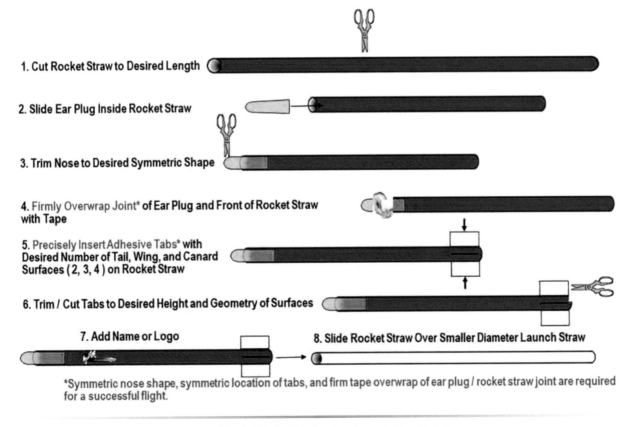

Fig. 7.80 Work Instructions for Soda Straw Rocket.

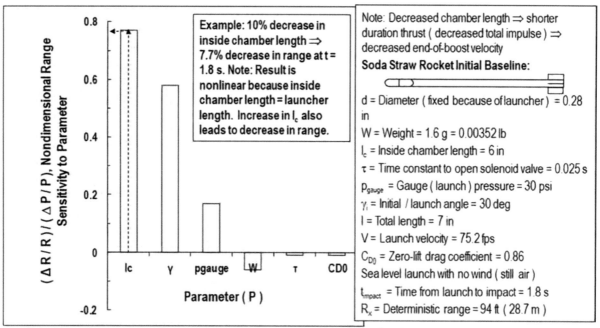

Fig. 7.81 Range Pareto Analysis Shows Fixed Diameter Soda Straw Rocket Baseline Range Is Driven by Length and Launch Angle.

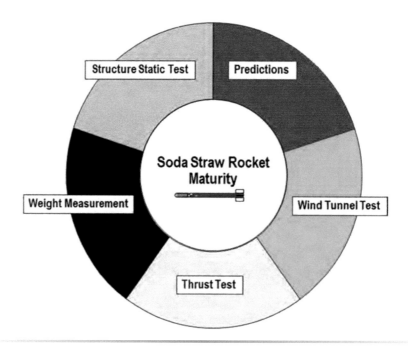

Fig. 7.82 Maturity Assessment of Soda Straw Rocket Is Based on a Comparison of Predictions and Tests.

P, Parameter	Baseline Value	($\Delta P/P$)	($\Delta R/R$)/($\Delta P/P$)	$\Delta R/R$
1. Chamber Length (l_c)	6 in	+/- 2%, 1σ	0.77	+/- 1.5%, 1σ
2. Initial (Launch) Angle (γ_i)	30 deg	+/- 3%, 1σ	0.58	+/- 1.7%, 1σ
3. Gauge Pressure (p_{gauge})	30 psi	+/- 3%, 1σ	0.17	+/- 0.5%, 1σ
4. Weight (W)	1.6 g	+/- 6%, 1σ	- 0.06	+/- 0.4%, 1σ
5. Solenoid Valve Time Constant (τ)	0.025 s	+/- 20%, 1σ	- 0.01	+/- 0.2%, 1σ
6. Zero Lift Drag Coefficient (C_{D_0})	0.86	+/- 20%, 1σ	- 0.01	+/- 0.2%, 1σ

♦ Customer Flight Range Requirement Is R > 90 ft, 3σ
♦ Total Flight Range Uncertainty for 30 psi Launch at 30 deg at Sea Level Standard Atmosphere in Still Air
  ♦ $\Delta R/R = [(\Delta R/R)_1^2 + (\Delta R/R)_2^2 + (\Delta R/R)_3^2 + (\Delta R/R)_4^2 + (\Delta R/R)_5^2 + (\Delta R/R)_6^2]^{1/2} = $ +/- 2.4%, 1σ
  ♦ R = 94 ft +/- 2.3 ft, 1σ ⇨ <u>87</u> ft < $R_{3\sigma}$ < 101 ft

Note: Primary Drivers of Uncertainty in Flight Range for Assumed Fixed Diameter Soda Straw Rocket Are Uncertainties in Chamber Length l_c and Launch Angle γ_i.

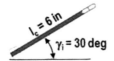

Note: Soda straw rocket diameter d = 0.28 in because constraint of specified launch straw diameter = 0.25 in.

Fig. 7.83 Because of Uncertainty, Soda Straw Rocket Initial Baseline Does Not Meet Range Requirement.

Concept	Tail Planform Area	Length	Weight	Unit Cost	Drag Coefficient @ 75 ft/s	Static Margin	3σ Range
Initial Baseline	0.250 in²	7.0 in	1.6 g	$0.34	0.86	6.00 diam	<u>87</u> to 101 ft (Fails ≥ 90 ft, 3σ Requirement)
Revised Baseline	0.0784 in²	7.0 in	1.2 g	$0.13	0.75	0.17 diam	93 to 107 ft (Meets ≥ 90 ft, 3σ Requirement)

Note: Revised baseline tails are made from light weight, low cost paper adhesive strip tabbing

Fig. 7.84 Revised Soda Straw Rocket with Smaller Tails (Lower Static Margin, Lighter Weight, Lower Drag) Meets Range Requirement.

Notional Example Design Space of Soda Straw Rocket Range / Dispersal / Weight

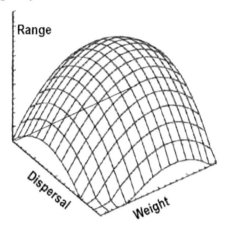

Broad Envelope of Possible Design Solutions, e.g. Extremes:
1. Longest Flight Range That Also Meets Dispersal, Weight and Cost Requirements
2. Lowest Dispersal That Also Meets Range, Weight, and Cost Requirements
3. Lightest Weight that Also Meets Range, Dispersal, and Cost Requirements
4. Lowest Cost that Also Meets Range, Dispersal, and Weight Requirements

Fig. 7.85a House of Quality Translates Customer Requirements into Engineering Emphasis.

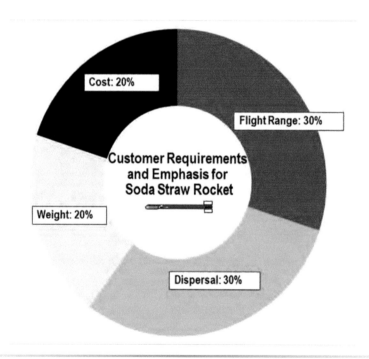

Fig. 7.85b House of Quality Translates Customer Requirements into Engineering Emphasis (cont).

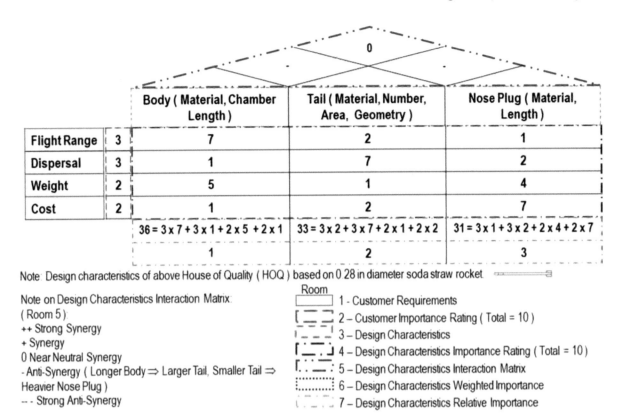

		Body (Material, Chamber Length)	Tail (Material, Number, Area, Geometry)	Nose Plug (Material, Length)
Flight Range	3	7	2	1
Dispersal	3	1	7	2
Weight	2	5	1	4
Cost	2	1	2	7
		36 = 3 x 7 + 3 x 1 + 2 x 5 + 2 x 1	33 = 3 x 2 + 3 x 7 + 2 x 1 + 2 x 2	31 = 3 x 1 + 3 x 2 + 2 x 4 + 2 x 7
		1	2	3

Note: Design characteristics of above House of Quality (HOQ) based on 0.28 in diameter soda straw rocket.

Note on Design Characteristics Interaction Matrix:
(Room 5):
++ Strong Synergy
+ Synergy
0 Near Neutral Synergy
- Anti-Synergy (Longer Body ⇒ Larger Tail, Smaller Tail ⇒ Heavier Nose Plug)
-- Strong Anti-Synergy

Room
1 – Customer Requirements
2 – Customer Importance Rating (Total = 10)
3 – Design Characteristics
4 – Design Characteristics Importance Rating (Total = 10)
5 – Design Characteristics Interaction Matrix
6 – Design Characteristics Weighted Importance
7 – Design Characteristics Relative Importance

Fig. 7.85c House of Quality Translates Customer Requirements into Engineering Emphasis (cont).

Design of Experiments (DOE) Approach	Number Required Designs	Designs Required for Simple Design (2 Parameters), 2 Levels	Designs Required for More Complex Design (5 Parameters), 2 Levels	Time Required for Each Design	Interactions / Confidence in Predicted Optimum
Parametric Study Using One Factor at a Time (OFAT)	$1 + k(l-1)$	☺ (3)	☺ (6)	☺	☹ (No Interaction)
Adaptive OFAT	$1 + k(l-1)$	☺ (3)	☺ (6)	☺	☹ to ☺ (Some)
Full Factorial	l^k	☺ (4)	☹ (32)	☹	☺ (Excellent)

☺ Superior ☺ Average ☹ Below Average

Note:
k = number of parameters, l = number of levels*
*Specifying l = 2 levels for a parameter assumes a small response from changing the parameter. However, if two levels of a parameter result in a large response, there is likely a nonlinear sensitivity and more than two levels are required.

Reference: Montgomery, D.G., *Design and Analysis of Experiments*, John Wiley & Sons, 2001

Fig. 7.86 Design of Experiments (DOE) Approaches Are a Tradeoff of Time Required vs. Confidence in Result.

Example of Parametric Study: 4 parameters, 2 Levels ⇨ 1 + k (l – 1) ⇨ 1 + 4 (2 – 1) = 5 DOE Designs

Design Number	Parameter (x_1)	Parameter (x_2)	Parameter (x_3)	Parameter (x_4)	Result (y)
1	Low	Low	Low	Low	y_1
2	High	Low	Low	Low	y_2
3	Low	High	Low	Low	y_3
4	Low	Low	High	Low	y_4
5	Low	Low	Low	High	y_5

One Factor at a Time (OFAT) Design of Experiments (DOE) Process:
1. Specify levels for each parameter
2. Change one parameter at a time, with all others at row 1 level
3. Compare row 1 with other rows to find best result (y_{Best}) for each x
   1) Compare y_1, y_2 ⇨ (x_1)$_{Best}$
   2) Compare y_1, y_3 ⇨ (x_2)$_{Best}$
   3) Compare y_1, y_4 ⇨ (x_3)$_{Best}$
   4) Compare y_1, y_5 ⇨ (x_4)$_{Best}$
4. DOE "best" design consists of the x for each best row result. y_{best} consists of ((x_1)$_{Best}$, (x_2)$_{Best}$, (x_3)$_{Best}$, (x_4)$_{Best}$)
5. Compare DOE "best" design with baseline and refine baseline for next iteration

Note: OFAT assumes linear sensitivity of each parameter without interactions between parameters.

Fig. 7.87 A DOE Parametric Study Changes One Factor at a Time, with All Others Held at Base Level.

Example of Adaptive OFAT Study with 4 parameters, 2 Levels ⇨ 1 + k (l – 1) ⇨ 1 + 4 (2 – 1) = 5 DOE Designs

Design Number	Parameter (x_1)	Parameter (x_2)	Parameter (x_3)	Parameter (x_4)	Result (y)
1	Low	Low	Low	Low	y_1
2	High	Low	Low	Low	y_2
3	Best	High	Low	Low	y_3
4	Best	Best	High	Low	y_4
5	Best	Best	Best	High	y_5

Adaptive One Factor at a Time (OFAT) Design of Experiments (DOE) Process:
1. Specify row 1 and row 2 levels for first parameter x_1
2. Compare results for first parameter x_1
3. Set new level for x_1 in row 3 based on row 1 vs row 2 that gives best result
4. Move on to next parameter (x_2) and repeat until all rows evaluated
5. Assume row that provides best y is y_{Best}
6. Compare y_{Best} with baseline and refine baseline for next iteration

Note: Adaptive OFAT provides some limited interactions between parameters.

Fig. 7.88 A DOE Adaptive OFAT Sequentially Selects the Values of Parameters, Based on Results.

Example of Design of Experiments Full Factorial Study with 4 parameters, 2 Levels ⇨ $l^k = 2^4 = 16$ DOE Designs

Design Number	Parameter (x_1)	Parameter (x_2)	Parameter (x_3)	Parameter (x_4)	Result (y)
1	Low	Low	Low	Low	y_1
2	High	Low	Low	Low	y_2
3	Low	High	Low	Low	y_3
4	High	High	Low	Low	y_4
5	Low	Low	High	Low	y_5
6	High	Low	High	Low	y_6
7	Low	High	High	Low	y_7
8	High	High	High	Low	y_8
9	Low	Low	Low	High	y_9
10	High	Low	Low	High	y_{10}
11	Low	High	Low	High	y_{11}
12	High	High	Low	High	y_{12}
13	Low	Low	High	High	y_{13}
14	High	Low	High	High	y_{14}
15	Low	High	High	High	y_{15}
16	High	High	High	High	y_{16}

Process: Select Row that Provides y_{Best}

Fig. 7.89 A DOE Full Factorial Evaluates All Combinations of Parameters, but Requires More Time.

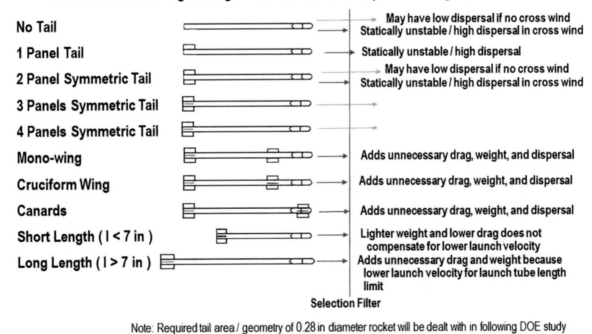

Example: Evaluation of Design of Experiments (DOE) Geometric Design Space for 0.28 in Diameter Soda Straw Rocket Flight Range to Screen the Most Important Design Parameters

No Tail — May have low dispersal if no cross wind / Statically unstable / high dispersal in cross wind
1 Panel Tail — Statically unstable / high dispersal
2 Panel Symmetric Tail — May have low dispersal if no cross wind / Statically unstable / high dispersal in cross wind
3 Panels Symmetric Tail
4 Panels Symmetric Tail
Mono-wing — Adds unnecessary drag, weight, and dispersal
Cruciform Wing — Adds unnecessary drag, weight, and dispersal
Canards — Adds unnecessary drag, weight, and dispersal
Short Length (l < 7 in) — Lighter weight and lower drag does not compensate for lower launch velocity
Long Length (l > 7 in) — Adds unnecessary drag and weight because lower launch velocity for launch tube length limit

Selection Filter

Note: Required tail area / geometry of 0.28 in diameter rocket will be dealt with in following DOE study

Fig. 7.90 A DOE Study Should Use Engineering Intuition in Searching the Broad Possible Design Space.

Example: 0.28 in Diameter Soda Straw Rocket Tail Geometric Design Space for Range and Dispersal

Engineering Characteristics Envelope	$n_{Tail\ Panels}$, Number of Tail Panels	S_T, Tail Planform Area, in^2
Lower Value	3	0.0392
Upper Value	4	0.1568

Note:
* Lower value of S_T, based on ½ revised baseline tail area. It is considered a reasonable lower value because it is probably statically unstable
* Upper value of S_T, based on 2x revised baseline tail area. It is considered a reasonable upper value because it probably has excess static stability
* For the same static margin, a 3-panel tail design will require larger individual tail area than a 4-panel tail design

Fig. 7.91 A Design of Experiments (DOE) Study Should Search the Broad Possible Design Space.

As an Example, Based on Slender Body Theory and Slender Wing Theory, Soda Straw Rocket Geometry Should Be Comparable to an Operational Unguided Rocket with Near-Neutral Static Stability (e.g., Hydra 70, M26 MLRS)

Concept	Sketch	l / d, Total Length / Diameter	b / d, Total Tail Span / Diameter	c / d, Tail Chord / Diameter
Revised Baseline		25	2.00	1
Hydra 70		15.1	2.66	1
M26 MLRS	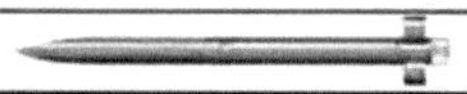	17.3	2.31	0.97

Note: A body-rectangular tail [$S_T = (b - d) c$, $S_{Ref} = (\pi / 4) d^2$] rocket with the center-of-gravity in the center of the rocket ($x_{CG} = l / 2$), slender body theory [($x_{AC})_B \approx 0$, ($C_{N_\alpha})_B = 2$ per rad], and slender surface theory [($x_{AC})_T \approx l$, ($C_{N_\alpha})_T = \pi A_{Exposed} / 2$ per rad, has neutral static stability if $b / d \approx 2$ and $c / d \approx 1$.

Fig. 7.92 Engineering Experience Should Guide Design of Experiments (DOE) Study.

DOE Parametric One Factor at a Time (OFAT) Study with k = 2, l = 2 ⇒ 1 + k (l – 1) = 3 Designs
Example: 0.28 in Diameter Soda Straw Rocket

Design Number	Number of Tail Panels (x_1)	Tail Planform Area (x_2)	Result (Range, Dispersal) (y)
1	3	0.0392 in^2	To Be Determined (TBD)
2	4	0.0392 in^2	TBD
3	3	0.1568 in^2	TBD

Design of Experiments (DOE) OFAT Process
1. Compare $y_1, y_2 \Rightarrow (x_1)_{Best}$
2. Compare $y_1, y_3 \Rightarrow (x_2)_{Best}$
3. DOE "best" design consists of the x for each best row result. y_{best} consists of (($x_1)_{Best}$, ($x_2)_{Best}$,)
4. Compare DOE best design with revised baseline design for next iteration

Note: OFAT assumes linear sensitivity of each parameter without interactions between parameters

Fig. 7.93 A DOE Parametric OFAT Study Provides a Quick Evaluation but Has Lowest Confidence.

DOE Adaptive OFAT Study Example with k = 2, l = 2 ⇨ 1 + k (l – 1) = 3 Designs
Example: 0.28 in Diameter Soda Straw Rocket

Design Number	Number of Tail Panels (x_1)	Tail Planform Area (x_2)	Result (Range, Dispersal) (y)
1	3	0.0392 in²	To Be Determined (TBD)
2	4	0.0392 in²	TBD
3	$(x_1)_3 = (x_1)_{Best}$	0.1568 in²	TBD

Design of Experiments (DOE) Adaptive One Factor at a Time (OFAT) Process
1. Compare y_1, y_2 ⇨ $(x_1)_{Best}$
2. Set level for $(x_1)_3 = (x_1)_{Best}$
3. Compare y_2, y_3 ⇨ $(x_2)_{Best}$
4. Compare DOE best design with revised baseline design for next iteration

Note: DOE Adaptive OFAT includes some limited interactions between parameters

Fig. 7.94 DOE Adaptive OFAT Provides a Quick Evaluation with Higher Confidence Than Parametric OFAT.

DOE Full Factorial Study Example with k = 2, l = 2 ⇨ l^k = 4 Designs
Example: 0.28 in Diameter Soda Straw Rocket

Design Number	Number of Tail Panels (x_1)	Tail Planform Area (x_2)	Result (Range, Dispersal) (y)
1	3	0.0392 in²	To Be Determined (TBD)
2	4	0.0392 in²	TBD
3	3	0.1568 in²	TBD
4	4	0.1568 in²	TBD

Note Row that Provides y_{Best}

Design of Experiments (DOE) Full Factorial Study Process to Determine x_1, x_2 for Optimum (max / min):
1. Assume $y = \beta_0 + \beta_1 x_1 + \beta_2 x_2 + \beta_{12} x_1 x_2$
2. Find values of $\beta_0, \beta_1, \beta_2, \beta_{12}$, by solving equations from results of 4 designs
3. Optimum (max / min) occurs if
   $\Delta y / \Delta x_1 = 0$
   $\Delta y / \Delta x_2 = 0$

Note: Full Factorial DOE includes interactions between parameters

Fig. 7.95 A Full Factorial Design of Experiments (DOE) Study Provides Highest Confidence but Requires More Time.

Chapter 8 Development Process

8.1 Missile Technology and System Development Process

Fig. 8.1 Chapter 8 Development Process—What You Will Learn.

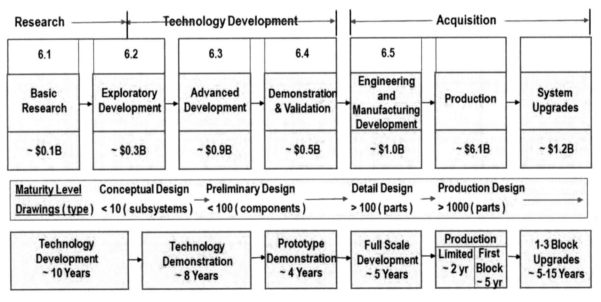

Note:
Total US Department of Defense (DoD) Research and Technology Development for tactical missiles ≈ $1.8 billion per year
Total US DoD Acquisition (EMD + Production + Upgrades) for tactical missiles ≈ $8.3 billion per year
Tactical missiles ≈ 11% of U.S. DoD RT&A budget
US Industry Independent Research & Development (IR&D) typically similar to US DoD 6.2 and 6.3A

Fig. 8.2 The US Has a Systematic Research, Technology Development, and Acquisition Process for Missile Development.

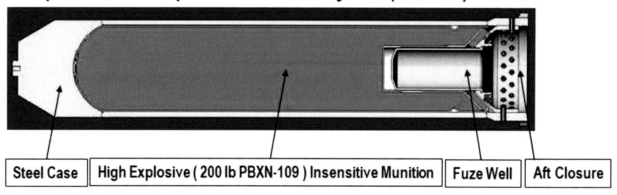

Fig. 8.3 Engineering and Manufacturing Development (EMD) Requires Subsystem Tests.
Note: See video supplement, GMLRS Warhead EMD Demonstration.

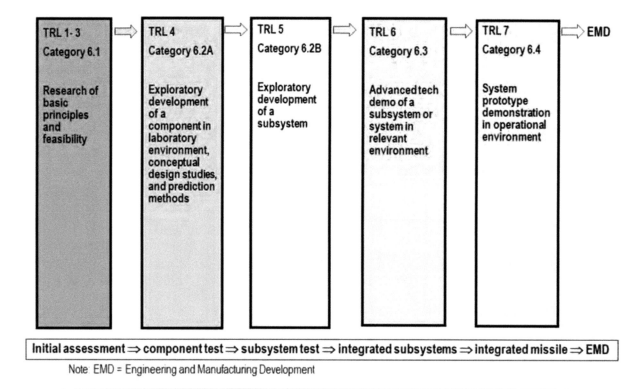

Fig. 8.4 Technology Readiness Level (TRL) Indicates the Maturity of Technology.

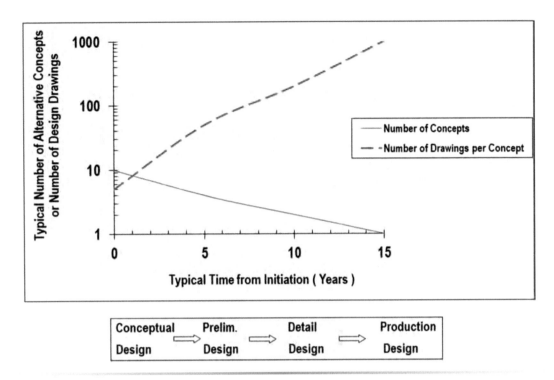

Fig. 8.5 Conceptual Design Requires Broad Range of Alternatives—Production Requires Configuration Management of Single Design.

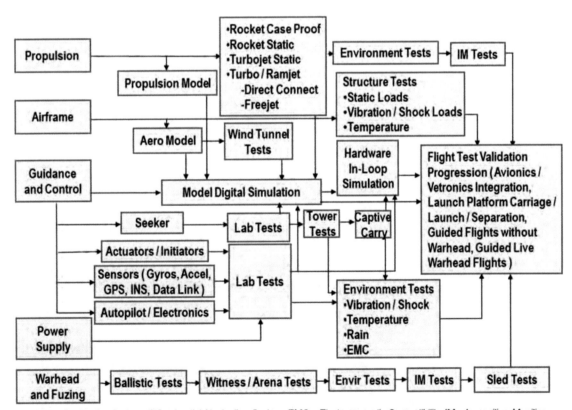

Note: GPS = Global Positioning System, INS = Inertial Navigation System, EMC = Electromagnetic Compatibility, IM = Insensitive Munition

Fig. 8.6 Missile Development, Integration, Tests, Design Verification, and Validation Is an Integrated Process.

- Airframe Wind Tunnel Test ....

- Propulsion Static Firing with Thrust Vector Control (TVC) ....

- Propulsion Direct Connect Test ....

- Propulsion Freejet Test ....

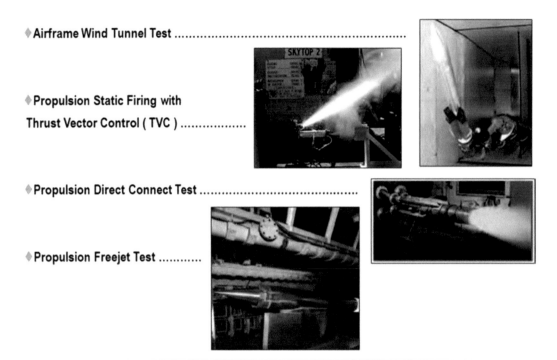

Fig. 8.7a Examples of Missile Development Tests and Facilities.

- Warhead Witness / Arena Test ....

- Warhead Sled Test ....

- Insensitive Munition Test ....

- Structure Load Test ....

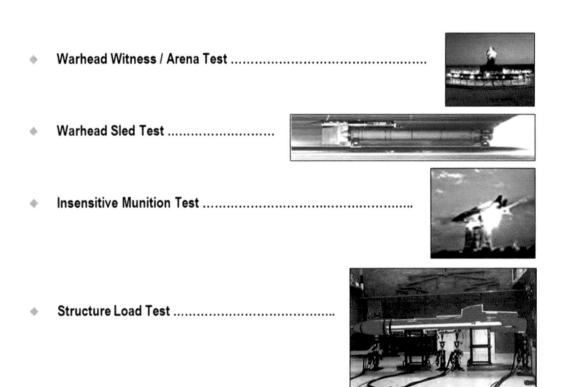

Fig. 8.7b Examples of Missile Development Tests and Facilities (cont).

- Seeker Test ....

- Guidance, Navigation, & Control (GN&C) Hardware-In-Loop

- Environmental Test ....

- Submunition Dispenser Sled Test ....

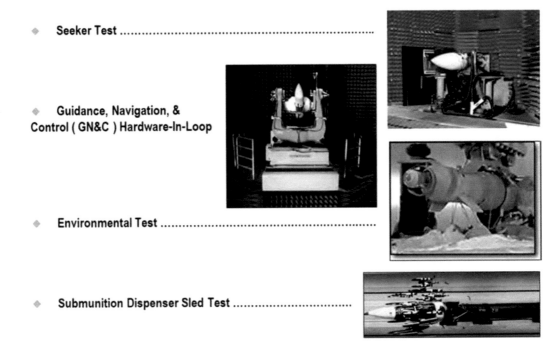

Fig. 8.7c Examples of Missile Development Tests and Facilities (cont).

- Radar Cross Section (RCS) Test..

- Avionics / Vetronics Integ

- Flight Test ....

- Video of Facilities and Tests

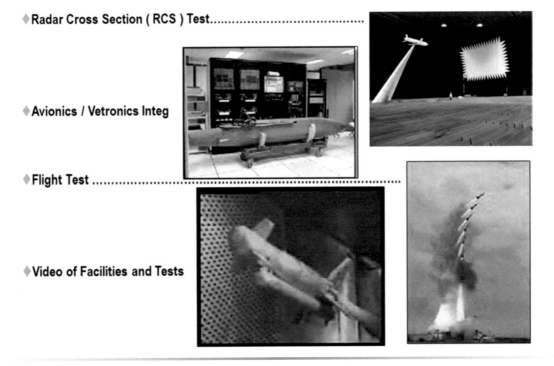

Fig. 8.7d Examples of Missile Development Tests and Facilities (cont).
Note: See video supplement, Facilities and Tests.

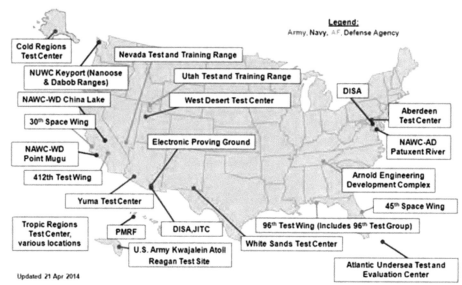

US Army, Navy, and Air Force Test Facilities

Considerations for Best Range to Conduct Missile Flight Test
- Range Size
- Geography
- Climate
- Instrumentation
- Environmental Impact
- Cost
- Availability

Fig. 8.7e Examples of Missile Development Tests and Facilities (cont).
Note: See video supplement, US Army, Navy, and Air Force Test Facilities.

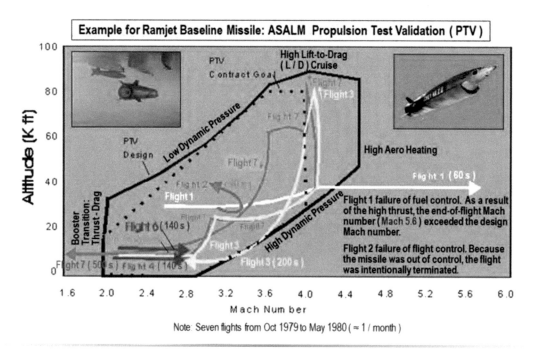

Fig. 8.8 Missile Development Flight Test Should Cover the Extremes and Corners of the Flight Envelope.

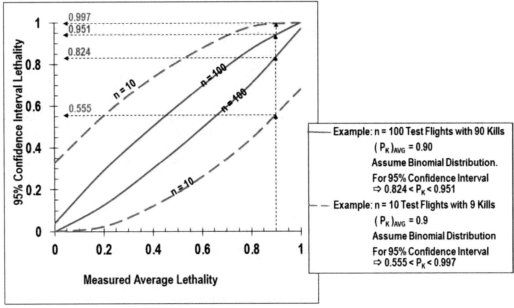

Fig. 8.9 A Relatively Large Number of Test Flights Are Required for a High Confidence Missile System.

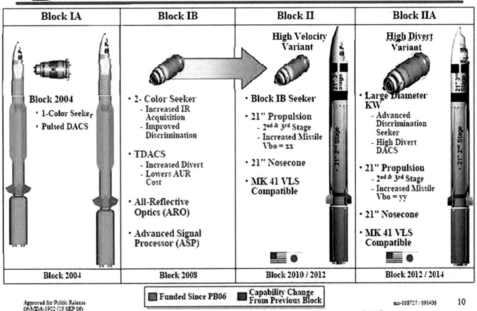

Note: SM-3 IIA first flight test (Feb 2017) was years later than original plan.

Nomenclature: BMD = ballistic missile defense, DACS = Divert and attitude control system, IR = Infrared, TDACS = Throttleable divert and attitude control system, AUR = Average unit round, VLS = Vertical Launch System, KW = Kill weapon, Vbo = Velocity at burnout

Fig. 8.10 Missile Block Upgrades Often Provide Improved Flight Performance, Guidance, Flight Control, and Propulsion.

332 Missile Design Guide

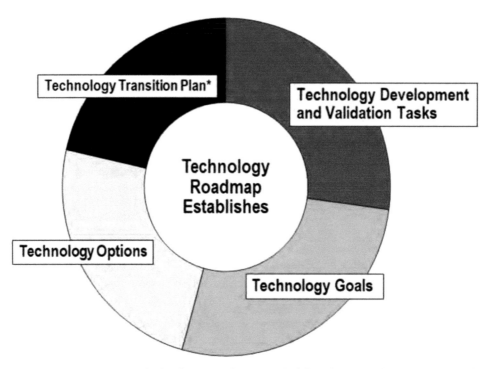

*Typical Technology Transition: **Advanced Technology Development (ATD)** ⇒ **Advanced Concept Technology Demonstration (ACTD)** ⇒ **Demonstration and Validation (DemVal)** ⇒ **Program Definition and Risk Reduction (PDRR)** ⇒ **Engineering and Manufacturing Development (EMD)**

Fig. 8.11a Technology Roadmap Is a Plan for the Development Process.

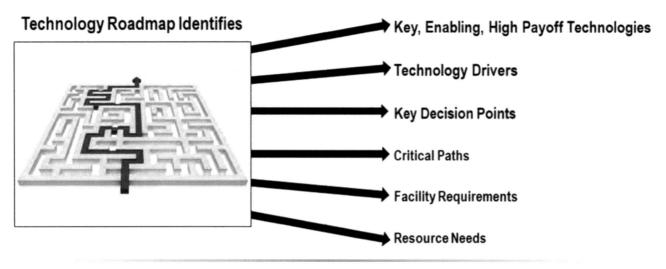

Fig. 8.11b Technology Roadmap Is a Plan for the Development Process (cont).

Example of Alternative Design Concepts	Example Constraints
Highest Performance Concept (s)	Cost and Risk Threshold / Standard Requirements
Lowest Cost Concept (s)	Performance and Risk Threshold / Standard Requirements
Lowest Risk Concept (s) (e.g., Derivative Concept)	Performance and Cost Threshold / Standard Requirements
Best Value (Balanced Design) Concept (e.g., Best Combination of Performance, Cost, Risk)	Performance, Cost, and Risk Standard / Goal Requirements

Example of Questions to be Answered before Picking a Preferred Concept / Approach
1. Preferred (best value) concept selected only after evaluating highest performance, lowest cost, and lowest risk concept alternatives
2. The preferred concept / approach strengths will lead to a successful program
3. The preferred concept / approach weaknesses are minor
4. Alternative concepts / approaches have more serious weaknesses

Conceptual design and system engineering studies are required to evaluate performance, cost, and risk of candidate concepts

Fig. 8.12 Conceptual Design and System Engineering Is Required to Explore Alternative Approaches.

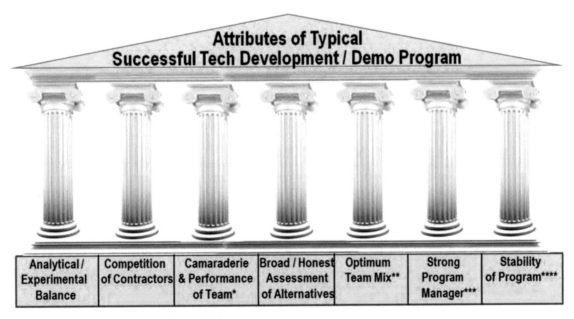

*Teams are often under pressure from rapid changes
** Optimum team size, background, creativity, and experience mix
***Willing to take risks and present politically incorrect information to upper management
****Stable funding / schedule (e.g., no peanut butter funding) and stable requirements (e.g., no requirements creep)

Fig. 8.13a Some Attributes of a Good Technology Development and Demonstration Program.

Fig. 8.13b Some Attributes of a Good Technology Development and Demonstration Program (cont).

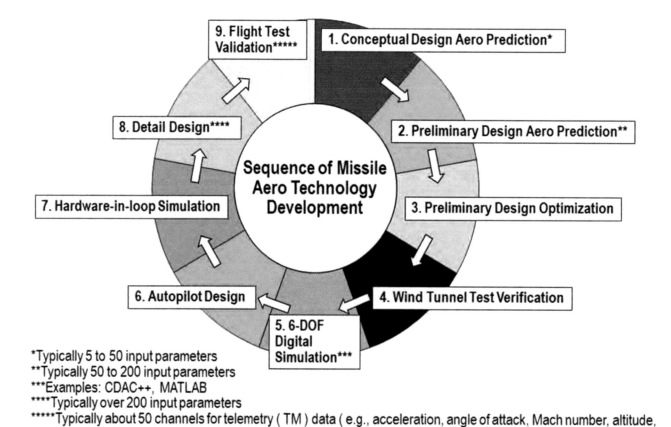

*Typically 5 to 50 input parameters
**Typically 50 to 200 input parameters
***Examples: CDAC++, MATLAB
****Typically over 200 input parameters
*****Typically about 50 channels for telemetry (TM) data (e.g., acceleration, angle of attack, Mach number, altitude, roll angle, flight control deflection, hinge moment, temperature)

Fig. 8.14a Example of a Typical Technology Development Process—Missile Aerodynamics.

Preliminary Design Aero Prediction Methods

Method	Contact	Attributes Include
Missile DATCOM	AFRL	Relatively Fast Setup, Low Cost
MISL3	Analytical Mechanics Associates / NEAR	Modeling Vortex Shedding
MISDL	Analytical Mechanics Associates / NEAR	Paneling Complex Geometry
AP09	Aeroprediction, Inc	Periodic Updates
FASTRAN	ESI / CFDRC	Computational Fluid Dynamics (CFD) Modeling Complex Shape, Jet Interaction, and Store Separation Using Personal Computer (PC)
FLUENT	ANYSYS, Inc	CFD Modeling Described Above. Relatively Fast Setup for Unstructured Mesh
CFD++	Metacomp Technologies	High Fidelity Aerodynamic, Navier-Stokes Solution of Combustion and Reacting Flows
CART3D	NASA	CFD Inviscid Modeling Complex Geometry Using Triangulation and Adjoint Driven Mesh

Fig. 8.14b Example of a Typical Technology Development Process—Missile Aerodynamics (cont).

Preliminary Design Optimization Methods

Method	Attributes Include
Response Surface Model	Meta-Model with Rapid Computation of Most Important Parameters
Probabilistic Analysis	Design Robustness
OTIS	Flight Trajectory Optimization
MATLAB GPOPS	Flight Trajectory Optimization
TAOS	Flight Trajectory Optimization

Fig. 8.14c Example of a Typical Technology Development Process—Missile Aerodynamics (cont).

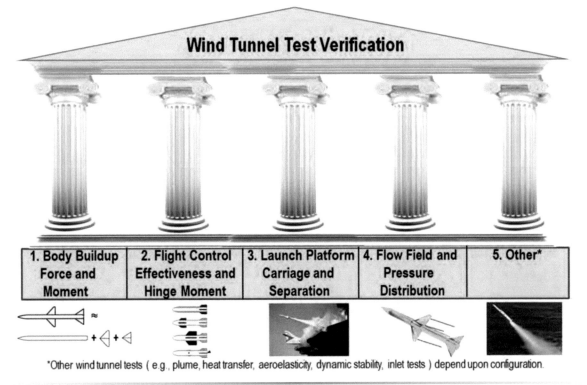

*Other wind tunnel tests (e.g., plume, heat transfer, aeroelasticity, dynamic stability, inlet tests) depend upon configuration.

Fig. 8.14d Example of a Typical Technology Development Process—Missile Aerodynamics (cont).

336 Missile Design Guide

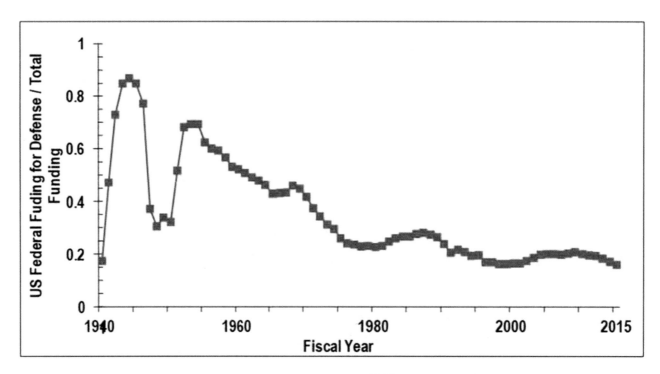

Source: US Office of Management and Budget, FY 2015 Budget, Table 6-1

Fig. 8.15 There Is Currently Relatively Low US Emphasis on Funding Defense Programs.

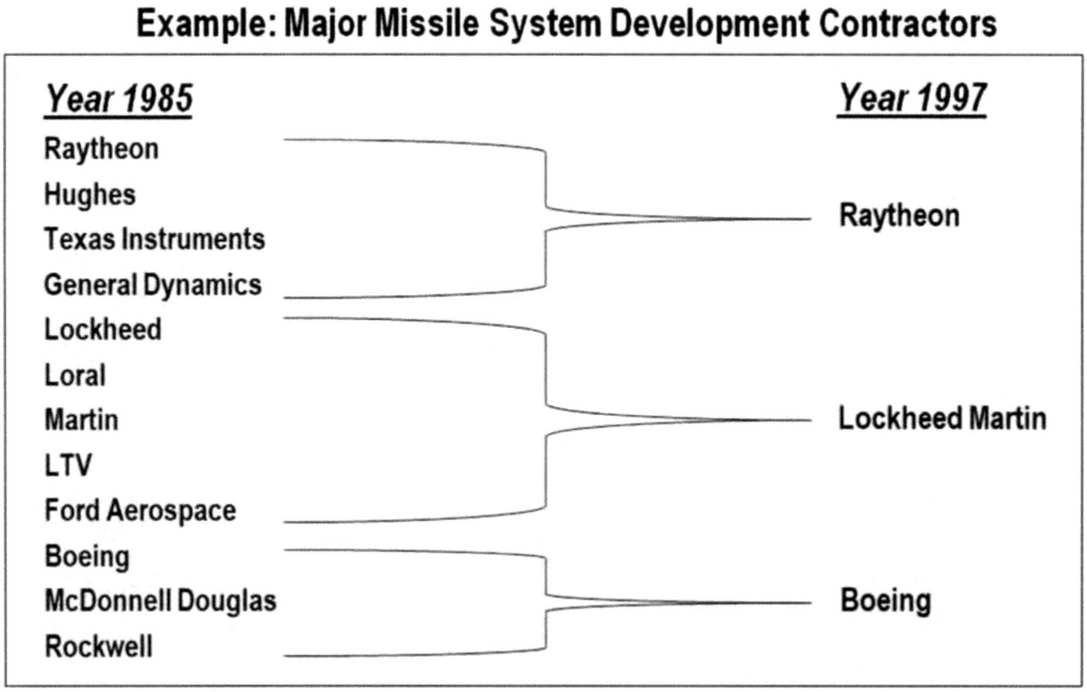

Fig. 8.16a There Is Currently Relatively Low Competition for US Missile Development.

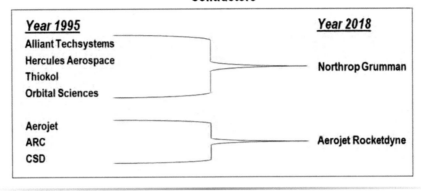

Fig. 8.16b There Is Currently Relatively Low Competition for US Missile Development (cont).

8.2 Examples of State-of-the-Art Advancement

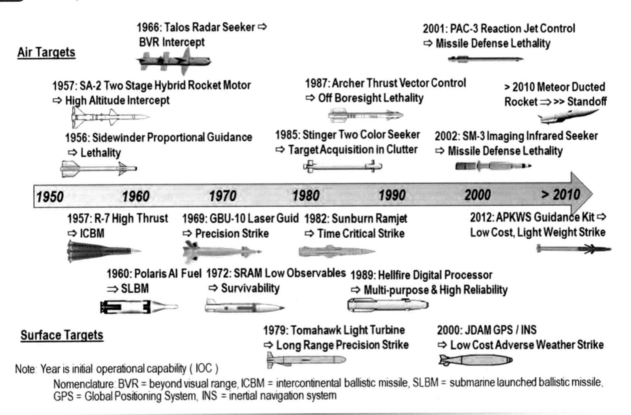

Fig. 8.17 Missile Technologies Have Transformed Warfare.

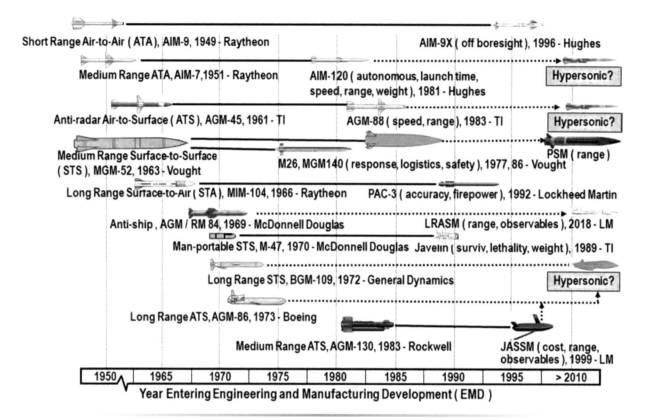

Fig. 8.18 US Tactical Missile Follow-On Programs Occur about Every 24 Years.

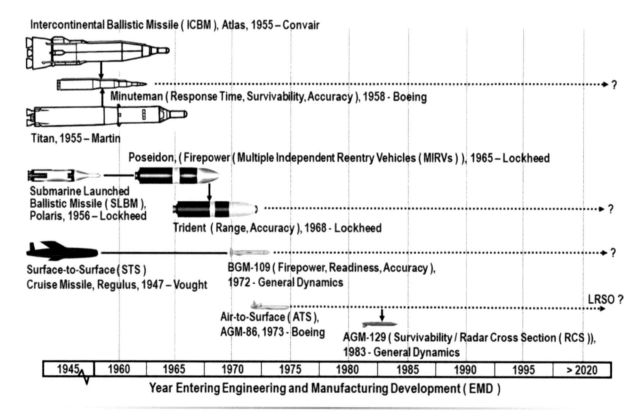

Fig. 8.19 US Strategic Missiles and Their Follow-on Programs Occurred During the Cold War.

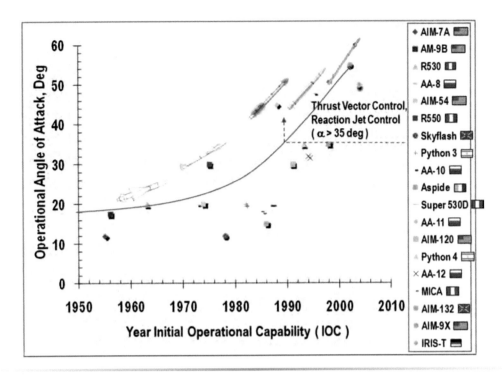

Fig. 8.20 Example of Missile Technology State-of-the-Art Advancement—Air-to-Air Missile Maneuverability.

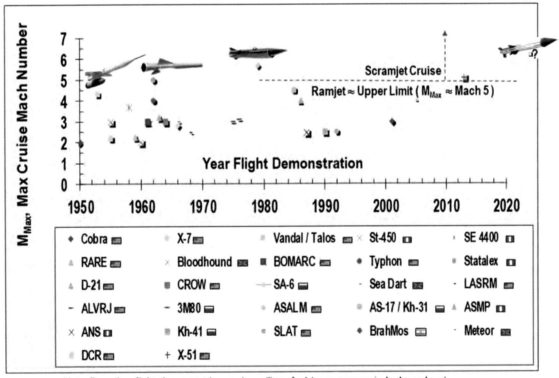

Note: Based on flight demonstrations using military fuel (non-cryogenic, hydrocarbon)

Fig. 8.21 Example of Missile Technology State-of-the-Art Advancement—Ramjet Propulsion Mach Number.

8.3 Enabling Technologies for Missiles

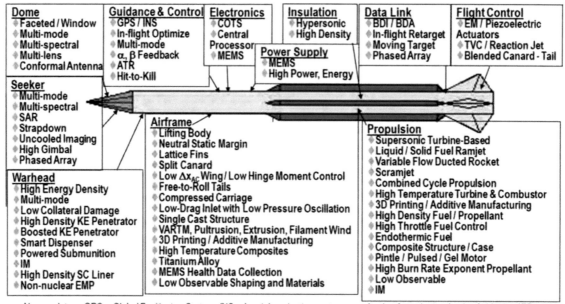

Fig. 8.22 There Are Many Enabling Technologies for Missiles.

Chapter 9 Lessons Learned

Government Aerospace Research Engineer

Industry Aerospace Program Manager

University Missile Design Teacher

Author of Missile Textbooks

Short Course Instructor for Missiles

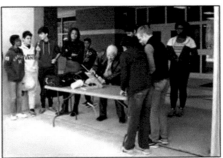
STEM Aerospace Education

Eugene L. Fleeman, E-mail: GeneFleeman@msn.com, Web Site: https://sites.google.com/site/eugenefleeman/home

Fig. 9.1 Lessons Learned Based on Career in Missile Design, Development, and System Engineering.

Fig. 9.2 Quickly Switch Back-and-Forth Focus from the Big Picture to a Detail Picture (See the Forest As Well As the Trees).

Fig. 9.3 Use Standards As a Guide, to Avoid Going Off Course.

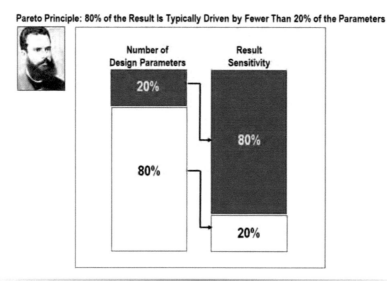

Fig. 9.4 Find the Driving Parameters for Most Important Measures of Merit.

Fig. 9.5 Select the Best Value for the Driving Measures of Merit.

Fig. 9.6 Start with a Good Baseline Design.

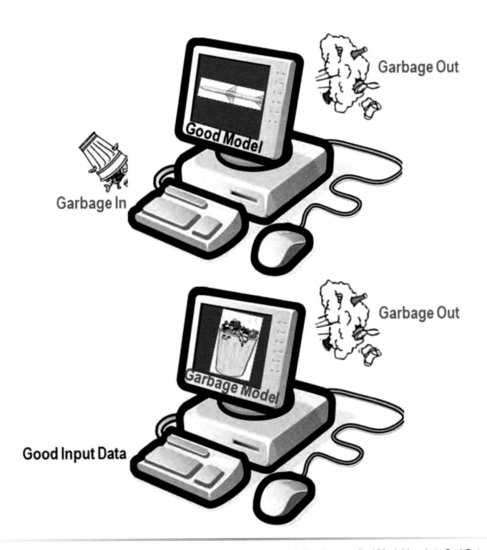

Fig. 9.7 Confirm Accuracy of Computer Input Data and Model (Bad Input or Bad Model Leads to Bad Output).

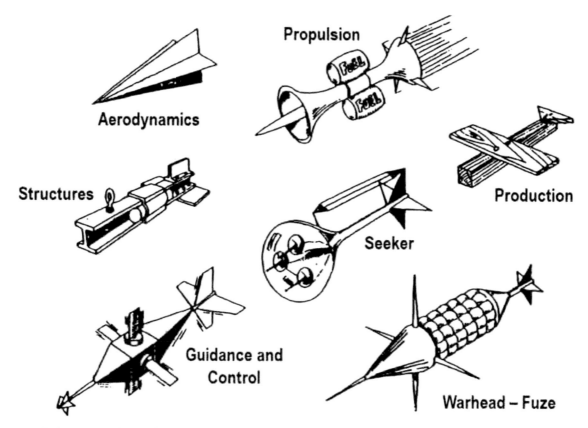

Reference: Anderson, K.D., 'Smart Munitions Technology,' Technology Training Corporation, July 1989

Fig. 9.8 Conduct Balanced, Unbiased Tradeoffs.

Example of System Integration: V-2 Tail Span Was Limited by a Railroad Tunnel

V-2 on Railroad Car

Mittelbau-Dora Tunnel in Kohnstein Mountain, Leading to / from Mittelwerk V-2 Factory

Fig. 9.9 Include System Integration.

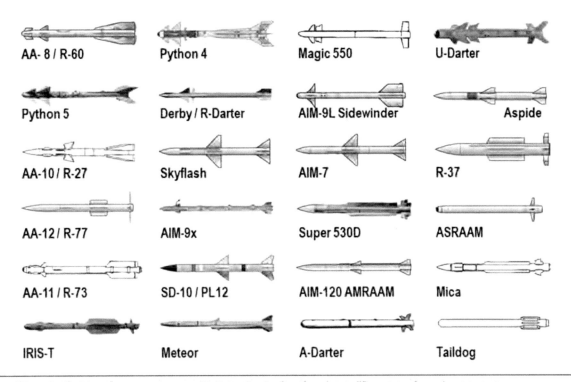

Fig. 9.10a Evaluate Many Alternatives.

Fig. 9.10b Evaluate Many Alternatives (cont).

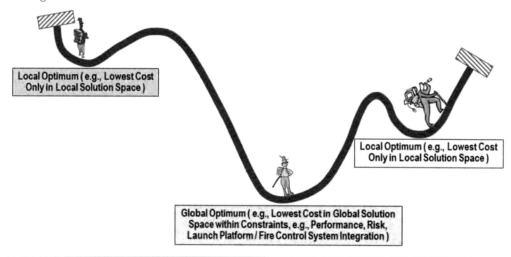

Fig. 9.11 Search a Broad Design Solution Space (Global Optimization vs. Local Optimization).

Fig. 9.12 Evaluate and Refine as Often as Possible for Conceptual Design—Maintain Strict Configuration Management for Production Design.

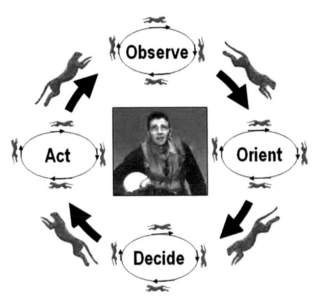

Fig. 9.13 Be Faster and More Accurate Than the Competition.

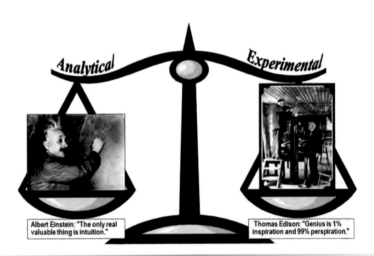

Fig. 9.14 Provide Balanced Emphasis of Analytical vs. Experimental.

Fig. 9.15 Consider Law of Unintended Consequences-Decisions May Have Bad Unforeseen Consequences.

Fig. 9.16 Keep Track of Assumptions and Develop Real-Time Documents.

CHAPTER 9 Lessons Learned 349

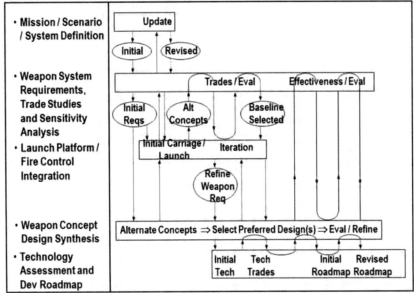

Fig. 9.17a Conduct Broad, Unbiased, Creative, Iterative, and Rapid Design Evaluations.

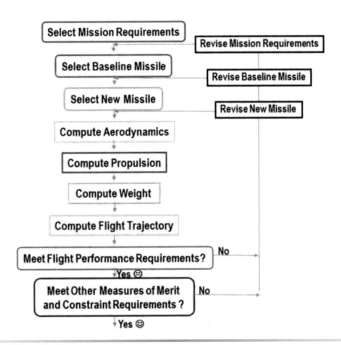

Fig. 9.17b Conduct Broad, Unbiased, Creative, Iterative, and Rapid Design Evaluations (cont).

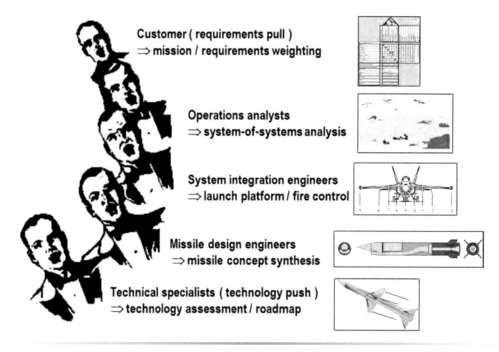

Fig. 9.18 Establish a Diverse Team for a Balanced Design.

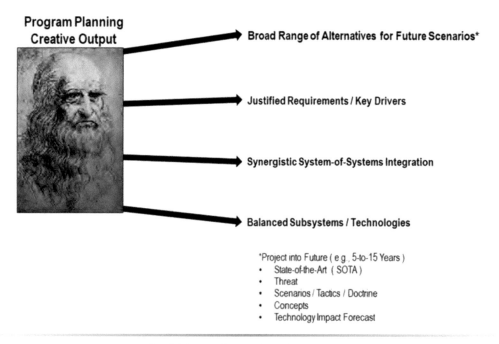

Fig. 9.19 Use Creative Skills for Program Planning.

CHAPTER 9 Lessons Learned 351

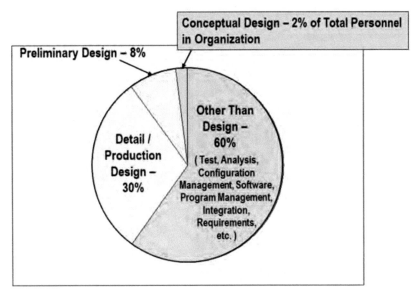

Fig. 9.20 Use Input from the Entire Organization.

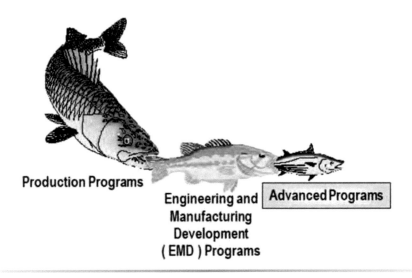

Fig. 9.21 Provide Balanced Investments for Near, Mid, and Long-Term Payoff.

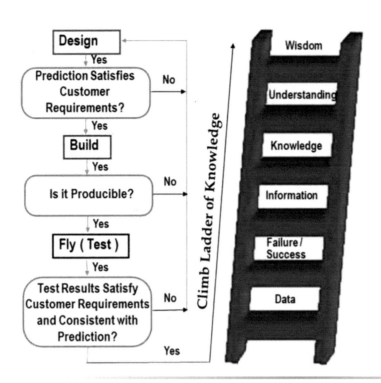

Fig. 9.22 Use the Design, Build, and Fly Process for Broader Knowledge and Understanding.

Fig. 9.23 Provide Broad Documentation.

Chapter 10 Summary

10.1 Missile Design Guidelines

Configuration Sizing Parameter	Aerodynamics Design Guideline
Body Fineness Ratio	$5 < l/d < 25$
Nose Fineness Ratio	$l_N/d \sim 2$ if $M > 1$
Boattail Angle or Flare Angle	$< \approx 10$ deg
Aero Flight Control Surface Stall	$(\alpha + \delta)_{stall} \approx 22$ deg

Nomenclature: l = body length, d = body diameter, l_N = nose length, M = Mach number, α = angle of attack, δ = flight control deflection

Fig. 10.1 Missile Conceptual Design and System Engineering Sizing Guidelines for Aerodynamics.

Configuration Sizing Parameter	Flight Control Design Guideline
Body Bending Frequency	$\omega_{BB} > \omega_{Flight\ Control\ Actuator}$
Trim Flight Control Effectiveness and Maneuverability	$\alpha/\delta > 1$
Neutral Static Stability for Body-Tail and Body-Flare	$b/d \approx 2$, $c/d \approx 1$ and $d_F/d > 1.4$
Wing Aerodynamic Center	Near Missile Center of Gravity ($\approx 50\%$ Missile Length)
Stability & Control Cross Coupling	$< 30\%$
Flight Control System Feedback Bandwidth	> 5-to-10x Guidance System Feedback Bandwidth

Nomenclature: ω_{BB} = body bending frequency, $\omega_{Flight\ Control\ Actuator}$ = flight control actuator frequency, α = angle of attack, δ = flight control deflection, b = total tail span, d = body diameter, c = chord, d_F = flare diameter,

Fig. 10.2 Missile Conceptual Design and System Engineering Sizing Guidelines for Flight Control.

Configuration Sizing Parameter	Propulsion Design Guideline
High Specific Impulse (I_{SP}) for Rocket	High Chamber Pressure and Optimum Nozzle Expansion
Inlet Sideslip	$\beta < 6$ deg
High I_{SP} for Ramjet / Ducted Rocket	High Temperature Insulation for Combustor
High I_{SP} for Turbojet	High Temperature Turbine Material
Oblique Shocks to Satisfy MIL-E-5007D Inlet	> 2 Oblique Shocks if M > 3 and > 3 Oblique Shocks if M > 3.5
Supersonic Inlet Flow Capture	Shock On Inlet Cowl Lip at M_{max} Cruise
Ramjet Minimum Cruise Mach Number	$M > 1.2\ M_{Inlet\ Start}$ and $M > 1.2\ M_{Max\ Thrust\ =\ Drag}$

Nomenclature: I_{SP} = specific impulse, β = angle of sideslip, M = Mach number, M_{max} = maximum Mach number, $M_{Inlet\ Start}$ = inlet start Mach number, $M_{max\ Thrust\ =\ Drag}$ = Mach number for maximum thrust = drag.

Fig. 10.3   Missile Conceptual Design and System Engineering Sizing Guidelines for Propulsion.

Configuration Sizing Parameter	Weight Design Guideline
Subsystems Packaging	Maximize Available Volume for Fuel / Propellant
Subsystems Density	≈ 0.05 lbm / in^3
Structure Design Factors of Safety	≈ 1.25 for Ultimate Flight Loads, ≈ 1.10 for Yield Flight Loads
Body Structure Weight	$\approx 22\%$ Missile Launch Weight
Solid Propellant Weight Fraction	$\approx 71\%$ Rocket Motor Weight
Ballistic Missile Staging for Minimum Weight	$\Delta V_{First\ Stage} \approx \Delta V_{Second\ Stage}$

Nomenclature: $\Delta V_{First\ Stage}$ = incremental velocity from first stage, $\Delta V_{Second\ Stage}$ = incremental velocity from second stage

Fig. 10.4   Missile Conceptual Design and System Engineering Sizing Guidelines for Weight.

Configuration Sizing Parameter	Flight Performance Design Guideline
Missile Maneuverability	$n_{Missile} / n_{Target} > 3$
Missile Heading Rate	$\dot{\gamma}_{Missile} > \dot{\gamma}_{Target}$
Missile Turn Radius	$(R_T)_{Missile} < (R_T)_{Target}$
Missile Homing Velocity	$V_M / V_T > \approx 1.5$ for Off-Boresight Target
Efficient Cruise / Glide Dynamic Pressure	$q < 1000$ psf

Nomenclature: $n_{Missile}$ = missile maneuver acceleration, n_{Target} = target maneuver acceleration, $\dot{\gamma}_{Missile}$ = missile heading rate, $\dot{\gamma}_{Target}$ = target heading rate, $(R_T)_{Missile}$ = missile turn radius, $(R_T)_{Target}$ = target turn radius, V_M = missile velocity, V_T = target velocity, q = dynamic pressure

Fig. 10.5   Missile Conceptual Design and System Engineering Sizing Guidelines for Flight Performance.

Configuration Sizing Parameter	Guidance Design Guideline
Missile Time Constant	$\tau < 0.2$ s, $\tau < \omega_{Target}^{-1}$
Proportional Guidance Ratio	$3 < N' < 5$
Target Span Resolution by Seeker	$< b_{Target}$

Nomenclature: τ = missile time constant, ω_{Target} = target frequency, N' = proportional guidance ratio, b_{Target} = target span

Fig. 10.6 Missile Conceptual Design and System Engineering Sizing Guidelines for Guidance.

Configuration Sizing Parameter	Other Measures of Merit (MOM) Design Guideline
Aging Lifetime	Temperature Spec > Lifetime Peak & Average Temperatures
Typical Environment Requirement	1% of World – Wide Extreme
Warhead Lethal Radius	> 3σ Miss Distance
Fire Control System Range	> ≈ 2x Missile Lethal Range
Blast / Frag Warhead Maximum Kinetic Energy	Explosive Charge Weight ≈ Total Fragment Weight
Shaped Charge Warhead Maximum Penetration	Large Diameter and Dense Liner
Insensitive Munition	Satisfy MIL-STD-2105C

Nomenclature: σ = standard deviation

Fig. 10.7 Missile Conceptual Design and System Engineering Sizing Guidelines for Lethality.

10.2 Wrap Up

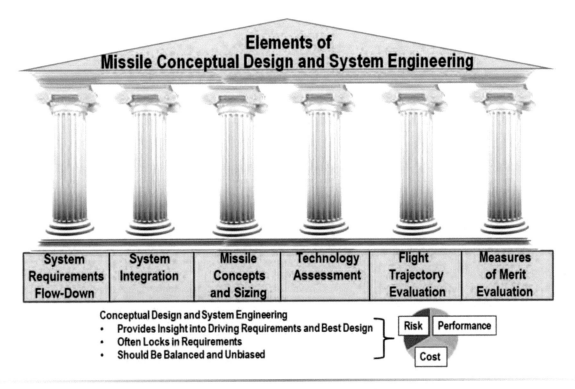

Fig. 10.8a Wrap Up (Part 1 of 3).

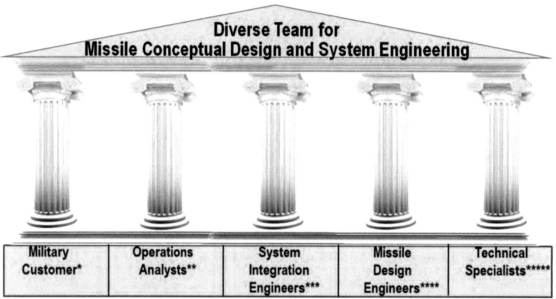

Fig. 10.8b Wrap Up (Part 2 of 3).

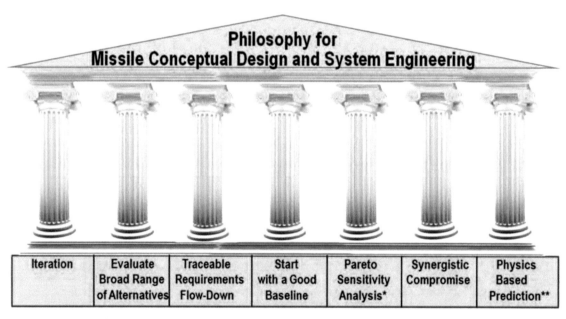

Fig. 10.8c Wrap Up (Part 3 of 3).

Chapter 11 References, Bibliography

11.1 References

1. Nicholas, T., and Rossi, R., "US Missile Data Book, 2011," Data Search Associates, 2011.
2. Alberts, D.S., Garstka, R.E., Hayes, D.A., and Signori, D.A., "Understanding Information Age Warfare", August 2001.
3. Pitts, W.C., Nielsen, J.N., and Kaattari, G.E., "Lift and Center of Pressure of Wing-Body-Tail Combinations at Subsonic, Transonic, and Supersonic Speeds," NACA Report 1307, 1957.
4. AIAA Aerospace Design Engineers Guide, American Institute of Aeronautics and Astronautics, 2012.
5. Bonney, E.A., et al., Aerodynamics, Propulsion, Structures, and Design Practice, "Principles of Guided Missile Design", D. Van Nostrand Company, Inc., 1956.
6. Jerger, J.J., Systems Preliminary Design, "Principles of Guided Missile Design", D. Van Nostrand Company, Inc., 1960.
7. Chin, S.S., Missile Configuration Design, McGraw-Hill Book Company, 1961.
8. Mason, L.A., Devan, L., and Moore, F.G., "Aerodynamic Design Manual for Tactical Weapons," NSWCTR 81-156, 1981.
9. Jorgensen, L.H., "Prediction of Static Aerodynamic Characteristics for Space-Shuttle-Like, and Other Bodies at Angles of Attack From 0 Degrees to 180 Degrees," NASA TND 6996, January 1973.
10. Hoak, D.E., et al., "USAF Stability and Control DATCOM," AFWAL TR-83-3048, Global Engineering, 1978.
11. "Nielsen Engineering & Research (NEAR) Aerodynamic Software Products," http://www.nearinc.com/near/software.htm
12. Brebner, G.G., "The Control of Missiles," AGARD-LS-98, February 1979.
13. Durham, F.P., Aircraft Jet Powerplants, Prentice-Hall, 1961.
14. Ashley, H., Engineering Analysis of Flight Vehicles, Dover Publications, Inc., 1974.
15. Roux, E., Turbofan and Turbojet Engines Database Handbook, 2007.
16. Anderson, J.D. Jr., "Modern Compressible Flow," Second Edition, McGraw Hill, 2004.
17. Kinroth, G.D., and Anderson, W.R., "Ramjet Design Handbook," CPIA Pub. 319 and AFWAL TR 80-2003, June 1980.
18. Ames Research Staff, "Equations, Tables, and Charts for Compressible Flow", NACA Report 1135, 1953.
19. Lyon, M., "Advanced Propulsion for Tactical Missiles", NDIA Conference on Armaments for Army Transformation, 2001.
20. "Internal Aerodynamics Manual," North American Rockwell Corporation, DTIC AD 723823, June 1970.
21. Oswatitsch, K.L., "Pressure Recovery for Missiles with Reaction Propulsion at High Supersonic Speeds", NACA TM-1140, 1947.
22. Beckstead, M., "Fundamentals of Combustion and Detonation", BYU Lecture, April 2000.
23. Blomshield, Fred S., "Lessons Learned in Solid Rocket Combustion Instability", AIAA 2007-5803, July 2007.
24. Stowe, R.A., "Performance Prediction of a Ducted Rocket Combustor", Defence Research Establishment Valcartier, DREV TR 2001-200, November 2001.
25. Bithell, R.A., and Stoner, R.C., "Rapid Approach for Missile Synthesis," AFWAL TR 81-3022, March 1982.
26. Cornelisse, J.W., Schoyer, H.F.R., and Wakker, K.F., *Rocket Propulsion and Spacecraft Dynamics*, Pitman, 1979.
27. Carslaw, H.S., and Jaeger, J.C., *Conduction of Heat in Solids*, Clarendon Press, 1988.
28. Grover, J.H., and Holter, W.H., "Solution of the Transient Heat Conduction Equation for an Insulated, Infinite Metal Slab," Journal of Jet Propulsion, Number 27, 1957.
29. Allen, J., and Eggers, A.J., "A Study of the Motion and Aerodynamic Heating of Ballistic Missiles Entering the Earth's Atmosphere at High Supersonic Speeds", NACA Report 1381, April 1953.
30. Atkinson, J.R., and Staton, R.N., "Missile Body Weight Prediction", SAWE 1497, May 1982.
31. Levens, A.S., *Nomography*, John Wiley & Sons, 1937.
32. Bruhn, E.F., Orlando, J.I., and Meyers, J.F., Analysis and Design of Missile Structures, Tri-State Offset Company, 1967.
33. AMCP 706-283, Engineering Design Handbook, Ballistic Missile Series, Aerodynamics, April 1965.
34. Klein, L.A., *Millimeter-Wave and Infrared Multisensor Design and Signal Processing*, Artech House, Boston, 1997.
35. Schneider, S.H., *Encyclopedia of Climate and Weather*, Oxford University Press, 2011.
36. Richardson, M.A., et al., *Surveillance & Target Acquisition Systems*, Brassey's, London, 2000.
37. Hudson, R.D., Infrared System Engineering, John Wiley & Sons, 1969.
38. Wolfe, W.I., and Zissis, G.J., The Infrared Handbook, Revised Edition, Office of Naval Research, 1985.
39. Khan, F., and Pi, J., "Millimeter-wave Mobile Broadband: Unleashing 3–300 GHz Spectrum," Proceedings of IEEE Wireless Communications and Networking Conference (WCNC), March 2011.
40. Heaston, R.J., and Smoots, C.W., "Precision Guided Munitions," GACIAC Report HB-83-01, May 1983.
41. Stimson, G.W., Introduction to Airborne Radar, SciTech Publishing, 1998.
42. Lawrence, A.L., Modern Inertial Technology, Springer, 2001.
43. Dow, R.D., *Fundamentals of Advanced Missiles*, John Wiley, 1958.
44. US Army Ordnance Pamphlet ORDP-20-290-Warheads, 1980.
45. Taylor, G.I., "The Formation of a Blast Wave by a Very Intense Explosion, Part I. Theoretical Discussion", Proceedings of the Royal Society of London, Vol. 201, March 1950.
46. Carleone, J. (Editor), *Tactical Missile Warheads*, "AIAA Vol. 155 Progress in Astronautics and Aeronautics," American Institute of Aeronautics and Astronautics, 1993.

47. Christman, D.R., and Gehring, J.W., "Analysis of High-Velocity Projectile Penetration Mechanics," Journal of Applied Physics, Vol. 37, 1966.
48. Walters, W., and Zukas, J.A. Fundamentals of Shaped Charges. John Wiley and Sons, New York, NY, 1989.
49. Przemieniecki, J.S., *Mathematical Methods in Defense Analysis*, American Institute of Aeronautics and Astronautics, 2000.
50. Eichblatt, E.J., *Test and Evaluation of the Tactical Missile*, American Institute of Aeronautics and Astronautics, 1989.
51. Bennett, R.R., and Mathews, W.E., "Analytical Determination of Miss Distances for Linear Homing Navigation Systems," Hughes Tech Memo 260, 31 March 1952.
52. Zarchan, P., *Tactical and Strategic Missile Guidance*, Vol. 5, American Institute of Aeronautics and Astronautics, 2007.
53. Crispin, J.W., Jr., Goodrich, R.F., and Siegel, K. M., "A Theoretical Method for the Calculation of the Radar Cross Section of Aircraft and Missiles," University of Michigan Report 2591-1-H, July 1959.
54. Military Specification Mil-A-8591H, "Airborne Stores, Suspension Equipment and Aircraft-Store Interface (Carriage Phase); General Design Criteria For," 23 March 1990.
55. Cengel, Y.A., *Heat and Mass Transfer: A Practical Approach*, McGraw-Hill, 2006.
56. Jones, D.B. et al., "Air-to-Ground Target Acquisition Source Book", Martin Marietta, 1974.
57. Jarvinen, P.O., and Hill, J.A.F., "Universal Model for Underexpanded Rocket Plumes in Hypersonic Flow", Proceedings of the 12th JANAF Liquid Propulsion Meeting, Nov. 1970.
58. Fleeman, E.L., and Donatelli, G.A., "Conceptual Design Procedure Applied to a Typical Air-Launched Missile," AIAA 81-1688, August 1981.
59. Teledyne Turbine Engines Brochure, "Model 370, J402-CA-400 Performance," 2006.
60. St. Peter, J., *The History of Aircraft Gas Turbine Engine Development in the United States: A Tradition of Excellence*, ASME International Gas Turbine Institute, 1999.
61. Hindes, J.W., "Advanced Design of Aerodynamic Missiles (ADAM)", October 1993.
62. Gregorion, G. "Advanced Design Code for Missile Aerodynamics", AIAA Paper 89-2171, 1989.
63. Fleeman, E.L., *Missile Design and System Engineering*, American Institute of Aeronautics and Astronautics, 2012.
64. Donatelli, G.A., and Fleeman, E.L., "Methodology for Predicting Miss Distance for Air Launched Missiles," AIAA Paper 82-0364, Jan. 1982.
65. Frits, A.P., et al., "A Conceptual Sizing Tool for Tactical Missiles", AIAA Missile Sciences Conference, November 2002.
66. Spears, S., Allen, R., and Fleeman, E.L. "First-Order Conceptual Design for Turbojet Missiles and Guided Bombs", SciTech Forum 2022.
67. Monta, William J., "Supersonic Aerodynamic Characteristics of an Air-to-Air Missile," NASA 1078, November 77.
68. Montgomery, D.G., *Design and Analysis of Experiments*, John Wiley & Sons, 2001.
69. Dixon, W.J., and Massey, F.J., *Introduction to Statistical Analysis*, McGraw-Hill, 1969.
70. Bruns, K.D., Moore, M.E., Stoy, S.L., Vukelich, S.R., and Blake, W.B., "Missile DATCOM," AFWAL-TR-91-3039, April 1991.
71. Moore, F.G., and Moore, L.Y., "The 2009 Version of the Aeroprediction Code", AIAA 2009-309, AIAA Aerospace Sciences Meeting, January 2009.
72. Anderson, K.D., "Smart Munitions Technology," Technology Training Corporation, July 1989.
73. Boyd, John R., "The Essence of Winning and Losing," Jan. 1996.
74. Nicolai, L.M., "Designing a Better Engineer," AIAA Aerospace America, April 1992.

11.2 Bibliography

System Design

"Missile Index," http://missile.index.ne.jp/en/
"DoD Index of Specifications and Standards," http://everyspec.com/
Defense Technical Information Center, http://www.dtic.mil/
NATO Research & Technology Organisation, https://www.sto.nato.int/Pages/default.aspx
"National Defense Industrial Association Conference Presentations," http://www.dtic.mil/ndia
"Missile System Flight Mechanics," AGARD CP270, May 1979.
Strickland, J.S., Missile Flight Simulation Surface-to-Air Missiles, Simulation Educators, 2010.
Rigby, K.A., Aircraft Systems Integration of Air-Launched Weapons, John Wiley, 2013.
Boord, W.J., and Hoffman, J.B., Air and Missile Defense Systems Engineering, CRC Press, 2016.
Ball, R.E., The Fundamentals of Aircraft Combat Survivability Analysis and Design, American Institute of Aeronautics and Astronautics, 1985.
Hogan, J.C., et al., "Missile Automated Design (MAD) Computer Program," AFRPL TR 80-21, March 1980.
Rapp, G.H., "Performance Improvements with Sidewinder Missile Airframe," AIAA Paper 79-0091, January 1979.
Nicolai, L.M., Fundamentals of Aircraft Design, METS, Inc., 1984.
Lindsey, G.H., and Redman, D.R., "Tactical Missile Design," Naval Postgraduate School, 1986.
Lee, R.G., et al., Guided Weapons, Third Edition, Brassey's, 1998.
Giragosian, P.A., "Rapid Synthesis for Evaluating Missile Maneuverability Parameters," AIAA Paper 1992-2615, 10th AIAA Applied Aerodynamics Conference, June 1992.
Fleeman, E.L. "Aeromechanics Technologies for Tactical and Strategic Guided Missiles," AGARD Paper Presented at FMP Meeting in London, England, May 1979.

Raymer, D.P., Aircraft Design, A Conceptual Approach, American Institute of Aeronautics and Astronautics, 1989 "Aircraft Stores Interface Manual (ASIM)," http://akss.dau.mil/software/1.jsp
"Aircraft Stores Interface Manual (ASIM)," AOP-12, May 1989.
Macfadzean, R.H., Surface-Based Air Defense System Analysis, Artech House, 1992.
"Advanced Sidewinder Missile AIM-9X Cost Analysis Requirements Description (CARD)," http://deskbook.dau.mil/jsp/default.jsp
Wertz, J.R., and Larson W.J., Space Mission Analysis and Design, Microprism Press and Kluwer Academic Publishers, 1999.
"Directory of U.S. Military Rockets and Missiles", http://www.designation-systems.net/
Fleeman, E.L., et al., "Technologies for Future Precision Strike Missile Systems," NATO RTO EN-018, July 2001.
"The Ordnance Shop", http://www.ordnance.org/portal/
"Conversion Factors by Sandelius Instruments", http://www.sandelius.com/reference/conversions.htm
"Defense Acquisition Guidebook", http://akss.dau.mil/dag/
"Weaponry", http://www.deagel.com/
Puckett, A.E., and Ramo, S., Guided Missile Engineering, McGraw-Hill book Company, 1959.
Aeroweb, US Missiles, http://www.bga.aeroweb.com/Missiles.html
'Defense Systems Information Analysis Center', https://www.dsiac.org/
"US Government Science Information," https://www.science.gov/
Fleeman, E.L., and Teague, J.R., "A Historical Overview of a Half Century of U.S. Missile Development", DSIAC Journal Volume 3, Number 3, 2016.
Crispin, J.W., and Siege, K.M., "Methods of Radar Cross-Section Analysis", Academic Press, 1968.
Fuhs, A.E., "Radar Cross Section Lectures", American Institute of Aeronautics and Astronautics, 1984.

Aerodynamics

"A Digital Library for NACA," http://naca.larc.nasa.gov/
Briggs, M.M., Systematic Tactical Missile Design, Tactical Missile Aerodynamics: General Topics, "AIAA Vol. 141 Progress in Astronautics and Aeronautics," American Institute of Aeronautics, 1992.
Briggs, M.M., et al., "Aeromechanics Survey and Evaluation, Vol. 1-3," NSWC/DL TR-3772, October 1977.
"Missile Aerodynamics," NATO AGARD LS-98, February 1979.
"Missile Aerodynamics," NATO AGARD CP-336, February 1983.
"Missile Aerodynamics," NATO AGARD CP-493, April 1990.
"Missile Aerodynamics," NATO RTO-MP-5, November 1998.
Nielsen, J.N., Missile Aerodynamics, McGraw-Hill Book Company, 1960.
Moore, F.G., Approximate Methods for Weapon Aerodynamics, American Institute of Aeronautics and Astronautics, 2000.
Mendenhall, M.R. et al., "Proceedings of NEAR Conference on Missile Aerodynamics," NEAR, 1989.
Nielsen, J.N., "Missile Aerodynamics—Past, Present, Future," AIAA Paper 79-1818, 1979.
Dillenius, M.F.E., et al., "Engineering-, Intermediate-, and High-Level Aerodynamic Prediction Methods and Applications," Journal of Spacecraft and Rockets, Vol. 36, No. 5, September-October, 1999.
Nielsen, J.N., and Pitts, W.C., "Wing-Body Interference at Supersonic Speeds with an Application to Combinations with Rectangular Wings," NACA Tech. Note 2677, 1952.
Spreiter, J.R., "The Aerodynamic Forces on Slender Plane-and Cruciform-Wing and Body Combinations", NACA Report 962, 1950.
Simon, J.M., et al., "Missile DATCOM: High Angle of Attack Capabilities, AIAA-99-4258.
Burns, K.A., et al., "Viscous Effects on Complex Configurations," WL-TR-95-3060, 1995.
Lesieutre, D., et al., "Recent Applications and Improvements to the Engineering-Level Aerodynamic Prediction Software MISL3," AIAA-2002-0274.
"1976 Standard Atmosphere Calculator," http://www.digitaldutch.com/atmoscalc/
"Compressible Aerodynamics Calculator," http://www.aoe.vt.edu/~devenpor/aoe3114/calc.html
John, J.E.A., Gas Dynamics, Second Edition, Prentice Hall, 1984.
Fleeman, E.L., and Nelson, R.C., Aerodynamic Forces and Moments on a Slender Body with a Jet Plume for Angles of Attack up to 180 Degrees, AIAA-1974-110, Aerospace Sciences Meeting, 1974.
Anderson, J.D., Modern Compressible Flow, Third Edition, McGraw Hill, 2002.
Cummings, R.M., et al., Applied Computational Aerodynamics, Cambridge University Press, 2015.

Propulsion

Sutton, G.P., and Biblarz, O., Rocket Propulsion Elements, Ninth Edition, John Wiley & Sons, 2017.
Jenson, G.E., and Netzer, D.W., Tactical Missile Propulsion, American Institute of Aeronautics and Astronautics, 1996.
Hill, P.G., and Peterson, C.R., Mechanics and Thermodynamics of Propulsion, Addison-Weshley Publishing Company, 1970.
Johns Hopkins Energetics Research Group, https://www.erg.jhu.edu/
Mahoney, J.J., Inlets for Supersonic Missiles, American Institute of Aeronautics and Astronautics, 1990.
"Tri-Service Rocket Motor Trade-off Study, Missile Designer's Rocket Motor handbook," CPIA 322, May 1980.
Humble, R.W., Henry, G.N., and Larson, W.J., Space Propulsion Analysis and Design, McGraw-Hill, 1995.
Bathie, W.W., Fundamentals of Gas Turbines, John Wiley and Sons, 1996.
Mattingly, J.D., et al., Aircraft Engine Design, American Institute of Aeronautics and Astronautics, 1987.
Davenas, A., Solid Rocket Propulsion Technology, Pergamon, 1992.

Yang, V., et al., Solid Propellant Chemistry, Combustion, and Motor Ballistics, AIAA, 2000.
Oates, G.C., Aerothermodynamics of Gas Turbine and Rocket Propulsion, AIAA, 1988.
TU Delft, https://www.tudelft.nl/en/
Fry, R.S., "A Century of Ramjet Propulsion Technology Evolution", Journal of Propulsion and Power, Jan 2004.

Materials and Heat Transfer

Budinski, K.G., and Budinski, M.K., Engineering Materials Properties and Selection, Prentice Hall, 1999.
"Matweb's Material Properties Index Page," http://www.matweb.com
"NASA Ames Research Center Thermal Protection Systems Expert (TPSX) and Material Properties Database", http://tpsx.arc.nasa.gov/tpsxhome.shtml
Harris, D.C., Materials for Infrared Windows and Domes, SPIE Optical Engineering Press, 1999.
Kalpakjian, S., Manufacturing Processes for Engineering Materials, Addison Wesley, 1997.
MIL-HDBK-5J, "Metallic Materials and Elements for Aerospace Vehicle Structures", Jan 2003.
"Metallic Material Properties Development and Standardization (MMPDS)", http://www.mmpds.org
Baker, A.A., and Scott, M.L., Composite Materials for Aircraft Structures, American Institute of Aeronautics and Astronautics, 2016.
Mallick, P.K., Fiber-Reinforced Composites: Materials, Manufacturing, and Design, Second Edition, Marcel Dekker, 1993.
Chapman, A.J., Heat Transfer, Third Edition, Macmillan Publishing Company, 1974.
Incropera, F.P., and DeWitt, D.P., Fundamentals of Heat and Mass Transfer, Fourth Edition, John Wiley and Sons, 1996.
Ashby, F.A., Materials Selection in Mechanical Engineering, Third Edition, Elsevier, 2005.
Rohsenow, W.M., and Hartnett, J.P., Handbook of Heat Transfer, 3rd Edition, McGraw Hill, 1998.

Guidance, Navigation, Control, and Sensors

"Proceedings of AGARD G&C Conference on Guidance & Control of Tactical Missiles," AGARD LS-52, May 1972.
Garnell, P., Guided Weapon Control Systems, Pergamon Press, 1980.
Locke, A.S., Guidance, Principles of Guided Missile Design, D. Van Nostrand, 1955.
Blakelock, J.H., Automatic Control of Aircraft and Missiles, 2nd Edition, John Wiley & Sons, 1991.
Siouris, G.M., Aerospace Avionics Systems, Academic Press, 1993.
Lecomme, P., Hardange, J.P., Marchais, J.C., and Normant, E., Air and Spaceborne Radar Systems, SciTech Publishing and William Andrew Publishing, 2001.
Wehner, D.R., High-Resolution Radar, Artech House, Norwood, MA, 1995.
Donati, S., Photodetectors, Prentice-Hall, 2000.
Jha, A.R., Infrared Technology, John Wiley and Sons, 2000.
Schlessinger, M., Infrared Technology Fundamentals, Marcel Decker, 1995.
Skolnik, M.I., Introduction to Radar Systems, McGraw-Hill, 1980.
Skolnik, Radar Handbook, Third Edition, McGraw-Hill, 2008.
Zipfel, P.H., Modeling and Simulation of Aerospace Vehicle Dynamics, 3rd Edition, American Institute of Aeronautics and Astronautics, 2014.
Hovanessian, S.A., Radar Detection and Ranging Systems, Artech House, 1973.
Shneydor, N.A., Missile Guidance and Pursuit—Kinematics, Dynamics and Control, Horwood Publishing, 1998.
Moir, I., and Seavridge, A., Military Avionics Systems, American Institute of Aeronautics and Astronautics, 2006.
Mahafza, B.A., Radar Systems Analysis and Design Using MATLAB®, 2nd Edition, Chapman & Hall / CRC, 2005.
Wolf, W.L., and Zissis, G.J., Infrared Handbook, Environmental Research Institute of Michigan; 1985.
James, D.A., Radar Homing Guidance for Tactical Missiles, Macmillan Education LTD, 1986.
Wolfe, W.I., and Zissis, G.J., The Infrared Handbook, Revised Edition, Office of Naval Research, 1985.
Rogalski, Infrared Detectors, 2nd Edition, CRC Press, 2011.
"Electronic Warfare and radar Systems Engineering Handbook". NAWCWPNS TP 8347, 1999.
Neri, F., Introduction to Electronic Defence Systems, 2nd Edition, SciTech Publishing, 2010.
Brooker, G., Introduction to Sensors for Ranging and Imaging, SciTech Publishing, 2009.
Miller, M.M., et al., "Navigation in GPS Denied Environments: Feature-Aided Inertial Systems", NATO RTO-EN-SET-116(2011).
Schmid, G.T., "INS/GPS Integration Architecture Performance Comparisons", NATO RTO-EN-SET-116(2011).
Schmid, G.T., et al., "INS/GPS Integration Architectures", NATO RTO-EN-SET-116(2011).
Barbour, N.M., et al., "Inertial MEMS Systems and Applications", NATO RTO-EN-SET-116(2011).
Barbour, N.M., "Inertial Navigation Sensors", NATO RTO-EN-SET-116(2011).
Schmid, G.T., "INS/GPS Technology Trends", NATO RTO-EN-SET-116(2011).
Bezick, S.M., et al., "Inertial Navigation for Guided Missile Systems", JOHNS HOPKINS APL TECHNICAL DIGEST, (2010).
Advances in Navigation Sensors and Integration Technology, NATO EN-SET-064, October 2003.
Braun, R.D., "Advances in Inertial Guidance Technology for Aerospace Systems", AIAA 2013-5123.
Siouris, G.M., Missile Guidance and Control Systems, Springer, 2004.
Logsdon, T., The Navstar Global Positioning System, Springer, 1992.
Ben-Asher, J.Z., Optimal Control Theory with Aerospace Applications, American Institute of Aeronautics and Astronautics, 2010.
Zaikang, Q.I. and Defu, L., "Design of Guidance and Control Systems for Tactical Missiles", CRC Press, 2020.

Chapter 12 Appendices

Appendix

Appendix A:	Chapter Problem Reviews	363
Appendix B:	Homework Problems/Classroom Exercises	369
Appendix C:	Example of Request for Proposal	375
Appendix D:	Nomenclature	381
Appendix E:	Acronyms/Abbreviations	390
Appendix F:	Conversion Factors	399
Appendix G:	Example Syllabus	401
Appendix H:	CEU Credit Quizzes	403
Appendix I:	Design Case Studies	421
Appendix J:	Summary of Tactical Missile Design Spreadsheet	422
Appendix K:	Soda Straw Rocket Science	426

Appendix A: Chapter Problem Reviews

Problem Review, Chapter 1 – Introduction
1. Typical missile guidance subsystems packaging is longitudinal, with high d_____.
2. The design team addresses mission / scenario definition, weapon requirements, launch platform integration, design synthesis, and tec_____ a_____.
3. The steps to evaluate missile flight performance require computing aerodynamics, propulsion, weight, and flight t_____.
4. Elements of missile system engineering include system requirements flow-down / specification, system integration, system analysis, and sim_____.
5. Air-to-air missiles have light weight, high speed, and high m_____.
6. Air-to-surface missiles are often versatile and often have mod____ warheads.
7. Surface-to-air missile measures of merit include accuracy and max al_____.
8. Four aeromechanics measures of merit are weight, range, maneuverability, and t___ to target.
9. The launch platform often constrains the missile span, length, and w_____.
10. An enabling capability for missiles is fast and accurate C4ISR t_____.
11. Unforeseen requirements may arrive from tech_____ surp____.
12. A baseline design enhances design accuracy and gives fast star___.
13. The Pareto Effect states that the design is driven by only a f__ parameters

Problem Review, Chapter 2 – Aerodynamics
1. Missile diameter tradeoffs include consideration of drag, warhead lethality, subsystem packaging, structural mode frequency, and see___ performance.
2. Tradeoffs of nose fineness include supersonic drag, radar cross section, and se____ performance.
3. Three contributors to drag are base drag, wave drag, and s___ f_____ drag.
4. A boattail angle for a subsonic missile should be less than about 12 deg and for a supersonic missile less than about 7 deg, to avoid flow s_____.
5. A lifting body is most efficient at about 700 psf d_____ p_____.
6. At low angle of attack the aerodynamic center of the body is on the n___.
7. Tail stabilizers have low drag, while lower aero heating and a relatively small shift in static stability are benefits of a fl____ stabilizer.
8. Examples of simple physics-based methods for predicting body normal force and aerodynamic center are sl_____ body theory and New_____ theory.
9. Subsonic missiles often have high aspect ratio wings, for longer ra___.
10. Examples of simple physics-based methods for predicting wing normal force and aerodynamic center are l_____ wing, sl_____ wing, and New_____ theory.
11. The aerodynamic center of the wing is between 25% and 50% of the m___ a_____ c____.
12. Hinge moment increases with the local flow angle due to control surface deflection and the a_____ o_ a_____.
13. Increasing the surface area increases the s___ f_____ d___.
14. Leading edge sweep reduces drag and r____ c____ s_____.
15. A missile with six control surfaces, four surfaces providing combined pitch / yaw control plus two surfaces providing roll control, has an advantage of good con____ ef_____.
16. A missile with two control surfaces providing only combined pitch / yaw control has advantages of lower c___ and good pac_____.

363

17. A tail flight control missile has larger trim normal force if it is statically u_____.
18. Lattice fins have low h____ m_____.
19. Split canards allow higher maximum angle of attack and higher m_____.
20. Aerodynamic flight control surfaces stall at a control surface local angle of attack of approximately __ degrees.
21. Two types of unconventional flight control are thrust vector control and r_____ j__ control.
22. The most common type of TVC for tactical missiles is j__ v___ control.
23. Four maneuver laws are skid to turn, bank to turn, rolling airframe, and d_____.
24. Bank to turn maneuvering is usually required for missiles with a single wing or with non-cr_____ inlets.
25. A missile is statically stable if the aero center is behind the c_____ o_ g_____.
26. A typical maneuverability guideline is the change in angle of attack with flight control deflection is $\alpha / \delta > \_$.
27. If the moments on the missile are zero the missile is in tr__.
28. Total normal force on the missile is approximately the sum of the normal forces on the planar surfaces (e.g., wing, tail, canard) plus the normal force on the b___.
29. Increasing the tail area increases the missile static st_____.

Problem Review, Chapter 3 – Propulsion

1. An advantage of a turbojet compared to a ramjet is thrust at sub_____ Mach number.
2. The thrust of a turbojet is often limited by the maximum allowable temperature of the t_____.
3. The specific impulse of a ramjet is often limited by the maximum allowable temperature of the combustor i_____.
4. Ducted rockets are based on a fuel-rich g__ g_____.
5. Advantages of solid propellant rocket propulsion over liquid propulsion include higher thrust / acceleration and less t_____.
6. A rocket boost to a supersonic take-over Mach number is required by ramjets and s_____.
7. Parameters that enable the long range of subsonic cruise turbojet missiles are high lift, low drag, available fuel volume, and high s_____ i_____.
8. In a turbojet the power to drive the compressor is provided by the t_____.
9. The compressor exit temperature is a function of the flight Mach number and the compressor p_____ r____.
10. Compressor exit temperature, fuel heating value, and fuel-to-air ratio determine the turbojet t_____ temperature.
11. A high compressor pressure ratio requires a large number of axial st____.
12. A long-range subsonic turbojet usually has higher compressor pressure ratio, with higher s_____ i_____.
13. A scoop inlet has higher efficiency while a flush inlet has lower r____ c____ s_____.
14. Mach number, fuel heating value, and fuel-to-air ratio determine the ramjet c_____ temperature.
15. A large inlet flow area is required for a ramjet with low combustion temperature, high drag, and high M___ n_____.
16. An example of a low-drag and lightweight ramjet is an i_____ r_____ ramjet.
17. Russia, France, China, United Kingdom, Taiwan, and India are the only countries with currently operational r_____ missiles.
18. 100% inlet capture efficiency for a supersonic missile occurs when the forebody shock waves intercept the i____ l__.
19. Excess air that does not flow into the inlet is called s_____ air.
20. Optimum pressure recovery across shock waves is achieved when the total pressure loss across each shock wave is e____.
21. The specific impulse and thrust of a ramjet are a function of the efficiency of the combustor, nozzle, and i____.
22. A high-density fuel has high payoff for a typical, v_____ limited missile.

23. The specific impulse of a ducted rocket with large excess fuel from the gas generator can approach that of a r_____.
24. A high-speed rocket requires a large weight fraction of p_____.
25. At the rocket nozzle throat, the flow area is minimum, sonic, and ch____ .
26. For an optimum nozzle expansion, the nozzle exit pressure is equal to the at_____ pressure.
27. High thrust and high chamber pressure are achievable through a large propellant b___ area.
28. Three approaches to rocket thrust magnitude control are pulse motor, pintle motor, and g__ motor.
29. Pulse motor thrust magnitude control may provide higher terminal m_____.
30. Three tradeoffs in selecting a solid propellant are safety, observables, and s_____ i_____.
31. An average temperature increase of 10 deg C decreases propellant lifetime by __%.
32. A low-cost motor case is usually based on steel or aluminum material while a lightweight motor case is usually based on graphite c_____ material.
33. Rockets with high chamber pressure or long burn time may require a tu_____ throat insert.
34. An under-expanded rocket nozzle may cause problems such as flow separation on the missile aft body, aero heating of nearby missiles, and aero heating of the launch platform due to its large diameter plu__.
35. A ducted rocket using carbon fuel has lower visual observables, while a ducted rocket using metal fuel typically has longer ra___.
36. Thrust magnitude control for a ducted rocket is often provided by a p_____ valve.

Problem Review, Chapter 4 – Weight

1. Lightweight missiles tend to be lower c___.
2. Ballistic missile launch weight is driven by maximum range, payload weight, propellant weight, and s_____ i_____.
3. For a multi-stage ballistic missile, maximum range is achieved when the incremental velocity from each stage is e____.
4. Missile weight is proportional to its vol___.
5. Missile subsystems d_____ is about 0.05 lbm / in^3
6. Missile structure factor of safety for free flight is usually about 1.25 for ultimate loads and about 1.10 for y____ loads.
7. Airframe manufacturing processes with low parts count include vacuum assisted resin transfer molding of composites and ca_____ of metals.
8. Other examples of composite airframe fabrication include vacuum bagging, pultrusion, thermal forming, compression molding, and fi_____ wi_____.
9. Other missile airframe manufacturing processes of metals include forming, sintering, welding, and ma_____.
10. Aluminum has lower hardness than steel, which makes it easier to m_____.
11. Increasing metal hardness increases its st_____.
12. Low cost missile airframe materials are usually based on metal (e.g., aluminum) while lightweight airframe materials are usually based on laminate c_____ materials.
13. Carbon fiber has high strength and high m_____ o_ e_____.
14. The recovery factors of 1.0, 0.9, and 0.85 are appropriate respectively for stagnation, turbulent boundary layer, and l_____ boundary layer.
15. The most popular types of missile insulation for temperatures greater than 4000 R are charring insulators based on ph_____ composites.
16. Missiles experience transient heating, and with increasing time the temperature approaches the r_____ temperature.
17. The inner wall temperature is nearly the same as the surface temperature for a t_____ t___ structure.
18. A thermally thick surface is a good in_____.
19. A low conductivity structure is susceptible to thermal st____.
20. A very thin wall structure is susceptible to localized b_____.
21. Ejection loads and flight control loads often result in large b_____ moment.
22. An approach to increase the missile propellant / motor weight fraction over the typical value of 65% would be a g_____ c_____ motor case.

23. The required rocket motor case thickness is often driven by the combustion chamber p_____.
24. For low speed missiles, a popular infrared dome material is z___ su_____.
25. A thermal battery provides high p____.
26. Flight control actuator design is driven by h____ moment.
27. The most popular type of flight control actuator for missiles is an e_____ actuator.

Problem Review, Chapter 5 – Flight Performance

1. Flight trajectory calculation requires input from aero, propulsion, and w_____.
2. Missile robust flight envelope can be characterized by the maximum effective range, minimum effective range, and large o__ b_____.
3. Limitations to the missile effective range include the fire control system, seeker, time of flight, closing velocity, and missile ma_____ capability.
4. A simulation that includes only axial force, thrust, and weight is called a _DOF simulation.
5. A 3-DOF simulation that models 3 aero forces is called a p____ m___ simulation.
6. A simulation that includes 3 aero forces (normal, axial, side) 3 aero moments (pitch, roll, yaw), thrust, and weight is called a _DOF simulation.
7. An aero control missile has high agility if it has high control effectiveness, low static margin, high dynamic pressure, and light w_____.
8. Missile turn radius should be s_____ than the target turn radius.
9. Cruise range is a function of velocity, specific impulse, L / D, and fu__ fraction.
10. Maximum glide range of a lofted boost-glide missile is driven by initial velocity, initial altitude, and lift-to-d___ ratio.
11. A lofted boost-glide flight trajectory has a lower apogee than a ballistic trajectory, which may delay threat r____ acquisition.
12. Thrust vector control provides higher maneuverability than aero control at low v_____.
13. Coast range at constant altitude is a function of initial velocity, weight, drag, and t___ of flight.
14. Range for a ballistic missile Is driven by burnout velocity, burnout angle, burnout altitude, and d___.
15. Boost increment velocity is a function of ISP, drag, and p_____ weight fraction.
16. An analytical model of an unguided ($\alpha = 0$ deg) weapon can be developed from the ba_____ flight phase.
17. An analytical model of a ramjet / turbojet in co-altitude, non-maneuvering flight can be developed by patching the flight phases of boost followed by cr____.
18. An off-boresight missile terminal velocity should be g_____ than target velocity.

Problem Review, Chapter 6 – Other Measures of Merit

1. IR signal attenuation is greater than 100 dB per km through a cl___.
2. GPS / INS enhances seeker lock-on in adverse weather and ground cl_____.
3. Data link update can enhance missile seeker lock-on against a mo____ target.
4. A larger diameter seeker has longer range and higher res_____.
5. One missile counter-counter measure to flares is an im_____ in_____ seeker.
6. Compared to a mid-wave IR seeker, a long wave IR seeker receives more energy from a c___ target.
7. Compared to a cmW seeker, a mmW seeker has higher resolution, but it is limited to shorter r____.
8. A strapdown seeker is lighter weight but a gimbaled seeker has a larger fi___ of re____.
9. Drivers for tactical INS are cost and w_____.
10. Drivers for strategic INS are alignment, drift, and au_____.
11. High fineness kinetic energy penetrators are required to defeat bu____ targets.
12. For the same lethality with a blast fragmentation warhead, a small decrease in miss distance allows a large decrease in the required weight of the w_____.
13. High PK requires warhead lethal radius greater than t____ times miss distance.
14. A conceptual design tool for a blast / frag warhead is the Gur___ equation.
15. For a blast / frag warhead, a charge-to-metal ratio of about one is required to achieve a high total fragment k_____ e_____.
16. A blast fragmentation warhead tradeoff is the number of fragments versus the individual fragment w_____.

17. Benefits of accurate guidance include higher lethality, lower required warhead weight, and lower col_____ damage.
18. Kinetic energy penetration is a function of the penetrator diameter, length, density, strength, and v_____.
19. Shaped charge warhead penetration is driven by diameter, ductility, and den____.
20. Flight control feedback should be faster than g_____ feedback.
21. Autonomous guidance improves launch platform sur_____.
22. Proportional homing guidance drives the line-of-sight angle rate equal to z___.
23. Aeromechanics contributors to missile time constant include flight control effectiveness, flight control system dynamics, and dome e____ s____.
24. Miss distance due to heading error is a function of missile navigation ratio, velocity, time to correct the heading error, and the missile t___ c_____.
25. Missile maneuverability is typically at least t____ times that of the target.
26. Proportional guidance requires high initial maneuverability for a h_____ e____.
27. Miss distance against a weaving target is large if the missile time constant is comparable to the inverse of the target weave fr_____.
28. Minimizing miss distance due to radar glint requires a high-resolution seeker, optimum time constant and optimum na_____ ra___.
29. Weapons on low observable launch platforms use i_____ carriage.
30. Weapons on low observable launch platforms use m_____ smoke propellant.
31. Alternative approaches to improve precision strike missile survivability include 1) high altitude cruise, 2) high speed, 3) mission planning / threat avoidance, 4) terrain masking, 6) maneuvering, and 7) low o_____.
32. Drivers for RCS are radar frequency, target shape, and target o_____.
33. The largest contributor to the ramjet baseline frontal RCS is its ra___ see___.
34. The hypersonic ramjet baseline missile has a high i_____ signature.
35. For an insensitive munition, burning is preferable to detonation because it releases less p____.
36. System reliability is enhanced by subsystem reliability and low p____ count.
37. High cost subsystems of missiles are sensors, electronics, and p_____.
38. Missile EMD cost is driven by the program duration and ri__.
39. Missile production cost drivers include # of units, learning curve, and w_____.
40. First level maintenance is conducted at a d____.
41. Store compatibility problems include aeroelasticity, hang-fire, and s_____.
42. A standard launch system for U.S. Navy ships is the V_____ L_____ S____.
43. Most air-launched missiles that are lightweight use rail launchers, while most heavyweight air-launched missiles use e_____ launchers.
44. Higher firepower is provided by com_____ carriage.
45. Missile / fire control system terminal guidance alternatives include command, semi-active, autonomous, and m____-m___.
46. Typical environmental requirement in MIL-HDBK-310 is _% world-wide extreme.
47. A missile exposed to the sun probably has greater than ambient t_____.

Problem Review, Chapter 7 – Sizing Examples and Sizing Tools

1. Required flight range is shorter for a head-on intercept and it is longer for a ta__ ch___ intercept.
2. The rocket baseline missile center-of-gravity moves f_____ with motor burn.
3. The rocket baseline missile has an al_____ airframe.
4. The rocket baseline motor case and nozzle are made of st___.
5. The rocket baseline missile flight range is driven by I_{SP}, propellant weight fraction, drag, and st____ ma____.
6. Configuration contributors to the maneuverability of the rocket baseline missile are its body, tail, and w___.
7. Although the rocket baseline missile has sufficient g's and turn rate to intercept a maneuvering aircraft, it needs a smaller turn r_____.

8. Compared to a co-altitude trajectory, the rocket baseline missile has extended range with a lofted boost g____ trajectory.
9. The ramjet baseline missile has a ch__ inlet.
10. The Mach 4 ramjet baseline missile has a ti_____ airframe.
11. The ramjet baseline combustor is a nickel-based super alloy, in_____.
12. The flight range of the ramjet baseline missile is driven by I_{SP}, thrust, zero-lift drag coefficient, and the weight fraction of fu__.
13. Extended range for the ramjet baseline missile would be provided by more efficient packaging of subsystems and the use of sl____ fuel.
14. The flight range of the turbojet baseline missile is driven by fuel weight, specific impulse, and d___.
15. Efficiency corrections to the turbojet prediction account for total pressure losses in the compressor, combustor, turbine, nozzle, and i____.
16. A turbojet cruises most efficiently at high a_____.
17. A small diameter centrifugal compressor has higher ro_____ sp___.
18. A conceptual design sizing code should be based on the simplicity, speed, and robustness of ph_____ based methods.
19. The House of Quality room for design characteristics has engineering characteristics that are most important in meeting cu_____ re_____.
20. A Pareto sensitivity identifies design parameters that are most im_____.
21. DOE provides a systematic evaluation of the broad possible de____ sp___.

Problem Review, Chapter 8 – Development Process

1. The levels of design maturity from the most mature to least mature are production, detail, preliminary, and c_____ design.
2. Technology transitions occur from basic research to exploratory development, to advanced development, to d_____ and v_____.
3. Approximately 11% of U.S. RT&A budget is allocated to t_____ m_____.
4. In the U.S., a new tactical missile follow-on program occurs about every __ years.
5. Compared to the AIM-9L, the AIM-9X has enhanced m_____.
6. Compared to the AIM-7, the AIM-120 has autonomous guidance, lighter weight, higher speed, and longer r____.
7. Compared to the PAC-2, the PAC-3 has h__ t_ k___ accuracy.
8. A best value proposal is a best value of cost, risk, and p_____.
9. Guidance & control is verified in the h_____ in l___ simulation.
10. Aerodynamic force and moment data are acquired in w___ t_____ tests.
11. Many flight tests are required to develop a high con_____ missile system.

Appendix B: Homework Problems/Classroom Exercises

These homework problems/classroom exercises are designed to complement the problems at the end of the chapters. The problems at the end of the chapters can be posed as verbal questions to the students, providing an indication that the students were listening to the lecture. The following homework problems/classroom exercises are designed to provide a more in-depth learning reinforcement. Points associated with each homework problem/classroom exercise indicate the degree of difficulty, as a consideration in assigning the student's grade for the problem.

1. **(20 points). Chapter 1 (Introduction).** Compare measures of merit for the following alternative concepts to destroy threat ballistic missiles during their boost phase:
   - Space-based interceptors
   - Space-based lasers
   - Airborne laser
   - Air-launched interceptors
   - Ship-launched interceptors
   - Ground-launched interceptors

2. **(10 points). Chapter 1 (Introduction).** Develop a state-of-the-art comparison of tactical missile characteristics with the current state-of-the-art (SOTA) of UCAVs. Show examples where missiles are driving technology. Also show examples where the missile is not driving technology.

3. **(10 points). Chapter 1 (Introduction).** For the examples shown of air-launched and surface-launched missiles, which have been applied to more than one mission?

4. **(10 points). Chapter 2 (Aerodynamics).** Based on the example, calculate the first mode body bending frequency for a missile weight of 367 lb.

5. **(10 points). Chapter 2 (Aerodynamics).** In the body-flare example, what is the required diameter of the flare to provide neutral stability at launch?

6. **(20 points). Chapter 2 (Aerodynamics).** What are the strengths and weaknesses of the China SD-10/PL-12 tail planform?

7. **(10 points). Chapter 2 (Aerodynamics).** Calculate the rocket baseline wing normal force coefficient $(C_N)_{Wing}$ at Mach 1.1, $\delta = 13$ deg, $\alpha = 9$ deg.

8. **(40 points). Chapter 2 (Aerodynamics).** Calculate C_{D_0}, C_N, and C_m for the ramjet baseline at Mach 2.5 end-of-cruise and Mach 4.0 end-of-cruise. Compare with the aerodynamic data of Chapter 7. Why is it difficult to accurately predict (or even obtain accurate data) for C_m?

9. **(20 points). Chapter 2 (Aerodynamics).** What are the dynamic pressures at cruise $(L/D)_{max}$ for a circular cross section missile and an a/b = 2 lifting body cross section missile if the weight W = 500 lb, cross sectional reference area $S_{Ref} = 0.5$ ft^2, length-to-diameter ratio l/d = 10, and zero-lift drag coefficient $C_{D_0} = 0.2$?

10. **(20 points). Chapter 2 (Aerodynamics).** Calculate hinge moment for the ramjet baseline at the initiation of a pitch-over dive. Flight conditions are Mach 4.0, h = 80k ft altitude, end-of-cruise, $\delta_e = -30$ deg control deflection, and $\alpha = \alpha_{max}$ angle of attack.

11. **(20 points). Chapter 2 (Aerodynamics).** Using the approach of the text example, calculate the required tail area of the rocket baseline to provide neutral static stability at launch.

12. **(20 points). Chapter 3 (Propulsion).** Using the approach of the text example, compare compressor exit temperature at Mach number M = 3, altitude h = sea level with the Mach 2, h = 60k ft compressor temperature result of the text example.

369

13. **(20 points). Chapter 3 (Propulsion).** For the turbojet text example ($T_4 = 3000$ R, $A_0 = 114$ in^2, RJ-5 fuel, $M = 2$, $h = 60$k ft) show the impact on maximum ideal thrust and specific impulse of $+/-$ 10% uncertainty in specific heat ratio.

14. **(30 points). Chapter 3 (Propulsion).** For an ideal turbojet, calculate thrust, specific impulse, equivalence ratio, and nozzle exit area for the following conditions/assumptions: liquid hydrocarbon fuel, free stream Mach number = 2, angle of attack = 0 deg, altitude = 60k ft, compressor pressure ratio for maximum thrust, turbine maximum temperature = 2000 R, and inlet capture area = 114 in^2. Compare with the text example.

15. **(20 points). Chapter 3 (Propulsion).** Using the approach of the text example of a centrifugal compressor, what is the rotation rate if the impeller tip Mach number $M_{ImpellerTip} = 1.2$?

16. **(20 points). Chapter 3 (Propulsion).** For an axial compressor stage with a pressure coefficient $c_p = 0.6$ and rotor entrance local Mach number $M_{entrance} = 1.4$, what is stage pressure ratio $p_{exit}/p_{entrance}$?

17. **(10 points). Chapter 3 (Propulsion).** For an assumed axial compressor single stage pressure ratio $p_{exit}/p_{entrance} = 2$, what is the overall compressor pressure ratio p_3/p_2 of a four-stage compressor?

18. **(20 points). Chapter 3 (Propulsion).** For an assumed turbine entrance temperature $T_4 = 4000$ R, compare the thrust T and specific impulse I_{SP} with the example in the text ($T_4 = 3000$ R).

19. **(20 points). Chapter 3 (Propulsion).** At what combustion temperature does dissociation of water become a significant contributor to real gas effects?

20. **(30 points). Chapter 3 (Propulsion).** Calculate the thrust, specific impulse, equivalence ratio, and nozzle exit area of an ideal ramjet for the following conditions/assumptions: liquid hydrocarbon fuel, free stream Mach number = 2.5, angle of attack = 0 deg, altitude = 60k ft, ramjet combustor maximum temperature 4000 R, and inlet capture area = 114 in^2. Compare with the ramjet baseline data of Chapter 7.

21. **(40 points). Chapter 3 (Propulsion).** The corrected specific impulse of a ramjet is a function of the individual efficiencies of the combustor, nozzle, and inlet. Assuming that the driving parameter for the efficiencies is the total pressure recovery, derive an expression for correcting theoretical specific impulse.

22. **(100+ points). Chapter 3 (Propulsion).** Derive the one-dimensional equations for thrust and specific impulse for an ideal scramjet. Calculate thrust, specific impulse, combustor area, and nozzle exit area for the following conditions/assumptions: hydrocarbon fuel, free stream Mach number = 6.5, angle of attack = 0 deg, altitude = 100k ft, Mach 3 initial combustion, thermal choking limit, combustor maximum temperature = 4000 R, and inlet capture area = 114 in^2. How is a scramjet similar to a ramjet? How is it different?

23. **(10 points). Chapter 3 (Propulsion).** In the example, what is the inlet start Mach number if the inlet throat area $A_{IT} = 0.4$ ft^2? Is there a problem in having a large area for the inlet throat?

24. **(30 points). Chapter 3 (Propulsion).** Calculate the total pressure ratios entering the combustor for the ramjet baseline cruising at Mach 2.5, sea level and at Mach 4.0, 80k ft. The inlet is a mixed compression type with a total of four compressions prior to the normal shock, consisting of 1) the shock wave on the conical nose, followed by 2) the shock wave on the ramp leading to the cowl, followed by 3) the shock wave on the cowl, and finally 4) a series of nearly isentropic internal contraction shock waves leading to a normal shock. Compare with the maximum available total pressure ratio from four optimum compressions.

25. **(100+ points). Chapter 3 (Propulsion).** Derive the one-dimensional equations for thrust and specific impulse for an ideal ducted rocket. Calculate thrust, specific impulse, equivalence ratio, inlet throat area, and diffuser exit area for the following conditions/assumptions: 40% boron fuel, 8% aluminum fuel, 27% binder fuel, 25% ammonium perchlorate oxidizer, free stream Mach number = 4, angle of attack = 0 deg, altitude = 80k ft, gas generator pressure = 1000 psi, combustor maximum temperature = 4000 R, combustor area = 287 in^2, and inlet capture area = 114 in^2.

26. **(10 points). Chapter 3 (Propulsion).** Compute the turbojet specific impulse I_{SP} of a 40% JP-10/60% boron carbide slurry fuel that has a heating value $H_f = 23,820$ BTU/lbm. Compare with the text example (JP-10 fuel with $H_f = 18.700$ BTU/lbm).

27. **(20 points). Chapter 3 (Propulsion).** Calculate the rocket baseline thrust at altitudes of sea level, 20k ft, and 50k ft. Compare results with Chapter 7.

28. **(20 points). Chapter 3 (Propulsion).** If the rocket baseline throat area were reduced by 50%, with the propellant, burn area, and nozzle expansion ratio the same, what would be the resulting boost/sustain chamber pressure, thrust, specific impulse, and propellant weight flow rate?

29. **(30 points). Chapter 3 (Propulsion).** Assume a propellant burn rate exponent n = 0.9. Also assume the same nominal propellant burn rate $r_{p_c=1000\,psi}$, propellant characteristic velocity c*, propellant density ρ, nozzle geometry, and thrust profile as the rocket baseline. Calculate the chamber pressures and burn areas for boost/sustain. Compare with the rocket baseline chamber pressures and burn areas.

30. **(20 points). Chapter 4 (Weight Considerations in Missile Design and System Engineering).** The rocket baseline propellant grain is a slotted tube with a propellant volumetric efficiency of 90%. For the same volume of the motor case, what is the boost propellant weight and end-of-boost velocity for a grain with a propellant volumetric efficiency of 80%?

31. **(20 points). Chapter 4 (Weight).** A typical strategic ballistic missile motor has a much larger propellant fraction than a typical tactical ballistic missile, resulting in longer range. Assume a strategic ballistic missile has a typical inert subsystems weight fraction of 0.1 of the propellant weight. Also assume a payload weight of 1000 lb. Neglecting drag and the curvature of the earth, calculate the maximum range of a three-stage 100,000 lb missile with specific impulse of I_{SP} = 250 s (Minuteman-type solid propellant). As a comparison, calculate the maximum range of a two-stage 100,000 lb missile with specific impulse of I_{SP} = 300 s (Titan-type liquid propellant). Discuss the trade-off of the number of stages vs type of propellant/specific impulse.

32. **(20 points). Chapter 4 (Weight).** Calculate the center-of-gravity and pitch/yaw moment-of-inertia of the rocket baseline if the propellant density were increased by 50%, assuming the same weights for subsystems (e.g., motor case) in Chapter 7. Compare with the data in Chapter 7.

33. **(30 points). Chapter 4 (Weight).** Based on an average heat transfer coefficient, estimate the rocket baseline airframe temperature at the end of the flight trajectory example of Chapter 7.1 (Mach 0.8 launch at 20k ft altitude, 3.26 s boost, 10.86 s sustain, 9.85 s coast).

34. **(30 points). Chapter 4 (Weight).** Calculate the ramjet baseline radome internal wall temperature and surface wall temperature after 10 s flight at Mach 3, sea level.

35. **(30 points). Chapter 4 (Weight).** Estimate the ramjet baseline internal insulation required thickness to maintain warhead temperature less than 160 °F for 10 m time of flight at Mach 4/80k ft. Assume an initial temperature of 70 °F.

36. **(40 points). Chapter 4 (Weight).** For the ramjet baseline, compare the weight of an aluminum airframe with external micro-quartz insulation to that of the baseline uninsulated titanium airframe.

37. **(30 points). Chapter 4 (Weight).** Using the ramjet baseline inlet geometry and material data of Chapter 7, estimate the required inlet thickness and the required inlet weight based on an inlet start at Mach 2.5, sea level altitude. Compare with the inlet weight of Chapter 7.

38. **(20 points). Chapter 4 (Weight).** Calculate motor case weight for the rocket baseline if the case were titanium, with a forward dome ellipse ratio of 3 and a cylindrical cross section aftbody. Compare with the example calculation for a steel motor case and the data in Chapter 7.

39. **(20 points). Chapter 4 (Weight).** Calculate required tail thickness and the resulting weight of the rocket baseline tail surfaces. Assume a flight condition of Mach 2, altitude = 20k ft, motor burnout, angle of attack = 9.4 deg, and an ultimate stress factor of safety = 1.5. Compare result with the data of Chapter 7.

40. **(20 points). Chapter 4 (Weight).** Calculate radome weight for the rocket baseline if the radome were silicon nitride with an optimum transmission thickness. Compare with the example calculation for a pyroceram radome and the data in Chapter 7.

41. **(20 points). Chapter 4 (Weight).** Calculate the required thickness and the resulting weight of the rocket baseline radome to withstand the load from Mach 2, 20k ft altitude, and angle of attack = 9.4 deg for an ultimate stress factor of safety = 1.5. Compare with the Chapter 4 example of optimum transmission thickness and the data of Chapter 7.

42. **(10 points). Chapter 4 (Weight).** Calculate actuation system weight for the rocket baseline if the actuators were electromechanical. Compare with the text example calculation and the data in Chapter 7.

43. **(10 points). Chapter 5 (Flight Performance).** Using the Breguet range equation, calculate the Mach 2.5, sea level cruise range of the ramjet baseline, using data from Chapter 7. Compare with the range in Chapter 7.

44. **(10 points). Chapter 5 (Flight Performance).** Calculate the steady state rate of climb and the climb flight path angle for the ramjet baseline at Mach 2.5, sea level, maximum thrust. Use data from Chapter 7.

45. **(20 points). Chapter 5 (Flight Performance).** Calculate turn radius and turn rate of the ramjet baseline for horizontal and pitch-over turns at Mach 4, h = 80k ft altitude, end of cruise, angle of attack α = 15 deg. Use data from Chapter 7.

46. **(20 points). Chapter 5 (Flight Performance).** Calculate the velocity and range of the rocket baseline after 10 s of coast at a flight path angle of +30 deg for an initial velocity of 2151 ft/s and an initial altitude of 20k ft.

47. **(20 points). Chapter 5 (Flight Performance).** Calculate booster burnout Mach number and ramjet acceleration capability following booster burnout at sea level for the ramjet baseline. Compare with the performance data of Chapter 7.

48. **(10 points). Chapter 5 (Flight Performance).** Assuming a non-accelerating target at Mach 0.8, h = 20k ft altitude, and 30 deg aspect, what is the required missile lead angle for a constant bearing Mach 3 fly-out at h = 20k ft?

49. **(10 points). Chapter 6 (Other Measures of Merit).** For fog cover of 20 m height over the target, what is the required look-down angle to achieve less than 5 dB attenuation of a passive IR seeker?

50. **(20 points). Chapter 6 (Other Measures of Merit).** Based on the example, calculate imaging IR seeker detection range and instantaneous field of view for rainfall at 4 mm/hr and optics diameter d_o = 2 in.

51. **(20 points). Chapter 6 (Other Measures of Merit).** Compare the performance of an MWIR seeker vs an LWIR seeker using the data of the text example, but with a target temperature of 500 K.

52. **(20 points). Chapter 6 (Other Measures of Merit).** Based on the example, calculate radar seeker detection range and 3-dB beam width for a target cross section σ = 1 m², mmW transmitter frequency f = 35 × 10⁹ Hz, transmitted power P_t = 100 W, and antenna diameter d = 4 in.

53. **(10 points). Chapter 6 (Other Measures of Merit).** Assume a missile with a velocity V = 300 m/s, strapdown seeker with a field of view θ = 20 deg, and seeker detection range R_D = 1 km. Assume that an off-board sensor (e.g., UAV) provides the missile with a target location error TLE = 10 m. Assume the target has a velocity V_T = 10 m/s, laterally to the flight path of the missile. If the update time from the off-board sensor is equal to the target latency time (i.e., t_{Update} = $t_{Latency}$), what is the required update time from the off-board sensor?

54. **(10 points). Chapter 6 (Other Measures of Merit).** For the text example rocket baseline warhead, what is the blast overpressure Δp at a distance from the center of explosion of r = 5 ft and an altitude h = sea level?

55. **(20 points). Chapter 6 (Other Measures of Merit).** Calculate the maximum miss distance requirement of a 5 lb warhead with C/M = 1 to achieve a lethality of 0.5 for a typical air target vulnerability (overpressure Δp = 330 psi, fragments impact energy = 130k ft-lb/ft²).

56. **(10 points). Chapter 6 (Other Measures of Merit).** Assume a revised rocket baseline warhead that has reduced collateral damage, with a warhead metal case mass M_m = 0.4 slug. For a warhead charge mass M_c = 1.207 slug, what is the total kinetic energy KE of the warhead?

57. **(10 points). Chapter 6 (Other Measures of Merit).** Compare the text example of penetration through concrete with the penetration of the same penetrator through granite of density ρ = 0.0897 lb/in³ and ultimate strength σ = 20,000 psi.

58. **(30 points). Chapter 6 (Other Measures of Merit).** Assume a missile defense interceptor has a time constant of τ = 0.05 s, an effective navigation ratio of N' = 4, and the target glint noise bandwidth is B = 2 Hz. What is the miss distance from glint for the imaging IR seeker example in Chapter 2?

59. **(10 points). Chapter 6 (Other Measures of Merit).** In the text example for survivability through high altitude flight and low radar cross section (RCS), if the flight altitude h = 80k ft, what is the required RCS to avoid detection by a threat radar that has a transmitted power P_t = 10⁶ W and a wavelength λ = 0.01 m?

60. **(20 points). Chapter 6 (Other Measures of Merit).** Calculate the frontal RCS of the rocket baseline.

61. **(20 points). Chapter 6 (Other Measures of Merit).** Using data from the example in the text, calculate the frontal radiant intensity of the ramjet baseline for long duration flight at Mach 2.5/sea level.

62. **(10 points). Chapter 6 (Other Measures of Merit).** Using data from the example in the text, calculate the frontal IR detection range if the if the radiant intensity is reduced by 50%.

63. **(10 points). Chapter 6 (Other Measures of Merit).** Using data from the text example for reduced frontal RCS, calculate detection range and exposure time for +/− 4 dB, 1 σ uncertainty in RCS.

64. **(10 points). Chapter 6 (Other Measures of Merit).** Assuming a reliability of 98%, what is the missile circular error probable (CEP) that is required to provide a 95% probability of kill for a warhead lethal radius of 5 ft?

65. **(10 points). Chapter 6 (Other Measures of Merit).** Calculate a typical system reliability of a missile that combines the autopilot, navigation sensors and computer as a single subsystem and has no seeker. Compare with the text example.

66. **(10 points). Chapter 6 (Other Measures of Merit).** For a seven-year development program of a 300 lb missile with a learning curve of 0.7, calculate the following for a total buy of 10,000 missiles:
    - development cost
    - total production cost
    - average unit production cost
    - unit production cost of missile number 10,000

67. **(20 points). Chapter 6 (Other Measures of Merit).** Using the rocket baseline motor example of the text, what is the inner surface temperature of the motor if the initial temperature is 70 °F and it is subjected to an ambient temperature of −60 °F for 1 h?

68. **(20 points). Chapter 6 (Other Measures of Merit).** Using the rocket baseline motor example of the text, what is the maximum temperature of a titanium motor case?

69. **(10 points). Chapter 7 (Sizing Examples and Sizing Tools).** What are the visual detection and recognition ranges of a small UCAV target that has a presented area $A_P = 10$ ft^2 and a contrast $C_T = 0.1$?

70. **(30 points). Chapter 7 (Sizing Examples and Sizing Tools).** Compare the rocket baseline motor case (see Chapter 7) with a higher strength motor case made of 4130 heat treated steel with an ultimate tensile strength of 250,000 psi. Discuss design considerations such as brittleness, fracture sensitivity, material cost, motor case manufacturing cost (e.g., machining, welding), required case thickness, and required case weight. Is a higher strength 4130 motor case a good idea?

71. **(60 points). Chapter 7 (Sizing Examples).** Using the rocket baseline air-to-air standoff range example, compare the results based on the 1DOF analytical equations of motion of Chapter 5 with results from a numerical solution of the time marching equations of motion.

72. **(30 points). Chapter 7 (Sizing Examples and Sizing Tools).** For the rocket baseline, what is the required wing area for 40 g maneuverability at Mach 3, burnout, and 20k ft altitude?

73. **(30 points). Chapter 7 (Sizing Examples and Sizing Tools).** Conduct a Design of Experiment (DOE) to define the wind tunnel model configuration alternatives for the harmonized rocket baseline. Specify the nose, body, wing, and tail geometry options for a light weight/small miss distance missile.

74. **(40 points). Chapter 7 (Sizing Examples and Sizing Tools).** For the ramjet baseline, compare the inlet mass flow rate at Mach 2.5, sea level altitude with the mass flow rate at Mach 4.0, 80k ft altitude.

75. **(10 points). Chapter 7 (Sizing Examples and Sizing Tools).** Based on the ramjet baseline data in Chapter 7 and the Breguet range equation, calculate the cruise range for Mach 4/80k ft altitude. Assume that all of the fuel is available for cruise.

76. **(80 points). Chapter 7 (Sizing Examples and Sizing Tools).** Size an extended range ramjet that has 30% greater range than the ramjet baseline. Assume launch at Mach 0.8/sea level with cruise at Mach 2.3/sea level. Size the design by extending the missile length while maintaining constant missile diameter and static margin. Compare the new body length, tail area, launch weight, and the individual weights of fuel, fuel tank, boost propellant, booster case, and tails with the ramjet baseline.

77. **(20 points). Chapter 7 (Sizing Examples and Sizing Tools).** Calculate the frontal radar cross section RCS of the turbojet baseline.

78. **(30 points). Chapter 7 (Sizing Examples and Sizing Tools).** Calculate the Mach $M = 0.4$ zero-lift drag coefficient C_{D_0}, normal force coefficient C_N, and lift-to-drag ratio L/D of the turbojet baseline. Compare with the data of Chapter 7.

79. **(40 points). Chapter 7 (Sizing Examples and Sizing Tools).** From the Request for Proposal given in Appendix B, develop a House of Quality Customer Requirements/Most Important Requirements (MIRs) and an Importance Rating of the MIRs. Expand the rows and columns of the House of Quality. Give rationale for your selections and values.

80. **(20 points). Chapter 7 (Sizing Examples and Sizing Tools).** For the soda straw rocket baseline, specify a tail geometry and area that provides neutral static stability.

81. **(20 points). Chapter 7 (Sizing Examples and Sizing Tools).** For the soda straw rocket baseline, calculate the range neglecting drag ($C_{D_0} = 0$). Compare with the Chapter 7 range that includes the effect of drag.

82. **(30 points). Chapter 7 (Sizing Examples and Sizing Tools).** For the soda straw rocket baseline, what is the zero-lift drag coefficient C_{D_0} if we assume a laminar boundary layer?

83. **(80 points). Chapter 7 (Sizing Examples and Sizing Tools).** Design, build, and fly a soda straw rocket that is optimized for maximum range at an assumed launch condition of 30 psi launch pressure. The rocket design must be compatible with a launch platform constraint of a "Super Jumbo" straw launcher of 0.25 in diameter and 6 in available launch length. You may base your rocket on the materials provided in class, or you may use your own materials as long as you satisfy the launch straw constraint (0.25 in diameter, 6 in available length). Provide the following information for your design:
    - Geometric, weight, center-of-gravity, aerodynamic, and thrust-time characteristics and the rationale for their values.
    - Dimensioned drawing.
    - Velocity during boost as a function of time and distance.
    - Post-boost flight trajectory height and horizontal range as a function of time.
    - Effect of $+/- 10\%$ (1σ) prediction uncertainty of the drag coefficient on horizontal range.
    - Effect of $+/- 10$ ft/s (1σ) horizontal head/tail wind velocity on horizontal range.
    - Comparison of predicted range with flight test.

84. **(10 points). Chapter 7 (Sizing Examples and Sizing Tools).** For the soda straw rocket house of quality of the text example, what is relative ranking of engineering design parameters if the customer emphasis is changed to 70% emphasis on light weight/30% emphasis on long range?

85. **(40 points). Chapter 8 (Development Process).** For the ASALM PTV flight envelope boundaries, what are the values of the booster transition thrust – drag, high dynamic pressure, high aero heating, high L/D cruise, and low dynamic pressure.

86. **(20 points). Chapter 8 (Development Process).** For the ramjet baseline, compute the Mach number when thrust equals drag at an altitude h = 40k ft if the equivalence ratio $\phi = 1$. Compare with the ASALM PTV flight test result.

87. **(20 points). Chapter 8 (Development Process).** What is the 95% confidence interval of an operational test program with 45 kills out of 50 flights?

88. **(20 points). Chapter 8 (Development Process).** Give an example of the typical sequence of events for flight trajectory modeling development of a typical tactical missile system.

89. **(20 points). Chapter 8 (Development Process).** Give an example of the typical sequence of events for propulsion system development of a typical tactical missile system.

90. **(20 points). Chapter 8 (Development Process).** Give an example of the typical sequence of events for structure development of a typical tactical missile system.

91. **(30 points). Chapter 8 (Development Process).** Show a history of the events in state-of-the-art SOTA advancement in the reduction in weight of synthetic aperture radar (SAR).

92. **(30 points). Chapter 8 (Development Process).** Show a history of the events in state-of-the-art SOTA advancement in the size of missile infrared seeker focal plane array.

93. **(30 points). Chapter 8 (Development Process).** Show a history of the events in state-of-the-art SOTA advancements in energy per weight and power per weight of missile power supply.

94. **(60 points). Appendix B.** Develop a technology roadmap for the Request for Proposal given in Appendix C.

95. **(60 points). Design Case Studies.** Select one of the design case study presentations from the web site in support of the textbook and conduct a review of the design case study. Provide a scoring/evaluation of the presentation, including rationale. The review should address the areas of technical content (35% weighting), organization and presentation (20% weighting), originality (20% weighting), and practical application and feasibility (25% weighting).

Appendix C: Example of Request for Proposal

Introduction

The Missile Systems Technical Committee (MSTC) of the American Institute of Aeronautics and Astronautics (AIAA) sponsors the graduate team missile design competition in order to further the understanding of the sciences associated with the design and development of missile systems. The competition is designed to allow a team of graduate level college engineering students to gain hands-on design and development experience in a real life competitive environment.

Statement of Work

Multi-Mission Cruise Missile (MMCM) Design and Analysis Study

1.0 Statement of Need

This study addresses the opportunity of an advanced cruise missile to counter time critical targets (TCTs) as well as long range surface targets of the year 2015 time frame. Threat TCTs include (1) mobile theater ballistic missile (TBM) launchers, (2) surface-to-air (SAM) missile systems, (3) major command, control and communication (C3) sites, (4) storage and support sites for weapons of mass destruction, and (5) other strategic targets such as bridges and transportation choke points. Mobile threat TBMs and SAMs are of particular interest because the stationary dwell time may be less than 10 minutes.

The Tomahawk subsonic cruise missile is a current baseline approach used by the U. S. Navy to counter long range surface threats. However, because the location of TCTs is often uncertain or their appearance is sudden, the response capability of a subsonic cruise missile is often insufficient. A multi-mission cruise missile (MMCM) system with the combined capabilities of 1) fast response against time critical targets, 2) subsonic cruise and extended range against other targets, and 3) loitering while waiting for retargeting would have operational and logistical advantages.

An enabling synergistic capability for MMCM is the anticipated advancement of the command, control, communication, computers, intelligence, surveillance, reconnaissance (C4ISR) network projected for the year 2015 time frame. It is anticipated that near real time, accurate targeting will be available from overhead tactical satellite and overhead unmanned air vehicle (UAV) sensors. Figure 1 illustrates an example of a ground station, overhead satellite sensors and satellite relays, and overhead UAV sensor platform elements of the C4ISR architecture. The assumed C4ISR of the year 2015 is projected to have a capability for a target location error (TLE) of less than 1 meter (1 sigma) and an off-board sensor-to-shooter connectivity time of less than 2 minutes (1 sigma).

A second example of an enabling capability for MMCM is low cost precision guidance available from GPS/INS. The assumed GPS/INS of the year 2015 is projected to have a guidance navigation error of less than 3 meters circular error probable (CEP).

A third example of an enabling capability is the application of high bandwidth data link technology to a cruise missile, allowing target position updates, retargeting, and precision command guidance.

A fourth example of an enabling capability for MMCM is the recent cost reductions in standoff missiles, due in part to manufacturing processes such as castings that reduce parts count, the "spin on" application of commercial technologies (in areas such as electronics and materials) and the application of procurement reform. Examples include JDAM, JASSM and Tactical Tomahawk. Low cost, combined with high

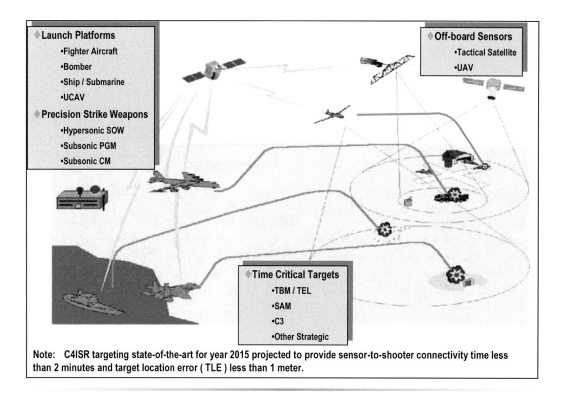

Fig. 1. Example of C4ISR Architecture for Hypersonic Standoff Missile Systems

performance and operational flexibility, has the potential to allow MMCM to be used not only in TCT missions, but also other more traditional standoff missions. It could provide a neck-down benefit of a more simplified missile logistics system with fewer types of missiles.

Finally, a fifth example of an enabling capability for a multi-mission missile is the projected advances in turbine-based propulsion. It is anticipated that future turbine-based propulsion systems such as turbojet, turbo ramjet, and air turborocket will be capable of higher compressor pressure ratio and higher turbine temperature and will be capable of operating at higher Mach number. By the year 2015 it is anticipated a turbine-based propulsion system could be operational that could provide the required specific impulse and thrust necessary to operate across a Mach number range from subsonic to about Mach 3 to 5.

2.0 Objective

The objective of this study is to evaluate alternative concepts and technologies for a subsonic to hypersonic standoff missile system that would be used against TCTs and other surface targets. The baseline requirements in sizing the missile concept and the selection of subsystems/technologies are given below.

- Maximum Mach number of 3 (threshold), 4 (standard), and 5 (goal). An objective is a counterforce response capability against short dwell targets such as TBMs within the projected state-of-the-art of cost-effective technologies for the year 2010.
- Maximum subsonic cruise range against non-time-critical surface targets greater than 300 nm (threshold), 600 nm (standard), and 1,000 nm (goal).
- Response time against TCTs less than 30 minutes (threshold), 15 minutes (standard), and 5 minutes (goal). This includes the command and control (C&C) decision time.
- Subsonic loiter time greater than 45 minutes (threshold), 60 minutes (standard), and 90 minutes (goal).
- Maximum post-loiter dash range after retargeting greater than 20 nm (threshold), 50 nm (standard), and 100 nm (goal).
- Weight less than 3,400 lb. The basis is Vertical Launch System (VLS) carriage on cruisers and destroyers.
- Length less than 256 in for compatibility with VLS.
- Cross section less than 22 in by 22 in for compatibility with VLS.
- Electrical compatibility with VLS.

- Launch exhaust gas compatibility with VLS.
- Multipurpose (blast/frag/penetrator) warhead payload with a weight of 250 lb.
- Average unit production cost less than $1,000 K, based on 4,000 units produced over a ten-year time frame.
- Wooden round with no shipboard maintenance and 15 year depot life.
- Satellite and UAV data link communication payload for retargeting and guidance.
- Guidance navigation error less than 3 meters CEP.

3.0 Tasks

A trade study will be conducted to evaluate alternative turbine-based missile concepts, their associated subsystems, and their associated technologies. The study will be based on projecting the current state-of-the-art (SOTA) to a technology readiness level (TRL) of 7 for Fiscal Year (FY) 2010. TRL 7 represents the Program Definition and Risk Reduction (PDRR) phase of a full-scale prototype in a full-scale flight environment. PDRR is comparable to demonstration/validation (demval). Following PDRR, a baseline assumption is that system development and demonstration (SDD) will begin in FY 2010. The desired initial operational capability (IOC) is FY 2015.

Tasks to be performed under this conceptual design study are as follows:

- Alternative Baseline Concepts and Subsystems Definition
- Alternative Baseline Concepts and Subsystems Evaluation
- Recommended Concept Refinement
- Mission Verification
- Technology Roadmap
- Documentation

Task 1. Alternative Baseline Concepts and Subsystems Definition

At least three alternative turbine-based concepts and their appropriate subsystems will be defined based on the requirements of Section 2.0. The characteristics of the alternative baseline concepts will be determined using the Tactical Missile Design (TMD) methods of Georgia Tech or similar conceptual design tools. A "house of quality" based on an integrated product and process development (IPPD) will be developed with

Customer involvement for the relative weighting of the requirements. The turbine-based concepts will include at least the following:

- Turbojet
- Turbo ramjet
- Air turborocket

Task 2. Alternative Baseline Concepts and Subsystems Evaluation

The alternative baseline concepts and their subsystems will be evaluated against the requirements of Section 2.0, using the Georgia Tech TMD methods or similar conceptual design and analysis tools. Physics based models will be developed that show the transparent prediction of specific impulse and thrust. The robustness of the design subsystems / technologies uncertainties and risks will be evaluated. Based on the capability against the requirements and the design robustness, a recommended MMCM and its subsystems will be selected.

Task 3. Recommended Concept Refinement

The MMCM recommended concept and subsystems will be refined based on considerations of alternative technologies. More sophisticated preliminary design and analysis tools will be used and results compared with the previous results from the conceptual design methods. Probabilistic/robust design, cost estimation, and optimization will be conducted. Design refinement will include:

- Turbine materials
- Compressor pressure ratio
- Inlet
- Fuels
- Insulation materials
- Airframe materials

- Aerodynamic configuration
- Forebody angles in front of the inlet
- Tail sizing
- Power supply
- Guidance, navigation, and control
- Data link
- Other subsystems and components
- Radar cross section and infrared signature

Task 4. Mission Verification

The MMCM concept will be evaluated in a three degree-of-freedom (3DOF) digital computer simulation. The flight trajectory and thrust profile will be optimized for maximum range, maximum loiter, and minimum time-to-target scenarios.

Task 5. Technology Roadmap

A technology roadmap will be developed for MMCM. The technology roadmap will address key enabling technologies that are driven by the requirements but will require additional development and demonstration to provide a required TRL of 7 to support a required SDD decision. The technology roadmap will include the major milestones, alternative approaches, risk mitigation plan, exit criteria for each TRL phase, and exit plan for failures. The technology development and demonstration activities will consider the following technology readiness levels (TRLs), leading to a TRL of 7 and an assumed SDD start in FY 2010:

- TRL 4 laboratory test demonstration of a component in a representative environment, but not full scale (6.2A exploratory development category funding)
- TRL 5 laboratory development or demonstration of a subsystem. The subsystem usually has full-scale components and is tested in a representative, but not full-scale environment. The category of funding is 6.2B exploratory development category.
- TRL 6 laboratory or flight Advanced Technology Demonstration (ATD) of a full-scale subsystem in a full-scale environment, with integration of some of the other subsystems and some of the other new technologies of a follow-on missile system. The category of funding is 6.3 advanced technology development.
- TRL 7 flight demonstration of either an ACTD or a PDRR full-scale prototype in a full-scale environment. The category of funding is 6.4 demonstration/validation (demval). All of the new critical technologies and subsystems are demonstrated. A driving consideration is to assure the system engineering and management confidence in the technologies. TRL 7 is normally performed where the technology and/or subsystem application is critical and high risk. The assumed Technology Availability Date (TAD) of year 2010 is based on successfully completing TRL 7.

Task 6. Documentation.

The MMCM solution, rationale, and supporting technology will be documented. The documentation will be provided in a Kickoff Meeting, Midterm Review, Final Review, and Final Report.

- Kickoff Meeting. A presentation at the Kickoff Meeting will be given to the Sponsor and Customer representatives. It will be held in early December 2003. The Kickoff Meeting will address (1) "house of quality" weighting of the requirements, (2) clarification of requirements, (3) alternative concepts, and (4) the proposed plan for the study. A copy of the viewgraphs from the Kickoff Meeting will be provided.
- Midterm Review. A Midterm Review will be presented to the Sponsor and Customer at the completion of Task 2. It will be held in March 2004. Approval of the recommended concept will be required prior to the initiation of Task 3. A copy of the viewgraphs from the Midterm Review will be provided.
- Final Review. A Final Review will be presented to the Sponsor and Customer in late May 2004, after completing the study tasks. It will address the final results of the study. A copy of the viewgraphs from the Final Review will be provided.
- Final Report. Six copies of the Final Report will be delivered to the Sponsor and Customer for review and scoring no later than June 1, 2004. The size of the Final Report will be no more than 100 pages. It will include the following information shown in Table 1:

Table 1. Final Report Information

- **Table of contents**, with sections consistent with the tasks of this statement of work (i.e., Tasks 1, 2, 3, 4, 5).
- Literature review **bibliography**.
- Tabulation of MMCM design **requirements**.
- "**House of Quality**" IPPD relative weighting of requirements
- **Justification of MMCM concept,** including a comparison against the requirements, results of the design tradeoffs, robustness of the design and its technologies to uncertainties and risk, criteria used for selection, and advantages compared to alternative concepts.
- **Three-view drawing** of MMCM concept, with a layout showing the inboard profile of subsystems, including dimensions of the major subsystems. The center-of-gravity location at launch and burnout events will be shown on the drawing.
- Sketches of **alternative concepts**
- Mission **flight profiles** of MMCM showing altitude, range, Mach number, weight, flight path angle, angle-of-attack, time, and major events such as booster ignition, surface deployments, booster burnout, booster separation, engine start, climb, cruise, descent, loiter, dash, dive, engine burnout, and impact.
- Operational **flight profiles** defining performance boundaries including altitude, speed, and maneuver g limits.
- **Aerodynamic, propulsion,** and **thermal characteristics** as a function of Mach number, angle of attack, angle of roll or sideslip, control surface deflection, and altitude.
- Airframe **structure** and **material** characteristics.
- **Sensitivity** of the system and subsystem parameters to requirements, with typical uncertainties.
- **Weight** and **balance** statement with subsystem weight, subsystem location (x,y,z), launch/burnout center-of-gravity locations, and launch/burnout moments-of-inertia.
- Discussion of **prediction methods** used to size the missile (e.g., aerodynamic configuration, propulsion system, guidance system, flight control system, structure, data link, power supply, other subsystems) and the methods used to predict the performance, cost, and constraint compliance. When practical, results will be verified as reasonable and consistent with other methods, available data, prior practice, and theory.
- Traceable **flow-down of system requirements** to subsystem and technology performance, cost, and constraints.
- Estimated **unit production cost** with units produced, production rate, learning curve, and basis for the learning curve.
- Estimated **SDD development cost** and schedule of SDD activities leading to production.
- Discussion of **VLS** integration.
- Discussion of the **year 2015 architecture** to support the areas of targeting, fire control, C4ISR, and GPS.
- **Technology roadmap** with a time phased schedule of exploratory development and advanced development programs and milestones required to demonstrate key enabling technologies by the year 2010.
- Discussion of other **life cycle** considerations with any operational, environmental, social, and technological issues that may affect the fitness of the concept over the system life cycle.
- Discussion of **manufacturing**, with considerations of reduced parts count, ease of manufacturing, life cycle cost, risk reduction, and compatibility with the logistics for the launch platforms.

Scoring/Evaluation

1. Technical Content (35 points)
   This addresses the correctness of theory, validity of reasoning used, apparent understanding and grasp of the subject, etc. Are all the major factors considered and a reasonably accurate evaluation of these factors presented?

2. Organization and Presentation (20 Points)
   The description of the design as an instrument of communication is a strong factor. Organization of written design, clarity, and inclusion of pertinent information are major factors.

3. Originality (20 points)
   The design proposal should avoid standard textbook information, and should show the independence of thinking or a fresh approach to the project. Does the method and treatment of the problem show imagination? Does the method show an adaptation or creation of automated design tools?

4. Practical Application and Feasibility (25 points)
   The proposal should present conclusions or recommendations that are feasible and practical and not merely lead the evaluators into further difficult or insolvable problems. Is the project realistic from the standpoints of cost, hazardous materials, and commonality with future systems?

Award

Based on a review and scoring of the final report, a trip and an award plaque may be provided by the Sponsor. The trip may consist of attending a missile launch, facility tour, or other aerospace related event. The Sponsor may award up to $5000 toward travel expenses.

Additional Information and Clarifications

Questions regarding this competition can be referred to the AIAA MSTC Sponsor members presented below.

Jacqueline Murdock
MSTC Chair
Atlantic Research Corporation (ARC)
5945 Wellington Rd.
Gainesville, VA
Phone: (703) 754-5337
E-mail: murdock@arceng.com

Eugene L. Fleeman
Georgia Institute of Technology
School of Aerospace Engineering
P.O. Box 150
Atlanta, GA 30332-0150
Phone: (703) 697-2187
E-mail: Eugene.Fleeman@asdl.gatech.edu

Appendix D: Nomenclature

The figures in this text are designed as a stand-alone set of information. In most cases it is possible to use a figure without referring to other figures. When practical the symbols are defined in the figures. As a complement, a list of symbols is provided in this section.

a	speed of sound
a_0	free stream speed of sound
A	target aspect angle, aspect ratio, or area
A_0	free stream air flow area into inlet
A_1	inlet throat area
A_2	diffuser exit area
A_3	combustor flame holder entrance area
A_4	combustor exit area
A_5	nozzle throat area
A_6	nozzle exit area
A_b	propellant burn area
A^*	cross sectional area of inlet or nozzle with Mach 1 flow
A_C	inlet capture area
A_d	detectors total area
A_e	nozzle exit area or aspect ratio of exposed planform
A_{EF}	engine frontal area
A_{IT}	inlet throat area
Al	aluminum
a_M	acceleration of missile
A_o	optics aperture area
A_P	presented area of threat or target
A_{Ref}	reference area
A_t	throat area
A_V	target vulnerable area
b	span
B	bandwidth, Boron, body, bending, billion, or gyro bias drift
B_0	sound pressure level
$(b_t)_{res}$	target span resolution by seeker
c	specific heat or thermal capacitance, type of loading, warhead charge weight, or speed of sound
C	canard, contrast, explosive charge weight, coefficient, chord length, carbon, control, climb, or cost
C_1	cost of 1st unit
C_{1000th}	cost of 1000th missile
c^*	characteristic velocity
C_A	axial force coefficient
C/C_T	actual contrast-to-threshold contrast ratio
c_d	discharge coefficient
C_d	cadmium
C_D	drag coefficient
C_{D0}	zero-lift drag coefficient
cg	center-of-gravity

cg_{BO}	center-of-gravity at burnout
cg_{Launch}	center-of-gravity at launch
C_i	initial contrast
Cl	chlorine
C_l	rolling moment coefficient
C_L	lift coefficient
C_{L_α}	Lift coefficient derivative from angle of attack
C_{l_p}	rolling moment coefficient derivative from roll damping
C_{l_β}	rolling moment coefficient derivative from sideslip
$C_{l_{\delta a}}$	rolling moment coefficient derivative from roll control deflection
$C_{l_{\delta r}}$	rolling moment coefficient derivative from yaw control deflection
C_{l_ϕ}	rolling moment coefficient derivative from roll angle
C_m	pitching moment coefficient
c_{mac}	mean aerodynamic chord
C_{m_α}	pitching moment coefficient derivative from angle of attack
$C_{m_{\delta e}}$	pitching moment coefficient derivative from pitch control deflection
C_n	yawing moment coefficient
C_N	normal force coefficient
C_{N_C}	normal force coefficient from control deflection
C_{N_α}	normal force coefficient derivative from angle of attack
$C_{N_{\delta e}}$	normal force coefficient derivative from pitch control deflection
C_{n_β}	yawing moment coefficient derivative from angle of sideslip
$C_{n_{\delta a}}$	yawing moment coefficient derivative from roll control deflection
$C_{n_{\delta r}}$	yawing moment coefficient derivative from yaw control deflection
C_{Nc}	normal force coefficient on control surface
$C_{N_{Trim}}$	normal force coefficient at trim
c_p	specific heat at constant pressure or pressure coefficient
cp	center-of-pressure
c_R	root chord
C_{SDD}	cost of system development and demonstration
c_{skin}	specific heat ratio of skin (airframe) material
c_T	tip chord or threshold contrast
C_x	cost of unit x
d	diameter
D	drag force, descent, or detection
D_0	zero-lift drag
dB	decibel
d_{BT}	diameter of boattail
d_{EF}	engine frontal diameter
d_{hemi}	diameter of hemisphere
d_o	optics diameter
d_p	pixel diameter
d_{Ref}	reference diameter
d_{spot}	spot diameter
dt	incremental time
D_t	target diameter
dz/dt	velocity in z-direction
d^2z/dt^2	acceleration in z-direction
D*	specific detectivity
E	Young's modulus of elasticity, energy, or 10
E_c	energy per unit mass of charge
erf	error function
erfc	complementary error function
E_T	total energy

f	frequency, fragment, or focal length
f_N	resonance frequency of the Nth mode (where N = 1, 2, 3, ...)
f/a	fuel-to-air ratio
F	force, flare, or noise factor
F_{CR}	critical force
f-number	aperture length/diameter
F-Pole	standoff range at missile intercept
fps	feet per second
F_T	tensile force
F_{TU}	ultimate tensile force
g	acceleration of gravity (32.2 ft/s² or 9.81 m/s²), gram
g_c	gravitational constant (32.2 ft lbm/lbf/s² or 32.2 lbm/slug)
GHz	10^{12} Hz
G_R	gain of receiver antenna
G_T	gain of transmitter antenna
h	altitude or inlet height
h	convection heat transfer coefficient
HCl	hydrogen chloride
h_a	apogee altitude
h_e	entry altitude
H_f	heating value of fuel
h_i	initial altitude
h_f	final altitude
H_g	mercury
h_L	launch altitude
h_{mask}	mask altitude
$h_{obstacle}$	height of obstacle
I	moment-of-inertia, radiant intensity
I_{sp}	specific impulse
I_{Total}	total impulse
$(I_T)_{\Delta\lambda}$	target radiant intensity between λ_1 and λ_2
I_y	moment-of-inertia about y-axis (pitch)
j	integer
J	Joule
k	thermal conductivity, number of parameters, Boltzmann's constant = 1.38×10^{-23} J/K
K	thousands or thickness constant
kg	kilogram
km	kilometer
l	length, number of levels
L	lift force, missile-target lead angle, learning curve coefficient, loss factor, length, or lifetime
$\dot{L}$	change in lead angle with time
l_b	length of body
lb	pound
lbf	pound force (unit of force)
lbm	pound mass (unit of mass)
l_c	inside chamber length
l_{case}	case length
L_{CB}	missile-target lead angle for constant bearing flight trajectory
$(l_{comb})_{min}$	minimum efficient combustor length
l/d	length/diameter
L/D	lift/drag
l_N	length of nose
l_N/d	nose fineness ratio
l_{Ref}	reference length

L_T	target length
L_λ	spectral radiance (Plank's Law)
m	meter
M	Mach number, moment, missile aim point, warhead metal fragments total weight, or molecular weight
$\dot{m}$	mass flow rate
M_0	free stream Mach number
$(M_3)_{TC}$	combustor entrance Mach number with thermal choking
$M_{\wedge_{LE}}$	Mach number perpendicular to leading edge
M_∞	free stream Mach number
Max	maximum value
M_B	bending moment
M_C	mass of charge of warhead
M_g	magnesium
M_i	impact Mach number or initial Mach number
M_{IE}	inlet entrance Mach number
M_L	launch Mach number
M_m	mass of case liner of blast fragmentation warhead
M_{TC}	combustor entrance Mach number for thermal choking
M_{wh}	mass of warhead
n	number of pulses integrated, integer, type of warhead geometry, or burn rate exponent
N	normal force, noise, navigation ratio, Newton, engine rotational speed, number of parts, or number of resolution line pairs
N'	effective navigation ratio
$n_{fragments}$	number of fragments
n_{hits}	number of hits
nm	nautical miles
n_M	missile maneuver acceleration g
N_{NU}	Nusselt number
n_T	target maneuver acceleration g
n_w	number of wings ($n_w = 1$ is planar, $n_w = 2$ is cruciform)
N_w	normal force on wing
n_x	acceleration g in longitudinal direction
n_z	acceleration g in vertical direction
O	oxygen
p	pressure, propellant, or pixel
P	penetration, load, or power
p_0	free stream static pressure
p_2	compressor entrance pressure
p_3	compressor exit pressure
p_4	combustor exit pressure
p_{5_t}	turbine exit total pressure
P_a	Pascal
$P_{B,C}$	parameter of baseline, corrected (based on actual data)
$P_{B,U}$	parameter of baseline, uncorrected
p_{blast}	blast pressure
p_{burst}	burst pressure
p_c	chamber pressure
$P_{CD,C}$	parameter of conceptual design, corrected
$P_{CD,U}$	parameter of conceptual design, uncorrected
P_D	probability of detection
p_e	exit pressure
p_{gauge}	gauge pressure
P_K	probability of kill

$P_{K_{SS}}$	probability of kill, single shot
P_{KE}	penetration of kinetic energy warhead
P_r	Prandtl number
P_R	power received
psf	pounds per square foot
psi	pounds per square inch
p_t	total pressure
P_T	power transmitted
p_{t0}	total pressure of free stream
p_{t2}	total pressure after normal shock
q	dynamic pressure $= \frac{1}{2}\rho V^2$, pitch rate
q'	local dynamic pressure
$q\cdot$	heat transfer input flux
Q	heat transfer rate per unit area
r	radius, recovery factor, propellant burn rate, or yaw rate
r_M	radius of missile
r_T	radius of target
R	range, radius, dome error slope, reliability, or gas constant
R_0	radar reference range or radius of earth
R_C	climb incremental range
R_D	detection range or descent incremental range
R_e	Reynolds number
R_F	flight range
R_{FCS}	fire control system range
$R_{F\text{-Pole}}$	standoff range at missile intercept
R_L	launch range
R_{LOS}	line-of-sight range
R_{max}	maximum range
$R_{obstacle}$	range to obstacle
R_R	recognition range
R_T	turn radius
R.T.	room temperature
RT&A	research, technology, and acquisition
R_{TM}	radius of turn of missile
R_{TT}	radius of turn of target
R_x	range in x-direction
R_y	range in y-direction
S	area or signal
$S_{control}$	flight control surface area
S_{hemi}	cross sectional area of hemisphere
S/N	signal-to-noise ratio
$(S/N)_D$	signal-to-noise ratio for detection
S_p	presented area
S_{Ref}	reference area
S_T	tail area
S_W	wing area
S_{wet}	wetted area
t	time or thickness
T	thrust
T	temperature, torque, target, target position, or tail
t_0	total time of flight
T_0	free stream temperature
$t_{90\%}$	time to achieve 90%
t_0/τ	number of time constants for intercept

T_1	inlet entrance temperature
T_2	compressor entrance temperature
T_3	compressor exit temperature
T_4	combustor exit temperature or turbine entrance temperature
T_{4_t}	combustion total temperature
T_5	turbine exit static temperature
T_{5_t}	turbine exit total temperature
t/t_0	fraction of time-to-go
t_B	rocket motor boost time
T_B	boost thrust
t_{Burst}	warhead burst time
t_C	coast time
t/c	thickness-to-chord ratio
t_{comb}	time required for combustion
T_e	telluride
t_{exp}	exposure time
t_f	time of flight
THz	10^{15} Hz
t_i	initial time
T_i	initial temperature
Ti	titanium
t_{mac}	maximum thickness of mean aerodynamic chord
t_{react}	reaction time
t_S	rocket motor sustain time
T_S	stall torque of single actuator
t_{SDD}	time duration of system development and demonstration
T_T	target temperature, total temperature
t_{wall}	thickness of wall material
T_Δ	temperature difference
u	velocity
V	velocity or vanadium
$\dot{V}$	time rate of change in velocity
V_0	orbit velocity near the earth's surface
V_{BC}	velocity at begin of coast
V_{BO}	velocity at burnout
V_C	velocity of climb or closing velocity
V_{comb}	combustion velocity
V_{CW}	velocity of cross wind
V_D	velocity of descent
V_e	exit velocity or entry velocity
V_{EB}	velocity at end of boost
V_{EC}	velocity at end of coast
V_f	fragment velocity
V_i	initial velocity
V_L	launch velocity
V_M	velocity of missile
V_T	velocity of target
V_x	velocity in x-direction
V_y	velocity in y-direction
V_∞	free stream velocity
w	load per unit length or wing
$\dot{w}$	weight flow rate of fuel or propellant
$\dot{w}_a$	weight flow rate of air
$\dot{w}_f$	weight flow rate of fuel

W	weight, wing, glint noise spectral density, width, or Watt
W_{BC}	weight at begin of cruise
W_{BO}	weight at burnout
W_c	weight of warhead charge
W_E	weight per unit energy
W_f	weight of fuel
$\dot{W}_f$	fuel flow rate
W_i	initial weight
W_L	launch weight
W_m	weight of warhead metal case (fragments)
W_p	propellant weight or weight per unit power
$\dot{W}_p$	propellant weight flow rate
W_T	weight per unit torque
W_s	structure weight
x	longitudinal distance, independent variable
x_{AC}	aerodynamic center location
x_{cp}	center-of-pressure location
x_{CG}	center-of-gravity location
x_{HL}	hinge line location
x_{mac}	longitudinal location of leading edge of mean aerodynamic chord
y	lateral distance, dependent variable
Y_1	first year
y_{cp}	outboard center-of-pressure location
y_{offset}	lateral offset distance
z	warhead scaling parameter or vertical distance
z_{skin}	skin thickness
z_{max}	maximum value of z
α	angle of attack, thermal diffusivity, coefficient of thermal expansion, or atmospheric extinction coefficient
α'	local effective angle of attack
$\ddot{\alpha}$	angle of attack angular acceleration
$\dddot{\alpha}$	angle of attack angular jerk
α_{CW}	angle of attack of cross wind
α_{Trim}	trim angle of attack
α/δ	change in angle of attack with control deflection
$\alpha/\dot{\gamma}$	angle of attack sensitivity to turn rate
β	angle of sideslip, bi-static radar angle, or accelerometer misalignment
δ	control deflection, body angle, or seeker angle
$\dot{\delta}$	control surface defection rate
δ_{LE}	leading edge section angle
δ_{trim}	control deflection for trim
ε	nozzle expansion ratio (exit area/throat area), strain, dielectric constant, seeker angle error, track angle error, or emissivity coefficient
ϕ	equivalence ratio (actual fuel-to-air ratio compared to stochiometric ratio), missile roll angle, or non-dimensional range
ζ	damping coefficient
γ	specific heat ratio (≈ 1.4 for ambient air, ≈ 1.2 for rocket motor), radius of body-on-body impact, or flight path angle
γ_0	free stream specific heat ratio or aim bias
γ_1	specific heat ratio at inlet entrance
γ_2	specific heat ratio at compressor entrance
γ_3	specific heat ratio at compressor exit or combustor entrance
γ_4	specific heat ratio at turbine entrance or combustor exit
γ_5	specific heat ratio at turbine exit

$\dot{\gamma}$	Rate of change in flight path angle
γ_C	Flight path climb angle
γ_D	Flight path dive or descent angle
γ_i	initial flight path angle
γ_M	Initial heading error of missile
η	efficiency ($\eta = 1$ is 100% efficient) or fraction of maximum solar radiation
η_a	atmospheric transmission
λ	taper ratio (tip chord/root chord), wavelength, or surface geometry coefficient
μ	Mach angle, aggregate thermal resistance coefficient, viscosity, or mean value
θ	missile pitch attitude, beam width, shock angle, radial angle, warhead vertical spray angle, fuzing angle, or surface angle
$\ddot{\theta}$	pitch attitude angular acceleration
θ_{3dB}	1/2 power beam width
θ_a	angular resolution
θ_{BT}	boattail angle
θ_F	vision fovea angle of human eye that provides highest resolution
θ_i	incidence angle or initial angle
ρ	density or reflectivity
ρ_∞	free stream density
ρ_M	missile average density
ρ_p	density of penetrator warhead
ρ_T	density of target
ρ_{wall}	density of wall material
σ	radar cross section, stress, standard deviation, atmospheric attenuation coefficient or miss distance standard deviation
σ_{Buckle}	buckling stress
σ_{glint}	standard deviation of miss distance from glint
σ_{HE}	standard deviation of miss distance from heading error
$(\sigma_{HE})_{Max}$	envelope of local maximum values of miss distance from heading error
σ_{MAN}	standard deviation of miss distance from target maneuver acceleration
$(\sigma_{MAN})_{Max}$	envelope of local maximum values of miss distance from maneuvering target
σ_{Max}	maximum stress
σ_t	tensile stress
σ_{TU}	target ultimate stress
σ_{TS}	thermal stress
σ_{yield}	yield stress
π	3.1416...
π_K	pressure sensitivity to temperature
τ	time constant or pulse duration
τ_δ	time constant from control effectiveness
$\tau_{\dot{\delta}}$	time constant from deflection rate limit
τ_{Dome}	time constant from radome error slope
ω	angular velocity
ω_1	first mode wing bending frequency
$\omega_{Actuator}$	actuator bandwidth
ω_{BB}	first mode body bending frequency
ω_{Drift}	INS drift rate
ω_{SP}	flight dynamics short period frequency
ψ	yaw angle
Δ	incremental
Δf_p	pixel detector bandwidth
Λ	sweep angle, solar absorptivity
Σ	summation
$!$	factorial

| | absolute value
≈   approximately equal to
∼   similar to
<   less than
>   greater than
⊥   perpendicular to
number
2a   major axis of ellipse
2b   minor axis of ellipse

Appendix E: Acronyms/Abbreviations

1D	one dimensional
1-DOF	one-degree-of-freedom
2D	two dimensional
2-DOF	two-degrees-of-freedom
3D	three dimensional
3-DOF	three-degrees-of-freedom
4-DOF	four-degrees-of-freedom
5-DOF	five-degrees-of-freedom
6-DOF	six-degrees-of-freedom
6.1	basic research
6.2	exploratory development
6.3	advanced development
6.4	demonstration & validation
6.5	engineering and manufacturing development
AAA	anti-aircraft artillery
AAM	air-to-air missile
AARGM	Advanced Anti-radiation Guided Missile
A/B	air-breathing
AC	aerodynamic control
A/C	aircraft
ACM	Advanced Cruise Missile
ACTD	Advanced Concept Technology Demonstration
A/D	analog-to-digital conversion
ADAM	Advanced Design of Aerodynamic Missiles
ADN	ammonium dinitramide
ADS	air defense system
AEDC	Arnold Engineering Development Center
AESA	Active Electronically Steering Antenna
AFB	Air Force Base
AGM	air-to-ground missile
AIAA	American Institute of Aeronautics and Astronautics
AIM	air intercept missile
A/J	anti-jam
ALCM	Air Launched Cruise Missile
ALON	aluminium oxynitride
AMRAAM	Advanced Medium Range Air-to-Air Missile
ANS	Anti-Navire Supersonique
AoA	angle of attack or assessment of alternatives
AP	ammonium perchlorate
AP09	Aeroprediction Incorporated Code
APC	antenna pitch control or autopilot control
APKWS	Advanced Precision Kill Weapon System
ARH	anti-radiation homing
ARM	anti-radiation missile
ARO	all reflective optics

Appendix E: Acronyms/Abbreviations

ARRMD	Affordable Rapid Response Missile Demonstrator
ASALM	Advanced Strategic Air Launched Missile
ASCM	anti-ship cruise missile
ASM	air-to-surface missile
ASMP	Air-Sol Moyenne Portée
ASP	advanced signal processing
ASRAAM	Advanced Short range Air-to-Air Missile
ATA	air-to-air or automatic target acquisition
ATACMS	Army Tactical Missile System
ATD	Advanced Technology Demonstration
ATM	atmosphere
ATR	automatic target recognition or air turbo rocket
ATS	air-to-surface
AUR	all up round
AVG	average
AWACS	Airborne Warning and Control System
B	billion
BAe	British Aerospace
BAT	Brilliant Anti-armor Technology
BC	beginning of cruise
BDA	battle damage assessment
BDI	battle damage indication
BIT	built in test
BL	baseline or body line
BMD	ballistic missile defense
BMDO	Ballistic Missile Defense Organization
BO	burn out
BRU	bomb rack unit
BTT	bank-to-turn
BTU	British Thermal Unit
BVR	beyond visual range
C2	command and control
C3	command, control, and communication
C3I	command, control, communication, intelligence
C4ISR	command, control, communication, computers, intelligence, surveillance, reconnaissance
CAD	computer aided design
CALCM	Conventional Air Launched Cruise Missile
CAOC	Combined Air Operation Center
CAS	close air support
CBU	cluster bomb unit
CCM	counter-countermeasures
CDR	critical design review
CEB	Combined Effects Bomblet
CEP	circular error probable
CET	Commander Evaluation Test
CEU	Continuing Education Unit
CFD	computational fluid dynamics
cg	center of gravity
CL	centerline
CLS	canister launch system
CL Tk	centerline tank carriage
CLU	command and launch unit
CM	cruise missile, countermeasure

cmW	centimeter wave
CNC	Computer Numerical Control
CO_2	carbon dioxide
CONOPS	concept of operations
CONUS	continental United States
COTS	commercial off the shelf
cp	center of pressure
CPIA	Chemical Propulsion Information Agency
CPU	central processor unit
CSIP	Center for Signal and Image Processing
CTS	captive trajectory simulation
CVD	chemical vapor deposition
CW	continuous wave
D3	dully, dirty, or dangerous
DACS	divert and attitude control system
DAGR	Direct Attack Guided Rocket
DARPA	Defense Advanced Research Projects Agency
DATCOM	DATaCOMpendium
dB	decibel
DBF	design-build-fly
DCR	Dual Combustor Ramjet-scramjet
DemVal	Demonstration and Validation
DIRCM	Directed Infrared Countermeasure
DMLS	direct metal laser sinter
DoD	Department of Defense
DOE	design of experiments
DOF	degree of freedom
DOS	disk operating system
DPICM	Dual Purpose Improved Conventional Munition
DSMAC	Digital Scene Matching Area Correlation
DSP	Defense Support Program
DT&E	development test & evaluation
ECCM	electronic counter-countermeasures
ECM	electronic countermeasures
EDM	electrical discharge machining
EF	engine frontal
EFP	explosively formed projectile
EKV	Exo-atmospheric Kill Vehicle
ELINT	electronic intelligence
EM	electro-mechanical or electro-magnetic
EMC	electro-magnetic compatibility
EMD	Engineering and Manufacturing Development
EMI	electro-magnetic interference
EMP	electro-magnetic pulse
EMR	electro-magnetic radiation
EO	electro-optical
EOCM	electro-optical countermeasure
EPDM	ethylene propylene diene monomer
ESA	electronically steered antenna
ESD	electrostatic discharge
ESSM	Evolved Sea Sparrow Missile
EW	electronic warfare
FAC	Forward Air Controller
FAR	false alarm rate

FCS	fire control system
FDTD	finite difference time domain
FEM	finite element model
F-Kill	firepower kill
FLIR	forward looking infrared
FM	frequency modulation
FMCW	frequency modulation continuous wave
FMRAAM	Future Medium Range Air-to-Air Missile
FOG	fiber optic gyro
FOR	field of regard
FOS	factor of safety
$(FOS)_{Ultimate}$	design ultimate load or predicted ultimate load
$(FOS)_{Yield}$	design yield load or predicted yield load
FOV	field of view
FPA	focal plane array
F-Pole	distance from launch aircraft to target at missile intercept
fps	feet per second
ft	foot
FW	fiscal year
g	gram
G&C	guidance & control
GAP	glycidyl azide polymer
GATR	Guided Advanced Tactical Rocket
GBI	Ground Based Interceptor
GBS	Global Broadcast System
GBU	guided bomb unit
GEM	Guidance Enhanced Missile
GEO	geosynchronous earth orbit
GFE	government furnished equipment
GFP	government furnished property
GHz	Giga Hertz (10^9 Hertz)
GLONASS	Global Navigation Satellite System
GMD	Ground-Based Midcourse Defense
GMLRS	Guided Multiple Launch Rocket System
GN&C	guidance, navigation, and control
GNSS	Global Navigation Satellite System
GPS	Global Positioning System
GTD	geometric theory of diffraction
HARM	High Speed Anti-Radiation Missile
HAWK	Homing-All-the-Way-Killer
HDBK	handbook
HDPE	high density polyethylene
HE	high explosive or heading error
HELLFIRE	Heliborne Launched Fire and Forget
HEO	highly elliptical orbit
HiBEX	High Boost Experiment
HIMARS	High Mobility Artillery Rocket System
HIP	hot isostatic press
HL	hinge line
HM	hypersonic missile or hinge moment
HMMWV	High Mobility Multipurpose Wheeled Vehicle
HMX	Her Majesty's Explosive
HOJ	home on jam
HOQ	house of quality

HOT	**H**aut subsonique **O**ptiquement **T**éléguidé
HPM	high power microwave
HS	Hamilton Sundstrand
HSAD	High-Speed Anti-radiation missile Demonstration
HSCM	high speed cruise missile
HSSM	high speed stand-off missile
HTK	hit-to-kill
HTPB	hydroxyl-terminated polybutadiene binder
HUD	head up display
HWL	hardware-in-loop
Hz	Hertz (cycles per second)
IC	integrated circuit
ICBM	intercontinental ballistic missile
IE	inlet entrance
IF	intermediate frequency
IFF	identification friend or foe
IFOV	instantaneous field of view
IM	insensitive munition
IMU	inertial measurement unit
in	inch
INS	inertial navigation system
IOC	initial operational capability
IPN	isopropyl nitrate
IPPD	integrated product and process development
IR	infrared
IR&D	independent research & development
IRBM	intermediate range ballistic missile
I^2R	imaging infrared
IIR	imaging infrared
IRCM	infrared countermeasure
IRNSS	Indian Regional Satellite System
IRR	integral rocket-ramjet
ISAR	inverse synthetic aperture radar
IT	inlet throat
ITCV	Inter-Tropical Convergence Zone
JAGM	Joint Air-to-Ground Missile
JASSM	Joint Air-to-Surface Standoff Missile
JDAM	Joint Direct Attack Munition
JHMCS	Joint Helmet Mounted Cueing System
JI	jet interaction
JSOW	Joint Standoff Weapon
JSTARS	Joint Surveillance Target Attack Radar System
JTIDS	Joint Tactical Information Distribution System
KE	kinetic energy
K-Kill	catastrophic kill
kg	kilogram
km	kilometer (1000 meter)
KV	kill vehicle
KW	kill weapon
LADAR	laser detection and ranging
LAR	launch acceptable range
LASER	light amplification by stimulated emission of radiation
LAU	launch adapter unit
LCC	life cycle cost

LCS	laser cross section or littoral combat ship
LGB	laser guided bomb
LE	leading edge
LE	Lethality Enhancer
LEO	low earth orbit
LO	low observable
LOAL	lock on after launch
LOBL	lock on before launch
LOCAAS	Low Cost Autonomous Attack System
LOGIR	Low-Cost Guided Imaging Rocket
LOS	line of sight
LOSAT	Line-of-Sight Anti-Tank
LPI	low probability of intercept
LRAP	Long Range Attack Projectile
LRIP	low rate initial production
LRSM	Long Range Standoff Missile
LSCM	low speed cruise missile
LWIR	long wave infrared
m	meter
M	million or molecular weight
mac	mean aerodynamic chord
MAC	Metal Augmented Charge
MALD	Miniature Air-Launched Decoy
MALI	Miniature Air-Lunched Interceptor
MANPADS	man portable air defense system
M&S	modeling and simulation
MBB	Messerschmitt-Boelkow-Blohn
MCTR	Military Technology Control Regime
MEOP	maximum expected operating pressure
MD	miss distance
MDA	Missile Defense Agency
MEADS	Medium Extended Air Defense System
MEMS	micro-machined electro-mechanical systems
MEPHISTO	Multi-Effect Penetrator High Sophisticated and Target Optimised
MHTK	Miniature Hit-to-Kill
MIDS	Multifunctional Information Distribution System
MIL STD	military standard
MIRs	most important requirements
MIRV	multiple independent reentry vehicle
Mk	Mark
M-Kill	mobility kill
MLRS	Multiple Launch Rocket System
MMIC	monolithic microwave integrated circuit
mmW	millimeter wave
MOE	measure of effectiveness
MoM	method of moments
MOM	measure of merit
MRAAM	medium range air-to-air missile
MRBM	medium range ballistic missile
MSE	Missile Segment Enhancement
MSTC	Missile Systems Technical Committee
MTBF	mean time between failure
MW	molecular weight
MWIR	medium wave infrared

N/A	not applicable
NASA	National Aeronautics and Space Administration
NATO	North Atlantic Treaty Organization
NCTID	non-cooperative target identification
NEAR	Nielsen Engineering and Research
Nd	neodymium
nm	nautical miles
NOAA	National Oceanic and Atmospheric Administration
NPT	Noel Penny Turbines
NSWC	Naval Surface Warfare Center
NTW	Navy Theater Wide
NWC	Naval Weapons Center
OFAT	one factor at a time
OODA	observe-orient-decide-act
OP	operating pressure
OPS	operations
O&S	operations and support
OT&E	operational test & evaluation
P	parameter
Pa	Pascal
PAC-2	Patriot Advanced Capability-2
PAC-3	Patriot Advanced Capability-3
PGMM	Precision Guided Mortar Munition
P^3I	pre-planned product improvement
PBX	plastic bonded explosive
PC	personal computer
PDF	probability density function
PDR	preliminary design review
PDRR	Program Definition and Risk Reduction
PEK	polyether ether ketone
PEEK	polyether ether ketone
PGM	precision guided munition
PE	program element
PFR	preliminary flight rating
PK	probability of kill
PPE	polyphenylene ether
PRF	pulse repetition frequency
PS	power supply
pt	point
PTD	physical theory of diffraction
PTV	propulsion test validation
QRS	quartz rate sensor
QWIP	quantum well infrared photodetector
QZSS	Quasi-Zenith Satellite System
RA	rolling airframe
rad	radian
RADAR	radio detection and ranging
RAM	Rolling Airframe Missile or radar absorbing material
R&M	reliability and maintainability
RAS	radar absorbing structure
RCS	radar cross section
R&D	research and development
RDT&E	research, development, test and evaluation
RDX	Research Department Explosive (also known as Royal Demolition Explosive)

RES	radome error slope
RF	radio frequency
RFCM	radar frequency countermeasure
RFP	request for proposal
RJ	ramjet
RJ-4,-5,-6	ramjet fuel 4, 5, or 6
RJC	reaction jet control
RLG	ring laser gyro
RMS	root mean of squares
ROK	Republic of Korea
rpm	revolutions per minute
RSS	root sum of squares
RT&A	research, technology, and acquisition
RTM	resin transfer molding
RV	reentry vehicle
RWR	radar warning receiver
s	second
S&A	safe and arm
S&C	stability and control
SADARM	Sense and Destroy Armor Munition
SAL	semi-active laser
SAM	surface-to-air missile
SAR	synthetic aperture radar
SATCOM	satellite communications
SBAS	Satellite-Based Augmentation System
SBIRS	Space Based Infrared System
SC	shaped charge
SCM	subsonic cruise missile
SCR	signal-to-clutter ratio
SDB	Small diameter Bomb
SDD	system development and demonstration
SDL	Self Defense Launcher
SEAD	suppression of enemy air defense
SED	scramjet engine demonstration
sfc	specific fuel consumption
SFW	Sensor Fuzed Weapon
SL	sea level
SLAM	Standoff Land Attack Missile
SLAM-ER	Standoff Land Attack Missile-Expanded Response
SLAMRAAM	Surface Launched AMRAAM
SLBM	submarine launched ballistic missile
SLS	Space Launch System
SM	static margin or Standard Missile
S/N	signal-to-noise ratio
SOTA	state of the art
SOW	statement of work
SPICE	Small Precise Improved Cost Effective Guidance Kit
SPL	sound pressure level
SRAM	Short Range Attack Missile
SRBM	short range ballistic missile
SRR	system requirements review
SSM	surface-to-surface missile
STA	station or surface-to-air
STANAG	standardization agreement

STD	standard
STS	surface-to-surface
STT	skid-to-turn
SV	space vehicle
SW	Sidewinder
SWIR	short wave infrared
TAD	technology availability date
TALD	Tactical Air-launched Decoy
TBD	to be determined
TBM	theater ballistic missile
TC	thermal choking
TCAE	Teledyne Continental Aviation and Engineering
TCT	time critical target
TD	target detection
TDACS	throttle-able divert and attitude control system
TDD	target detection device
TE	trailing edge or target error
TEL	transporter, erector, launcher
TERCOM	terrain contour matching
THAAD	Theater High Altitude Area Defense
TK	tank
TLE	target location error
TM	telemetry
TMC	thrust magnitude control
TMD	Tactical Missile Design or theater missile defense
TNT	TriNitroToluene
TOF	time of flight
TOW	Tube Launched, Optically Tracked, Wire command link Guided
TRD	technical requirements document
TRL	technology readiness level
TSRM	Third Stage Rocket Motor
TURB	turbulent
TV	television
TVC	thrust vector control
TVM	track via missile
TWT	traveling wave tube
UAV	unmanned air vehicle
UCAV	unmanned combat air vehicle
UHF	ultra high frequency
UK	United Kingdom
US	United States
USD	United States dollars
UV	ultraviolet
VARTM	vacuum assisted resin transfer molding
VLS	Vertical Launch System
V&V	verification and validation
w	with
WAGE	Wide Area GPS Enhancement
WAN	wide area network
W/H	warhead
WI	Williams International
XLDB	cross-linked double base
w/o	without
YAG	yttrium aluminum garnet

Appendix F: Conversion Factors

Table F.1 Conversion of English to Metric Units

Parameter	English Unit	Metric Unit	Multiply by
Acceleration	ft/s^2	m/s^2	0.3048
Area	ft^2	m^2	0.09294
Area	in^2	m^2	6.452E-04
Density	lb m/in^3	kg/m^3	2.768E+04
Density	lb m/in^3	kg/l	27.67
Density	lb m/ft^3	kg/m^3	16.02
Density	lb f-s^2/in^4	kg/m^3	1.069E+07
Density	slug/ft^3	kg/m^3	515.4
Energy	BTU	J or N-m	1055
Energy	ft-lbf	J or N-m	1.3557
Force	lbf	N	4.4484
Heat capacity	BTU/lbm/°F	J/kg/°C	4188
Heating value	ft-lbf/lbm	J/kg, N-m/kg	2.990
Heating value	BTU/lbm	kJ/kg	2.236
Heat transfer coefficient	BTU/hr/ft^2/°F	W/m^2/°C	5.6786
Heat transfer coefficient	BTU/s/ft^2/°F	W/m^2/°C	2.044E+04
Heat transfer rate	BTU/ft^2/s	W/m^2	1.135E+04
Length	ft	m	0.3048
Length	in	m	0.0254
Length	mil (0.001 in)	micron	25.4
Length	mile	m	1609
Length	nm	m	1852
Mass	lbm	kg	0.4535
Mass	lbf-s^2/in	kg	1200
Mass	slug	kg	14.59
Mass flow rate	lbm/h	kg/s	1.260E-04
Mass flow rate	lbm/s	kg/s	0.4535
Mass flow rate	slug/s	kg/s	14.59
Moment-of-inertia	ft-lbf-s^2 (slug)	kg-m^2	1.3557
Moment-of-inertia	in-lbf-s^2	kg-m^2	0.113
Power	BTU/h	W	0.2931
Power	hp	W	745.71
Power per unit area	BTU/h/ft^2	W/m^2	3.155
Pressure	lbf/ft^2	Pa or N/m^2	47.89
Pressure	lbf/in^2	Pa or N/m^2	6895
Pressure	std atm	Pa or N/m^2	1.013E+05
Specific heat	BTU/lbm/°F	J/kg/°C	4186
Specific impulse	s	Ns/kg or m/s	9.81

Temperature	°F	°C	(°F - 32) 0.5556
Temperature	R	K	0.5556
Thermal conductivity	BTU/hr/ft/°F	W/m/°C	1.7307
Thermal conductivity	BTU/s/ft/°F	W/m/°C	6231
Thermal diffusivity	ft²/s	m²/s	0.09290
Torque	ft-lbf	N-m	1.3557
Torque	in-lbf	N-m	0.113
Torque	slug-ft²/s	N-m	1.3557
Velocity	ft/s	m/s	0.3048
Velocity	knot or nm/h	km/h	1.852
Viscosity – absolute	lbf-s/ft²	N-s/m²	47.87
Viscosity – absolute	lbm-s/ft	N-s/m²	1.4881
Viscosity – kinematic	ft²/s	m²/s	0.09294
Volume	ft³	m³	0.02831
Volume	in³	m³	1.639E-05
Volume	US gallons	L	3.785
Volume	US gallons	m³	0.003785

Appendix G: Example Syllabus

An example syllabus for a two-semester course on Missile Design, Development, and System Engineering is given below. This is a lecture-lab course that includes lectures, classroom team homework, individual homework problems, exams, student team design study/studies, customer reviews, and documentation. The first semester is 3 credit hours, consisting of 2 hours each week in the classroom plus design team participation outside the classroom. The second semester is 2 credit hours, consisting primarily of the design team study/studies, customer reviews, and documentation.

Summary: This is a self-contained course on the fundamentals of missile design, development, and system engineering. It provides a system-level, integrated method for the missile aerodynamic configuration/propulsion system design and analysis. The course addresses the broad range of alternatives in meeting missile performance, cost and system measures of merit/constraint requirements. The methods presented are generally simple closed-form analytical expressions that are physics-based, to provide insight into the primary driving parameters. Configuration sizing examples are presented for rocket-powered, ramjet-powered, and turbojet-powered baseline missiles. Typical values of missile parameters and the characteristics of current operational missiles are presented. Also presented are the enabling subsystems and technologies for missiles and the current/projected state-of-the-art of missiles. During this course the students will size the configuration, size the propulsion system, estimate the weight, and estimate the flight performance for a conceptual missile design, in accordance with the requirements of a Request for Proposal (RFP). The students will also include first order system measures of merit/constraints (such as miss distance, lethality, cost, and launch platform integration) for the conceptual design and system engineering of a missile, in accordance with the requirements of the RFP. Computational methods used will progress from hand calculation to spreadsheet analysis to more sophisticated computer programs during the design maturation. Results will be presented to a customer review team at kickoff, midterm, and final reviews.

Textbook: Missile Design and System Engineering by Eugene L. Fleeman.

Prerequisites: Previous courses in flight vehicle aerodynamics, propulsion, structure, and performance.

Schedule:

Lectures: Weeks 1-15 of the first course, on Chapters 1-9 of the textbook.

Classroom Review of Textbook Problems: Upon completion of the lecture for each chapter of the textbook. The review will be conducted in the classroom. The problems are at the end of the chapter.

Classroom Homework Problems: A classroom homework assignment will be given each week, from Appendix A of the textbook. The following week, a selected two-student team will present their solution in the classroom.

Individual Homework Problems: Individual homework assignments will be given each week, from Appendix A of the textbook. Homework is due the following week after the assignment.

Midterm Exam: Upon completion of the Chapter 5 lecture (Flight Performance Considerations in Missile Design and System Engineering). It is anticipated that the midterm exam will be given during week 9 of the first course.

Final Exam: Upon completion of the Chapter 9 lecture (Summary and Lessons Learned). It is anticipated that the Final Exam will be given during week 15 of the first course.

Request for Proposal: The RFP provides the statement of work and deliverable requirements for the missile design study. An example of an RFP is shown in Appendix B of the textbook. The RFP will be provided prior to week 10 of the first course.

Student Design Team(s): The student design team(s) will be formed prior to week 10 of the first course. Each student design team will consist of about ten students. One of the students will be the program manager and one of the students will be the systems engineer. The other students will be responsible for functional areas such as propulsion, aerodynamics, structure, flight control, guidance and navigation, weight and balance, design layout, warhead, flight performance, cost, and launch platform integration.

Customer Reviews: Three reviews are held with the customer – A Kickoff Meeting during week 15 of the first course, a Midterm Review upon completion of a down-select to a preferred concept(s) (about week 8 of the second course), and a Final Review (during week 13 of the second course). Dry runs will be held prior to the customer reviews.

Documentation: Documentation from the student design team(s) will include a final technical report of about 100 pages and slides from the customer reviews. The draft final report will provided 1 week following the Final Review. A paper may also be presented later at a technical conference. Drafts of the documentation will be reviewed prior to submittal to the customer(s).

Grading: Grades will be based on the exam scores, classroom team homework scores, individual homework scores, student team participation, classroom participation, presentation skills, the overall quality of the student team presentations, and the overall quality of the student team documentation.

Appendix H: Multiple Choice Questions

Chapters 1–5

Chapter 1: Introduction

1. Which missile is more maneuverable?
    a. ALCM
    b. Archer
    c. Storm Shadow
    d. Brimstone
2. Which has the least impact on aerodynamic configuration sizing?
    a. fuzing
    b. propulsion
    c. wing geometry
    d. length
3. Which design activity typically follows the others?
    a. mission/scenario/system definition
    b. weapon system requirements, trade studies and sensitivity analysis
    c. launch platform integration
    d. technology assessment and development roadmap
4. System integration includes
    a. environment extremes
    b. launch platform
    c. targeting
    d. all of the above
5. Which air-to-air missile has the longest range?
    a. Meteor
    b. AMRAAM
    c. IRIS_T
    d. Sparrow
6. Which air-to-surface missile has the highest speed?
    a. Hellfire
    b. HARM
    c. Storm Shadow
    d. Tomahawk
7. Which surface-to-surface missile is the lightest weight?
    a. Javelin
    b. MGM-140
    c. BGM-109
    d. MLRS
8. Which surface-to-air missile has the longest range?
    a. FIM-92
    b. PAC-3
    c. GBI
    d. SM-3

9. Campaign model results typically include
   a. effectiveness of different mixes of weapons in the campaign
   b. cost per target kill
   c. C4ISR interface
   d. all of the preceding

Chapter 2: Aerodynamics

10. The rocket baseline missile flight range is least sensitive to which parameter?
    a. Zero-lift drag coefficient
    b. Static margin
    c. Inert weight
    d. Thrust

11. A typical value of tactical missile body fineness ratio is
    a. $l/d = 5$
    b. $l/d = 10$
    c. $l/d = 25$
    d. $l/d = 50$

12. Missile drag is proportional to
    a. diameter
    b. diameter$^{1.5}$
    c. diameter^2
    d. diameter^3

13. Radar seeker range is proportional to
    a. diameter
    b. diameter$^{1.5}$
    c. diameter^2
    d. diameter^3

14. The first mode body bending frequency should be
    a. $>0.1 \times$ flight control actuator bandwidth
    b. $>0.5 \times$ flight control actuator bandwidth
    c. $\approx$ flight control actuator bandwidth
    d. $>$flight control actuator bandwidth

15. Maximum effective boattail angle of a hypersonic missile is about
    a. 0 deg
    b. 2 deg
    c. 4 deg
    d. 12 deg

16. A disadvantage of a circular body cross section is it typically has
    a. higher RCS
    b. less efficient packaging of subsystems
    c. more difficult launch platform integration
    d. more complex aerodynamic/structural analysis

17. A disadvantage of a large wing is
    a. short range in subsonic flight
    b. larger seeker gimbal angle requirement
    c. less angle of attack requirement for maneuver
    d. none of the preceding

18. Which cruise missile has the smallest wing?
    a. JASSM
    b. Apache
    c. CALCM
    d. Harpoon

19. According to slender wing theory, wing normal force coefficient is independent of
    a. angle of attack
    b. aspect ratio
    c. Mach number
    d. none of the preceding
20. As Mach number increases from Mach 0 to 5, the wing aerodynamic center moves from about
    a. 25% to 35% of the mac
    b. 25% to 40% of the mac
    c. 25% to 45% of the mac
    d. 25% to 50% of the mac
21. A disadvantage of a delta wing is
    a. larger variation in aerodynamic center
    b. larger bending moment
    c. higher supersonic drag
    d. greater tendency for aeroelastic instability
22. The mean aerodynamic chord (mac) length of a delta wing is
    a. 1/3 root chord length
    b. 1/2 root chord length
    c. 2/3 root chord length
    d. 3/4 root chord length
23. Which missile has 8 control surfaces?
    a. Stunner
    b. Stinger
    c. JASSM
    d. Kitchen
24. A disadvantage of tail control for a statically stable missile is
    a. less efficient packaging of flight control actuators
    b. higher hinge moment/actuator torque
    c. less tail lift
    d. higher induced rolling moment
25. Which missile has tail control?
    a. Derby
    b. Mistral
    c. Mica
    d. U-Darter
26. Which missile has canard control?
    a. FSAS Aster
    b. Rapier 2000
    c. SD-10
    d. Python 5
27. An advantage of wing control is
    a. lower body rotation
    b. more efficient actuator packaging
    c. lower hinge moment/actuator torque
    d. lower induced rolling moment
28. Which missile has wing control?
    a. Sea Wolf
    b. Exocet
    c. Aspide
    d. Magic
29. The most popular type of TVC for tactical missiles is
    a. jet vane
    b. movable nozzle

c. jet tab
d. reaction jet

30. An advantage of a rolling airframe is
    a. higher maneuverability
    b. lower rate gyros
    c. less sensitive to thrust misalignment
    d. capability for two axis flight control

31. Which missile has a top inlet?
    a. Tomahawk
    b. CALCM
    c. RBS-15
    d. Storm Shadow

Chapter 3: Propulsion

32. The lower limit for scramjet cruise is about
    a. Mach 5
    b. Mach 7
    c. Mach 8
    d. Mach 10

33. A turbojet compressor is powered by the
    a. combustor
    b. inlet
    c. nozzle
    d. turbine

34. A popular material for a subsonic tactical missile turbojet turbine is
    a. rhenium alloy
    b. single crystal nickel aluminide
    c. nickel super alloy
    d. ceramic matrix composite

35. Typical pressure ratio for a turbojet missile single stage axial compressor is about
    a. $p_3/p_2 \approx 1.6$
    b. $p_3/p_2 \approx 4$
    c. $p_3/p_2 \approx 8$
    d. $p_3/p_2 \approx 13$

36. Typical pressure ratio for a turbojet missile with a centrifugal +2 stage axial compressor is about
    a. $p_3/p_2 \approx 2$
    b. $p_3/p_2 \approx 3$
    c. $p_3/p_2 \approx 4$
    d. $p_3/p_2 \approx 8$

37. Which turbojet component typically has the highest efficiency?
    a. compressor
    b. nozzle
    c. inlet
    d. combustor

38. For which flight condition is the inlet free stream flow area A_0 approximately equal to the inlet capture area A_C?
    a. subsonic cruise
    b. transonic cruise
    c. supersonic cruise
    d. hypersonic cruise

39. A ramjet does not have the which component?
    a. inlet
    b. compressor
    c. combustor
    d. nozzle
40. Stochiometric combustion of liquid hydrocarbon fuel occurs at a fuel-to-air ratio of about
    a. 5%
    b. 7%
    c. 9%
    d. 12%
41. For a supersonic ramjet with a combustor entrance Mach number $M = 0.1$, the ratio of the inlet throat area A_{IT} to the combustor entrance area A_3 should be equal to about
    a. 0.2
    b. 0.3
    c. 0.4
    d. 0.5
42. An example of a ramjet with higher cruise drag is
    a. integral rocket-ramjet
    b. aft drop-off booster ramjet
    c. podded ramjet
    d. forward booster ramjet
43. A country that does not have a supersonic air-breathing missile is
    a. United States
    b. India
    c. France
    d. Russia
44. Which supersonic inlet typically has the best compromise of low drag, low pressure oscillation, and low start Mach number?
    a. external compression
    b. internal compression
    c. mixed compression
    d. none of the above
45. A ramjet inlet with an isentropic turning angle of 60 degrees is consistent with a flight Mach number of about
    a. Mach 1.2
    b. Mach 1.8
    c. Mach 3.6
    d. Mach 5.9
46. The fuel with the highest volumetric performance is
    a. JP-10
    b. boron
    c. HTPB
    d. aluminum
47. A typical value of solid propellant characteristic velocity is
    a. $c^* = 4200$ ft/s
    b. $c^* = 5200$ ft/s
    c. $c^* = 6200$ ft/s
    d. $c^* = 7200$ ft/s
48. A typical value of burn rate exponent is
    a. $n = 0.05$
    b. $n = 0.1$

c. n = 0.3
d. n = 0.9

49. A disadvantage of a bi-propellant gel motor compared to a solid propellant motor is
    a. higher toxicity
    b. lower specific impulse
    c. less available duty cycle
    d. less explosive safety

50. Which of the following is inconsistent with a minimum smoke propellant?
    a. ADN
    b. HMX
    c. RDX
    d. AP

51. A solid propellant rocket motor stored for 1 year at a temperature of 35 deg Celsius will have a reduction on its lifetime of about
    a. 30%
    b. 40%
    c. 50%
    d. 60%

52. A disadvantage of a graphite composite motor case is
    a. greater explosive sensitivity
    b. higher cost
    c. greater susceptibility to corrosion
    d. heavier weight

53. Which nozzle throat material would have the greatest erosion?
    a. silica/phenolic
    b. tungsten
    c. rhenium
    d. molybdenum

Chapter 4: Weight

54. Weight density of missiles is about
    a. 0.02 lbm/in^3
    b. 0.03 lbm/in^3
    c. 0.04 lbm/in^3
    d. 0.05 lbm/in^3

55. Which typically requires the highest design factor of safety?
    a. captive carriage loads
    b. free flight loads
    c. pressure bottle
    d. thermal loads

56. Which is a machining manufacturing process?
    a. forge
    b. mill
    c. weld
    d. cast

57. A manufacturing process that typically results in a larger number of parts is
    a. filament wind
    b. pultrusion
    c. machine
    d. cast

58. A disadvantage of a steel motor case is
    a. greater susceptibility to buckling
    b. greater difficulty in joining to other materials
    c. higher cost
    d. lower maximum temperature
59. Which material has the highest modulus of elasticity?
    a. Kevlar fiber
    b. S-glass fiber
    c. carbon fiber
    d. steel
60. High strength graphite composites have
    a. finely ground particles
    b. long oriented fibers
    c. chopped fibers
    d. flakes
61. Which structural material has the highest allowable temperature?
    a. steel
    b. titanium
    c. graphite polyimide
    d. Inconel
62. Which material has the best insulation efficiency?
    a. silicone rubber
    b. teflon
    c. porous ceramic
    d. carbon/carbon
63. Thermal radiation is proportional to
    a. $temperature^1$
    b. $temperature^2$
    c. $temperature^3$
    d. $temperature^4$
64. Which is least important in thermal stress?
    a. modulus of elasticity
    b. pressure load
    c. coefficient of thermal expansion
    d. temperature difference
65. Which has the lowest aerodynamic heating?
    a. nose tip
    b. surface leading edge
    c. leeward (upper) body
    d. flare
66. Body structure weight fraction of tactical missile launch weight is about
    a. 4%
    b. 11%
    c. 22%
    d. 45%
67. A military standard procedure for captive loads calculation is
    a. MIL-A-8590
    b. MIL-A-8591
    c. MIL-A-8592
    d. MIL-A-8593

68. For a typical tactical missile, solid propellant weight fraction of the total rocket motor weight is about
    a. 65%
    b. 83%
    c. 92%
    d. 96%
69. Compared to a composite motor case, a steel motor case is about
    a. 1.2 × heavier
    b. 1.5 × heavier
    c. 2 × heavier
    d. 7 × heavier
70. Infrared bandpass is best for which dome material?
    a. diamond
    b. zinc sulfide
    c. ALON
    d. quartz
71. Which dome material has the lowest dielectric constant?
    a. pyroceram
    b. polyimide
    c. silicon nitride
    d. zinc sulfide
72. Compared to a lithium battery, advantages of a thermal battery include
    a. lighter weight for short time of flight
    b. higher acceleration limit
    c. longer storage life
    d. all of the above
73. Electromechanical flight control actuators are used by
    a. Stinger
    b. AMRAAM
    c. Javelin
    d. all of the above

Chapter 5: Flight Performance

74. A point mass flight trajectory modeling has
    a. 3 degrees of freedom
    b. 4 degrees of freedom
    c. 5 degrees of freedom
    d. 6 degrees of freedom
75. For a 500 nm ramjet, the ratio of fuel weight to total missile weight at the beginning of cruise is about
    a. 30%
    b. 40%
    c. 50%
    d. 60%
76. In steady level flight
    a. thrust ≈ lift
    b. thrust ≈ weight
    c. drag ≈ weight
    d. none of the above
77. Which is false? A TVC missile has higher turn rate at
    a. higher thrust
    b. higher velocity
    c. lighter weight
    d. higher angle of attack

78. A ballistic flight trajectory has
    a. zero flight path angle
    b. zero acceleration
    c. zero angle of attack
    d. none of the above
79. Proportional guidance divert distance is least sensitive to which parameter?
    a. seeker lock-on range
    b. effective navigation ratio
    c. missile velocity
    d. specific impulse
80. A constant bearing is easier for
    a. a missile with high velocity
    b. a missile with a high gimbal seeker
    c. a low-speed target
    d. all of the above

Chapters 6–8

Chapter 6: Other Measures of Merit

1. The global annual average cloud cover is about
    a. 29%
    b. 43%
    c. 61%
    d. 69%
2. The atmospheric attenuation of a 10 GHz signal for clear weather is about
    a. 0.005 db/km
    b. 0.02 db/km
    c. 0.08 db/km
    d. 0.19 db/km
3. A "greenhouse" gas that attenuates an infrared signal through the atmosphere is
    a. H_2O
    b. O_3
    c. CO_2
    d. all of the above
4. A disadvantage of a synthetic aperture radar seeker compared to an infrared seeker is
    a. less capability in adverse weather
    b. shorter range
    c. limited Mach number
    d. lower technical maturity
5. A disadvantage of GPS/INS/data link guidance is
    a. less capability against moving target
    b. less capability in adverse weather
    c. heavier weight
    d. lower technical maturity
6. A typical advantage of a strapdown seeker compared to a gimbaled seeker is
    a. lower cost
    b. larger field of regard
    c. less susceptible to body bending error
    d. longer range
7. A scanning imaging infrared seeker has which advantage over a staring focal plane array seeker?
    a. field of view
    b. weight

c. range
d. performance in clutter

8. A subsonic airframe has maximum thermal emission at a wavelength of about
   a. 1 micron
   b. 4 micron
   c. 12 micron
   d. 29 micron

9. A disadvantage of a mmW seeker compared to a cmW seeker is
   a. resolution
   b. atmospheric attenuation
   c. performance in clutter
   d. bandwidth

10. Assume that a missile has an external data link with a target location error of 10 m. The required data link update rate that provides less than 10 m of additional error for a 10 m/s moving target is about
    a. 1 s per update
    b. 2 s per update
    c. 5 s per update
    d. 10 s per update

11. Optimum cruise altitude for a Mach 4 ramjet is about
    a. sea level
    b. 40,000 ft
    c. 80,000 ft
    d. 120,000 ft

12. Which has the least variation with altitude?
    a. temperature
    b. speed of sound
    c. pressure
    d. density

13. Typical EMD uncertainty in flight performance is about
    a. $+/- 2\%, 1\sigma$
    b. $+/- 5\%, 1\sigma$
    c. $+/- 10\%, 1\sigma$
    d. $+/- 20\%, 1\sigma$

14. A GPS guidance counter-countermeasure is
    a. pseudolite
    b. differential
    c. integrated GPS/INS
    d. all of the above

15. Which missile has an ARH-mmW seeker?
    a. AARGM
    b. HARM
    c. ARMAT
    d. Armiger

16. For the same weight, which warhead is typically the least effective against a tank?
    a. shaped charge
    b. EFP
    c. blast frag
    d. KE

17. The warhead blast scaling parameter z is which function of distance from the center of explosion r and explosive charge weight c?
    a. $r/c^{1/4}$
    b. $r/c^{1/3}$

c. $r/c^{1/2}$
  d. $r/c^{2/3}$
18. Which warhead is least representative of a cylindrical distribution of fragments?
    a. preformed fragments
    b. continuous rod
    c. smooth case
    d. multi-p charge
19. Maximum fragment total kinetic energy occurs for a charge-to-metal ratio of about
    a. 0.5
    b. 1
    c. 2
    d. 5
20. Compared to TNT, HMX explosive energy $(2E_C)^{1/2}$ is greater by about
    a. 30%
    b. 50%
    c. 70%
    d. 90%
21. Kinetic energy is a function of
    a. Velocity
    b. Velocity^2
    c. Velocity^3
    d. Velocity^4
22. An approach to minimize collateral damage of a blast fragmentation warhead is
    a. high charge-to-metal ratio
    b. light weight fragments
    c. low density case
    d. all of the above
23. An equation for proximity fuze angle is
    a. $\sin^{-1}[V_{fragment}/(V_{closing})_{max}]$
    b. $\tan^{-1}[V_{fragment}/(V_{closing})_{max}]$
    c. $\cos^{-1}[V_{fragment}/(V_{closing})_{max}]$
    d. $\cot^{-1}[V_{fragment}/(V_{closing})_{max}]$
24. An equation for proximity fuze time delay is
    a. $\sigma/(V_{closing}\cos\theta)$
    b. $\sigma/(V_{closing}\sin\theta)$
    c. $\sigma/(V_{closing}\cot\theta)$
    d. $\sigma/(V_{closing}\tan\theta)$
25. An advantage of an unguided weapon is
    a. lower cost
    b. less susceptibility to countermeasures
    c. higher reliability
    d. all of the above
26. The probability of a 1σ miss distance assuming a normal distribution is about
    a. 43%
    b. 50%
    c. 57%
    d. 68%
27. Which increases the effective navigation ratio of proportional guidance?
    a. higher missile velocity
    b. lower missile lead angle
    c. lower closing velocity
    d. all of the above

28. The missile time constant τ is the elapsed time to complete which fraction of the commanded maneuver acceleration?
    a. 63%
    b. 75%
    c. 95%
    d. 99%

29. Radome error slope is lowest for a nose fineness ratio of
    a. $l_N/d = 0.2$
    b. $l_N/d = 0.5$
    c. $l_N/d = 1$
    d. $l_N/d = 5$

30. Miss distance contribution due to initial heading error is typically lowest for a proportional guidance effective navigation ratio of
    a. $N' = 1$
    b. $N' = 2$
    c. $N' = 4$
    d. $N' = 6$

31. For a missile with proportional guidance effective navigation ratio $N' = 2.5$ intercepting a target that has a step maneuver acceleration n_T, the missile maneuver acceleration n_M must be at least
    a. $n_M > 2\, n_T$
    b. $n_M > 3\, n_T$
    c. $n_M > 5\, n_T$
    d. $n_M > 8\, n_T$

32. The adjoint miss distance equation for a weaving target is given by
    a. $\sigma_{Weave} = gn_T \tau^2 (\omega\tau)^{N'-1}/[1+(\omega\tau)^2]^{N'/2}$
    b. $\sigma_{Weave} = gn_T \tau^2 (\omega\tau)^{N'-2}/[1+(\omega\tau)^2]^{N'/2}$
    c. $\sigma_{Weave} = gn_T \tau^2 (\omega\tau)^{N'}/[1+(\omega\tau)^2]^{N'/2}$
    d. $\sigma_{Weave} = gn_T \tau^2 (\omega\tau)^{N'+1}/[1+(\omega\tau)^2]^{N'/2}$

33. Glint contribution to miss distance is typically lowest for a proportional guidance effective navigation ratio of
    a. $N' = 3$
    b. $N' = 4$
    c. $N' = 5$
    d. $N' = 6$

34. A solid propellant rocket motor contrail (H_2O) occurs at an atmospheric temperature of about
    a. $-35°F$
    b. $-20°F$
    c. $0°F$
    d. $32°F$

35. Radar transmitter power required for target detection is which function of target radar cross section?
    a. σ^{-4}
    b. σ^{-3}
    c. σ^{-2}
    d. σ^{-1}

36. Which is usually not applicable for conceptual design prediction methods of RCS?
    a. target in free space (no ground effect)
    b. perfectly conducting target surfaces
    c. target dimensions less than wavelength
    d. target components add in random phase

37. The greatest contribution to the frontal RCS of the ramjet baseline missile is
    a. nose
    b. seeker

c. inlet
d. tails

38. Which is the characteristic rise time of a detonation?
    a. $\sim 2 \times 10^{-6}$ s
    b. $\sim 2 \times 10^{-5}$ s
    c. $\sim 2 \times 10^{-4}$ s
    d. $\sim 2 \times 10^{-3}$ s

39. Reliability is a function of
    a. parts count
    b. environment
    c. time of flight
    d. all of the above

40. Which is typically the lowest cost subsystem for a missile?
    a. seeker
    b. structure
    c. electronics
    d. propulsion

41. A typical duration of a moderate risk tactical missile EMD program is about
    a. 2 years
    b. 3 years
    c. 5 years
    d. 7 years

42. For production with an 80% learning curve (L = 0.8), doubling the number of units produced decreases the unit production cost by about
    a. 2%
    b. 5%
    c. 10%
    d. 20%

43. World-wide military/aerospace semiconductor sales are about which fraction of the total world-wide market?
    a. 1%
    b. 3%
    c. 8%
    d. 19%

44. A 50% reduction in parts count reduces production cost by about
    a. 8%
    b. 15%
    c. 21%
    d. 50%

45. Missiles in a wooden round container have reduced aging due to
    a. humidity
    b. temperature
    c. shock
    d. all of the above

46. The number of MEMS devices that are typically produced on a 5 in silicon wafer is about
    a. 30
    b. 300
    c. 3000
    d. 30000

47. Which launch platform has the greatest weight and size constraint for a guided weapon?
    a. tank
    b. helicopter

c. submarine
d. fighter aircraft

48. The rocket plume lateral boundary is which function of thrust T and free stream dynamic pressure q_∞?
    a. $(T/q_\infty)^{1/3}$
    b. $(T/q_\infty)^{1/2}$
    c. $(T/q_\infty)^{2/3}$
    d. (T/q_∞)

49. Which is an aircraft store compatibility wind tunnel test?
    a. flow field
    b. captive trajectory
    c. drop
    d. all of the preceding

50. An aircraft store suspension lug spacing of either 14 in or 30 in may be used for ejection launch of stores with weight between
    a. 101 to 250 lb
    b. 101 to 550 lb
    c. 101 to 1450 lb
    d. 101 to 2050 lb

51. Which missile uses the LAU-7 rail launcher?
    a. Maverick
    b. Sidewinder
    c. Hellfire
    d. Sparrow

52. Which launcher provides an adjustable pitch rate for AMRAAM?
    a. LAU-106/A
    b. LAU-142/A
    c. BRU-33/A
    d. BRU-69/A

53. The CBU-87/CEB Longshot kit provides compressed carriage through
    a. small span wings and tails
    b. folded wings and tails
    c. switch blade wings and folded tails
    d. wrap around wings and tails

54. Which is an aircraft carriage and fire control avionics interface?
    a. MIL-STD-1760
    b. MIL-STD-1553
    c. NATO STANAG 3837
    d. all of the above

55. Which missile uses laser command guidance?
    a. RBS-70
    b. Roland
    c. TOW
    d. BILL

56. Which missile does not use semi-active guidance?
    a. AMRAAM
    b. Sea Dart
    c. Laser Maverick
    d. Sparrow

57. How many VLS canister cells are there in a module?
    a. 4
    b. 8

c. 12
d. 16

58. A typical missile environmental maximum temperature specification is
    a. 120°F
    b. 130°F
    c. 160°F
    d. 180°F

59. A typical missile cold temperature specification is
    a. −30°F
    b. −40°F
    c. −50°F
    d. −60°F

60. Maximum solar radiation at "high noon" in a clear desert environment is about
    a. 155 BTU/ft²/h
    b. 255 BTU/ft²/h
    c. 355 BTU/ft²/h
    d. 455 BTU/ft²/h

Chapter 7: Sizing Examples and Sizing Tools

61. The F-pole is the distance from the launch aircraft to the target when the
    a. missile impacts the target
    b. missile begins active guidance
    c. target begins evasive maneuvers
    d. target has turned 90 deg

62. Visual threshold contrast for detection is about
    a. 2%
    b. 5%
    c. 10%
    d. 20%

63. The rocket baseline missile radome is made of which material?
    a. silicon nitride
    b. slip cast silica
    c. pyroceram
    d. zinc sulfide

64. The rocket baseline missile body airframe is which material?
    a. titanium
    b. aluminum
    c. steel
    d. composite

65. The rocket baseline rocket motor case is which material?
    a. aluminum
    b. graphite composite
    c. steel
    d. inconel

66. Which typically provides the greatest sensitivity to the rocket baseline missile flight range?
    a. propellant weight
    b. specific impulse
    c. zero-lift drag coefficient
    d. static margin

67. The ramjet baseline missile airframe is which material?
    a. aluminum
    b. steel

c. titanium
d. composite

68. Maximum specific impulse of the ramjet baseline occurs at about
    a. Mach 2
    b. Mach 2.5
    c. Mach 3
    d. Mach 3.5

69. Which fuel provides the longest cruise range for the ramjet baseline?
    a. RJ-5
    b. JP-10
    c. boron
    d. JP-10/boron slurry

70. The compressor for the turbojet baseline missile has
    a. two stages
    b. three stages
    c. four stages
    d. five stages

71. $(L/D)_{max}$ of the turbojet baseline missile occurs at an angle of attack of about
    a. 7 deg
    b. 10 deg
    c. 13 deg
    d. 16 deg

72. What is the maximum temperature of the turbojet baseline missile combustor?
    a. 1800 R
    b. 2000 R
    c. 2200 R
    d. 2400 R

73. The turbojet baseline missile flight range is most sensitive to which parameter?
    a. fuel weight
    b. specific impulse
    c. inert weight
    d. thrust

74. The number of movable fins of the baseline guided bomb is
    a. 2
    b. 3
    c. 4
    d. 6

75. Explosive material of the baseline guided bomb is
    a. Tritonal
    b. HMX
    c. RDX
    d. TNT

76. Compared to a preliminary design code, a conceptual design code typically has
    a. faster turnaround time
    b. physics-based methods
    c. more stable computation
    d. all of the above

77. The far left column of the House of Quality is
    a. customer importance rating
    b. customer requirements
    c. design characteristics
    d. design characteristics importance rating

78. The number of required designs for a DOE full factorial study with k parameters and l levels is
    a. $1+k(l-1)$
    b. $1+k(l+1)$
    c. $1+kl$
    d. l^k
79. A DOE parametric study with k = 4 parameters and l = 2 levels requires how many designs?
    a. 3
    b. 4
    c. 5
    d. 6

Chapter 8: Development Process

80. EMD occurs before
    a. ACTD
    b. DemVal
    c. PDRR
    d. none of the preceding
81. US DoD RT&A expenditure on tactical missiles is what fraction of the US DoD total RT&A budget?
    a. 11%
    b. 16%
    c. 23%
    d. 28%
82. Which category is 6.4 funding?
    a. basic research
    b. exploratory development
    c. prototype demonstration
    d. advanced technology demonstration
83. The introduction of which missile led to the space race?
    a. R-7
    b. Atlas
    c. Polaris
    d. Titan
84. Which mission had the first application of laser guidance?
    a. air-to-air
    b. air-to-surface
    c. surface-to-air
    d. surface-to-surface
85. Which was the most important consideration in replacing AGM-130 with JASSM?
    a. higher reliability
    b. longer range
    c. better accuracy
    d. better capability in adverse weather
86. From the year 1985 to 1997, the number of US missile development contractors was reduced from 12 to
    a. 6
    b. 5
    c. 4
    d. 3
87. A flight test program with 10 kills for 10 shots has a P_K lower limit of the 95% confidence interval of
    a. $P_K = 0.7$
    b. $P_K = 0.8$
    c. $P_K = 0.9$
    d. $P_K = 0.95$

88. Which is not a preliminary design prediction method for missile aerodynamics?
    a. OTIS
    b. DATCOM
    c. SUPL
    d. AP09

Appendix I: Design Case Studies

The following missile design case studies conducted by Georgia Tech aerospace engineering students for the AIAA Graduate Student Missile Design Competition are available.

- High Speed Standoff Missile, 2000
- Ship Launched Cruise Missile, 2001
- Ship Launched Ramjet Cruise Missile, 2002
- Future Target Delivery System, 2003
- Boost Phase Interceptor, 2005
- Multi-mission Cruise Missile, 2004
- Long Range Liquid Boost Target Vehicle, 2004
- Extended Range Air-Launched Target, 2005
- Foreign Missile Target Surrogate System, 2006
- Hybrid Target System, 2007
- Affordable Low Fidelity Target System, 2008
- Boost Phase Interceptor, 2009
- Long Range, High Speed, Precision Strike Weapon, 2010
- Hypersonic Advanced Missile for Multi-mission Execution and Rapid Response, 2011
- Sea Based Missile Defense Interceptor, 2012
- Terminal High Altitude Naval Theater Operations Shield, 2012
- Aerial Reconnaissance Tactical Engagement Missile, 2013
- High-Speed Advanced Long Range Air-to-Air Missile, 2014
- Swarm Intercepting Rapid Engagement Missile, 2015

Appendix J: Summary of Tactical Missile Design Spreadsheet

Help / User's / Installation Guide for the Tactical Missile Design Spreadsheet

Developed by:

Gene Fleeman
Andrew Frits
Jack Zentner
at the
Aerospace Systems Design Laboratory
School of Aerospace Engineering
Georgia Institute of Technology
Atlanta, GA 30332

Contact Information:

Eugene L. Fleeman
4472 Anne Arundel Court
Lilburn, GA 30047

Email: GeneFleeman@msn.com

Phone: (770) 925-4635

Installing/Opening:

Note that the Tactical Missile Design Spreadsheet (TMD) is NOT a stand-alone program. It is a file that is designed to be run via Microsoft Excel 97, or any later versions of Excel. Microsoft Excel does not come with this software package, and must be purchased, licensed, and installed by the user.

To operate the TMD spreadsheet:

First, open Microsoft Excel 97 by clicking on the appropriate icon in the start menu or run the excel executable: Excel.exe

Second, go to the 'file' pull-down menu, then 'open', browse through your computer's file structure until you find the appropriate file named: TMDv1_0.xls (it should be on the CD-ROM drive if the TMD CD is inserted in your machine)

Double click on this file to open it.

A warning message may pop up, saying "The workbook you are opening contains macros. Some macros contain viruses that may be harmful to your computer." You should click on "Enable Macros". The TMD spreadsheet will not run correctly if macros are not enabled.

When the spreadsheet opens, it should be at the "Master Input" page. Other pages can be viewed by simply clicking on the page tab at the bottom of the excel window. All input should be handled at the "Master Input" page.

IMPORTANT NOTE: This spreadsheet uses circular cell references (i.e., a cell that calls itself) for some of its computations. In order for the circular reference to work properly, your Excel system must be set up so that it will iterate to solutions. Upon loading the file, this should be done automatically. However, opening other files while the TMD spreadsheet is open may actually turn iterations off. To check to see if Excel is set up to handle iterations appropriately to run the TMD spreadsheet, do the following steps:

1. Go to the 'Tools' pull-down menu
2. Select 'Options'
3. A new window should appear. The window has tabs near the top, select the tab that is called: 'Calculation'
4. See that 'Iteration' is checked. If it is not checked, click on the white box to the left in order to activate it
5. Set a 'Maximum iterations'. This should be set to at least 50
6. Set the 'Maximum change'. 0.0001 is a good number for the maximum change.

These steps should ensure that the TMD spreadsheet will handle circular references correctly.

Using The Tactical Missile Design Spreadsheet:

All inputs should be made to the "Master Input" sheet. Inputs are entered into the blue-colored fields, with the appropriate units specified. Some small menus are included so that the user may select discrete values, such as engine type, fuel type, etc. Note that some combinations of items will not work. For example, selecting a Ramjet with a Mach number of 0.5 will not work. Warnings to this effect will be printed, or, in some cases, the output will have errors such as: #value, #div/0

Variable names listed in the "Master Input" sheet are not used in the spreadsheet (each individual analysis will often re-define the variable name in a text-form). The variable name on the "Master Input" page matches the text, and is used to help the user correlate the text-examples and equations with the TMD spreadsheet.

The values of two example systems are included, a baseline rocket and a baseline ramjet system. These numbers and results should match those given in Gene Fleeman's Text: "Tactical Missile Design". These numbers make a good starting point for any design.

Each analysis type is made on a separate Excel Sheet (structures, aerodynamics, etc.). Each page is linked to the Master Input page, and is linked to output from other pages via the Master Output page. For example, the aerodynamics sheet uses the body minor diameter and major diameter (linked from the Master Input Page) to calculate the missile reference area. This reference area is then sent to the Master Output page, where it is used by the trajectory and propulsion pages for additional calculations.

The three analysis sheets that are intricately linked together are the aerodynamics, propulsion, and trajectory. These three sheets work together to take an initial aircraft geometry, propulsion system, and initial conditions to fly an appropriately sized missile. The system assumes that it is a co-altitude flight (i.e., the missile does not change altitude). Further, there are three stages to the flight:

Boost

Sustain

Coast

The boost stage uses a rocket engine (as specified by the propulsion inputs), with a specified boost fuel weight and burn time. The sustain phase can use a number of engines: rocket, ramjet, scramjet, etc. Note, however, that currently only a ramjet or a rocket type engine can be analyzed. Any other type of engine simply uses a look-up table for Isp. For rockets, only a constant thrust solid rocket system can be used, there is no adjustable grain geometry/variable thrust profiles. However, the user can have one thrust level for the boost motor and a different thrust level for the sustain motor, essentially creating a two-level constant thrust system (this is the thrust profile of the baseline rocket used in the Tactical Missile Design text). The coast stage has no thrust, it simply calculates the drag for a co-altitude coast, stopping when the Mach number reaches the 'Terminal Mach Number' Criteria. This trajectory analysis allows the user to get a preliminary range and time-to-target for a given missile system.

The other sheets have output that is not linked to the trajectory module. These are stand-alone analyses that can be used to get estimates for motor case weights, temperature loads, warhead effectiveness, and other measures of merit. Again, the user should be aware that these outputs are not used by the integrated trajectory analysis. Important parameters for the trajectory analysis, including the empty weight (inputted as launch weight) must be input by the designer, though the designer may use some elements of the warhead/structural analysis to give him/her a better estimate of the empty weight.

Output may be viewed via three mechanisms.

First, the Master Output sheet has key output from many of the sheets. It also contains all of the values that are linked between sheets.

Secondly, the Summary sheet has a summary of the trajectory output. It gives the stage-by-stage range, weight, etc. It also has plots of Mach number and range versus time for each flight stage.

Thirdly, individual analysis sheets have a lot of important data. They have numbers that are not included in the limited numbers on the Master Output sheet. The individual analysis sheets also have a lot of additional plots and charts.

The individual analyses are done, whenever possible, such that they closely follow the methodology illustrated in the "Tactical Missile Design" text. Example cases from the textbook should give results that are similar to those given in the text.

A Note on Printing:

Most of the spreadsheets are wider than the width of paper. After printing the user will probably need to go to the computer screen as an aid in assembling the printout. The computer screen shows the edges of each page as a dashed line.

WARNING: Often you will receive the following message: 'Not enough system resources to display completely'. This is a common error that excel has when it runs this spreadsheet. If you get this warning, simply hit 'ok' and ignore it.

Making Changes:

All of the sheets, with the exception of the Master Input sheets, are write- protected. This allows the user to change the input parameters and look at the results of the individual sheets, without the fear of accidentally altering the spreadsheet. The user is free to make changes wherever he/she feels is appropriate, and tailor the TMD spreadsheet to meet individual user needs. However, the responsibility of the results from any changes in the spreadsheet rests with the user. A sheet can be unprotected by selecting the 'Tools' pull-down menu, selecting 'Protection', and then selecting 'Unprotect sheet'.

When altering or adding analysis sheets, note the way that the current system is set-up. Each analysis only has links to the 'Master Input' and 'Master Output' sheets. This prevents a lot of invisible cross-linking by making all of the important numbers flow through the same choke points. Also note that each sheet, generally on the left hand side, has a listing of all the variables used in the analysis. These variables are often nothing more than links to the 'Master Input' and 'Master Output', however, by having a copy of all the important numbers on the analysis page, it further assists in cutting down on the number of cross-links between different analysis sheets of the TMD spreadsheet.

Here is an example of the way an analysis sheet is generally set up:

INPUTS	CALCULATIONS
-linked to Master Input -linked to Master Output	Equations that use only the inputs on the left
OUTPUTS	FIGURES
Output values are copies from results of the Calculations -linked to Master Output	Figures are placed throughout the sheet

Question/Comments:

Please direct any questions or comments regarding the Tactical Missile Sizing Spreadsheet and the text to Mr. Gene Fleeman. See his contact information above.

Sincerely,
Andrew P. Frits
Jack Zentner
-Spreadsheet implementers

Tactical Missile Design Spreadsheet Enhancement*

This work builds on the original Tactical Missile Design (TMD) spreadsheet for a rocket baseline missile (the AIM-7 Sparrow, a medium-range semi-active radar homing air-to-air missile) and a ramjet missile (the Advanced Strategic Air Launched Missile, ASALM, a medium-range strategic missile. The original spreadsheet was updated to include baseline parameters for a turbojet missile and a guided bomb, as well as subject to verification and validation with an independent model. The turbojet baseline missile is based on the Harpoon missile – an all-weather, over-the-horizon, anti-ship missile. The guided bomb is based upon the Joint Direct Attack Munition, JDAM – a guidance kit that converts unguided bombs, or "dumb bombs," into all-weather precision-guided munitions. Additional flight modes that were added to the original spreadsheet include cruise, climb, descent, and ballistic.

Primary advancements made to the original spreadsheet include the introduction of baseline data for a turbojet engine and guided bomb; this data comes from publicly released information pertaining to the Harpoon missile and the MK-84 2,000 lb bomb with Joint Direct Attack Munitions (JDAM) attachment. Turbojet propulsion data comes from the *Teledyne Turbine Engines Brochure*, Ashley, *Engineering Analysis of Flight Vehicles*, and Jerger, *Systems Preliminary Design Principles of Guided Missile Design*. The spreadsheet operates by a configuration buildup in which aerodynamic parameters are initially considered. By inputting the geometry of the body, canards, wings, and control surfaces, the TMD spreadsheet can summate the aerodynamic effects acting on the system. This first-order estimation neglects the interactions between each winged section, but does account for normal force, drag force, and center of pressure. Computations can be performed at both subsonic and supersonic flight conditions using slender body theory, slender wing theory, and linear wing theory as applicable.

To improve the robustness of this tool, an alternative method was developed for calculating the wing and tail geometry. Before, a user defines the geometry and leading-edge position of each component, then the spreadsheet iteratively sizes the wing and tail fins to provide zero static margin corresponding to the specified Mach number. This entails the original design specification are being altered to produce a more statically stable configuration. While this function may be beneficial and provide insight to the initial design and stability of the projectile, keeping the original tail dimensions may be desired.

*Spears, Seth[1], Allen, Randall[2], and Fleeman, Eugene[3] "First-Order Conceptual Design for Turbojet Missiles and Guided Bombs", SciTech Forum 2022

Appendix K: Soda Straw Rocket Science

Soda Straw Rocket Science is a STEM aerospace program to demonstrate the physics of flight and development using a small air rocket.

The initial idea to design, build, and fly a small air rocket first came to me several years ago as an activity for my grandchildren, who at that time were ages 4 and 7. We built the rocket launcher from pipe fittings, an air pressure gauge, a gas valve, an air supply hose, and a water filter tank. The rockets were constructed from soda straws, ear plugs, and adhesive filing tabs. My grandchildren found that launching their rockets was not only fun, but also interesting to try different launch angles, launch pressures, and rocket designs. After that, word got around and I was asked to participate in local school classroom, Cub Scout, and science camp programs. Small air rockets are particularly suitable as aerospace outreach—the process of designing, building, and flying the air rockets is a good introduction to the aerospace environment.

As an aerospace outreach program for students and their teachers, the program introduces them to an aerospace environment. Attributes include:

- Introduces the design, build, and fly process, a fundamental process of aerospace systems development
- Illustrates the competitive nature of aerospace systems
- Is exciting and fun—capturing interest from kindergarten students to adults

Other potential alternatives to using small air rockets include Estes-type hot exhaust rockets and water rockets. The Estes-type rocket is more spectacular—with fire, smoke, noise, and higher performance. However, it can only be launched outside the classroom in a large space to be safe. Another disadvantage is the student has less insight into the physics of the rocket. A water rocket is also spectacular and is great fun to launch. However, it is messy and also has less insight into the physics of the rocket. The small air rocket of this program is safe, clean, and can be launched inside the classroom if necessary. The range can be directly controlled by the chamber pressure, launch angle, and rocket design. Students are introduced to the experimental and theoretical physics of thrust, total impulse, boost velocity, drag, vertical impact velocity, and flight trajectory. These are easily predicted with simple physics equations and can be easily confirmed with ground/flight test data. For these reasons, a soda straw air rocket was selected.

Georgia Tech Science Camp for Middle School Girls

Appendix K: Soda Straw Rocket Science 427

List of Figures

Chapter 1 Introduction

1.1 Overview

Fig. 1.1   Chapter 1: Introduction—What You Will Learn
Fig. 1.2   Emphasis of This Material Is on Physics-Based Conceptual Missile Design

1.2 Missile Characteristics Comparison

Fig. 1.3   Examples of State of the Art (SOTA) Drivers—Tactical Missiles: Cost, Strategic Missiles: Range
Fig. 1.4   Tactical Missiles Are Produced in Larger Quantity Than Strategic Missiles
Fig. 1.5   Missile Subsystems Are Analogous to a Human Boxer
Fig. 1.6   Typical Missile Subsystems—Packaging Is Longitudinal, with High Density
Fig. 1.7   Configuration Sizing Parameters Emphasized in This Material

1.3 Conceptual Design and System Engineering Process

Fig. 1.8   Conceptual Design Should Be Unbiased, Creative, and Iterative—with Rapid Evaluations
Fig. 1.9   Conceptual Design and System Engineering Is Required to Explore Alternative Approaches
Fig. 1.10   Missile Conceptual Design and System Engineering Requires Broad, Creative, Rapid, and Iterative Evaluations
Fig. 1.11   Pareto Effect—Only a Few Parameters Are Drivers for Each Measure of Merit
Fig. 1.12   A Balanced Missile Design Requires Harmonized Mission Requirements and Measures of Merit
Fig. 1.13   System Engineering Includes Modeling, Analysis, Integration, Requirements, and Flight Simulation
Fig. 1.14   Missile Design and System Engineering Require System Integration
Fig. 1.15   System Engineering Requirements Flow-down Provides Subsystem Components Specifications

1.4 System-of-Systems Comparison

Fig. 1.16a   Missile Development Should Be Conducted in a System-of-Systems Context
Fig. 1.16b   Missile Development Should Be Conducted in a System-of-Systems Context (cont)
Fig. 1.16c   Missile Development Should Be Conducted in a System-of-Systems Context (cont)
Fig. 1.16d   Missile Development Should Be Conducted in a System-of-Systems Context (cont)
Fig. 1.16e   Missile Development Should Be Conducted in a System-of-Systems Context (cont)
Fig. 1.16f   Missile Development Should Be Conducted in a System-of-Systems Context (cont)
Fig. 1.16g   Missile Development Should Be Conducted in a System-of-Systems Context (cont)
Fig. 1.16h   Missile Development Should Be Conducted in a System-of-Systems Context (cont)

1.5 Examples of State-of-the-Art Missiles

Fig. 1.17   Examples of Design Drivers/State-of-the-Art (SOTA) for Air-Launched Missile Missions/Types/Attributes
Fig. 1.18   Examples of State-of-the-Art Surface-Launched Missile Missions, Types, and Attributes

1.6 Examples of Alternatives in Establishing Mission Requirements

- Fig. 1.19 Example of Assessment of Alternatives to Establish Requirements for Future Mission
- Fig. 1.20 C4ISR Satellites and UAVs Have High Effectiveness for Time Critical Target Cueing
- Fig. 1.21 Example of System-of-Systems Analysis to Develop Future Stand-off Missile Requirements
- Fig. 1.22 Technological Surprise May Drive Immediate Mission Requirements

1.7 Use of a Baseline Missile

- Fig. 1.23 Starting with a Baseline Design Expedites the Design Process
- Fig. 1.24 Example of Missile Baseline Data—Chapter 7 Data of Ramjet Baseline Missile
- Fig. 1.25 Baseline Missile Design Data Allows Correction of Conceptual Design Computed Parameters

Chapter 2 Aerodynamics

2.1 Introduction

- Fig. 2.1 Chapter 2: Aerodynamics—What You Will Learn
- Fig. 2.2 Aerodynamic Forces Impact Missile Maneuverability and Flight Range
- Fig. 2.3 Aero Configuration Sizing and System Engineering Has High Impact on Mission Requirements and Measures of Merit
- Fig. 2.4 Conceptual Design Aerodynamic Methods of This Text Are Based on Aero Configuration Buildup
- Fig. 2.5 Conceptual Design Total Aerodynamic Force May Be Estimated by Summing Individual Contributors

2.2 Missile Diameter Tradeoff

- Fig. 2.6 Missile Diameter Is a Tradeoff
- Fig. 2.7 A Small Diameter Missile Has Lower Drag
- Fig. 2.8 Missile Fineness Ratio May Be Limited by Resonance of Body Bending Frequency with Flight Control

2.3 Nose Fineness and Geometry Tradeoffs

- Fig. 2.9 Nose Fineness and Geometry Is a Tradeoff
- Fig. 2.10 Faceted and Flat Window Domes Can Provide Low Distortion, Low Drag, and Low Radar Cross Section

2.4 Body Drag Prediction

- Fig. 2.11 Body Maximum Zero-Lift Drag Coefficient Occurs Near Mach 1
- Fig. 2.12 Supersonic Body Wave Drag Is Driven by Nose Fineness
- Fig. 2.13 Moderate Nose Tip Bluntness Causes a Relatively Small Increase in Supersonic Drag

2.5 Boattail Tradeoffs

- Fig. 2.14 A Boattail Decreases Base Pressure Drag Area
- Fig. 2.15 A Boattail Is More Effective for a Subsonic Missile

2.6 Body Normal Force and Lift-to-Drag Prediction

- Fig. 2.16 A Lifting Body Has Higher Normal Force

Fig. 2.17   Body Lift-to-Drag Ratio Is Impacted by Angle of Attack, Zero-Lift Drag Coefficient, Body Fineness, and Cross Section Geometry

Fig. 2.18   A Lifting Body Requires Flight at Relatively Low Dynamic Pressure to Achieve High Lift-to-Drag Ratio

Fig. 2.19   A Lifting Body Typically Has Higher Lift-to-Drag and Lower RCS, A Circular Cross Section Body Has Higher Volumetric Efficiency

2.7 Sign Convention of Forces, Moments, and Axes

Fig. 2.20   Sign Convention of Forces, Moments, and Axes
Fig. 2.21   Pitch, Yaw, and Roll Animation of Rocket Baseline Missile

2.8 Static Stability and Body Aerodynamic Center Prediction

Fig. 2.22   Pitch Moment Stability $\Delta Cm/\Delta \alpha$ and Static Margin ($X_{AC} - X_{CG}$) Define Pitch Static Stability
Fig. 2.23   Body Aerodynamic Center Is Driven by Angle of Attack, Nose Length, and Body Length

2.9 Flare Stabilizer Tradeoffs

Fig. 2.24a   An Aft Flare Increases Static Stability
Fig. 2.24b   An Aft Flare Increases Static Stability (cont)
Fig. 2.25   Tail Stabilizers Have Less Drag and Provide Flight Control, Flare Stabilizer Has Less Aero Heating and Less Variation in Stability

2.10 Wings Versus No Wings

Fig. 2.26a   Most Supersonic Missiles Do Not Have Wings
Fig. 2.26b   Most Supersonic Missiles Do Not Have Wings (cont)
Fig. 2.27a   Most Subsonic Cruise Missiles Have Relatively Large Wings
Fig. 2.27b   Most Subsonic Cruise Missiles Have Relatively Large Wings (cont)
Fig. 2.28   Examples of Guided Bombs That Have Wings for Extended Range

2.11 Normal Force Prediction for Planar Surfaces

Fig. 2.29   Definition of Planar Aerodynamic Surface Geometry Parameters
Fig. 2.30   Normal Force Coefficient of a Planar Surface (Wing, Tail, Canard) Is Higher at Low Mach Number
Fig. 2.31   High Normal Force for a Planar Surface Occurs at High Local Angle of Attack

2.12 Aerodynamic Center Location and Hinge Moment Prediction for Planar Surfaces

Fig. 2.32   Aerodynamic Center of a Planar Surface Moves Aft with Increasing Mach Number
Fig. 2.33   Hinge Moment Increases with Mach Number, Aerodynamic Center Distance from Hinge Line, and Angle of Attack

2.13 Planar Surface Drag and Lift-to-Drag Prediction

Fig. 2.34   Skin Friction Drag Is Lower for Small Surface Area
Fig. 2.35   Supersonic Drag of Planar Surface Is Smaller if Leading Edge Has Sweep and Small Section Angle
Fig. 2.36   Wing Subsonic Aero Efficiency Lift-to-Drag Is Driven by Angle of Attack, C_{D_0}, and Aspect Ratio

2.14 Surface Planform Geometry and Integration Alternatives

Fig. 2.37 Planar Surface (Wing, Tail, Canard) Panel Geometry Is a Tradeoff with Many Considerations
Fig. 2.38 Examples of Wing, Stabilizer, and Flight Control Surface Arrangements and Alternatives

2.15 Flight Control Alternatives

Fig. 2.39 Most Aero Flight Control Missiles Have Four Control Surfaces, with Pitch, Yaw, and Roll Control Integration
Fig. 2.40 There Are Many Flight Control Aerodynamic Configuration Alternatives
Fig. 2.41 Missile Flight Control Selection Is Typically Driven by Maneuverability, Accuracy, Weight, Volume, and Cost
Fig. 2.42 Tail Flight Control Is Efficient at High Angle of Attack
Fig. 2.43 Tail Flight Control Can Usually Operate at Higher Angle of Attack Than Canard Flight Control
Fig. 2.44 About 70% of Tail Flight Control Missiles Have Wings
Fig. 2.45 Tail Flight Control Alternatives - Conventional Balanced Actuation Fin, Flap, and Lattice Fin
Fig. 2.46 Lattice Fin Flight Control Has Advantages for Low Subsonic and High Supersonic Missiles
Fig. 2.47 Lattice Fin Choked Flow Is Driven by Lattice Section Thickness and Transonic Mach Number
Fig. 2.48 Conventional Canard Flight Control Is Efficient at Low Angle of Attack, but Stalls at High Angel of Attack with Induced Roll
Fig. 2.49 Most Canard Flight Control Missiles Are Wingless and Most Are Supersonic
Fig. 2.50 Examples of Aerodynamic Approaches that Enhance Maneuverability and Accuracy of Canard Flight Control
Fig. 2.51 Split Canard Flight Control Provides Maneuverability at High Angle of Attack with Lower Hinge Moment
Fig. 2.52 A Wing Flight Control Advantage Is Low Body Rotation. Disadvantages Are Larger Hinge Moment, Induced Roll, and Tendency to Stall
Fig. 2.53 Wings Are Susceptible to Strong Vortex Shedding
Fig. 2.54 Current Wing Flight Control Missiles Are Supersonic and Are Old Technology
Fig. 2.55 Missile Aerodynamic Flight Control Typically Stalls at a Surface Local Angle of Attack of Approximately 22 Deg
Fig. 2.56 Surface Maximum Lift (Stall) Deceases with Supersonic Mach Number
Fig. 2.57 TVC and Reaction Jet Flight Control Provide High Maneuverability at Low Dynamic Pressure
Fig. 2.58 Most Missiles with TVC or Reaction Jet Flight Control Also Use Aerodynamic Flight Control

2.16 Maneuver Law Alternatives

Fig. 2.59a Skid-to-Turn Is the Most Common Maneuver Law for Missiles
Fig. 2.59b Skid-to-Turn Is the Most Common Maneuver Law for Missiles (cont)
Fig. 2.60 Examples of Skid-to-Turn, Bank-to-Turn, Rolling Airframe, and Divert Maneuvering
Fig. 2.61a Non-Cruciform Inlets Require Bank-to-Turn Maneuvering
Fig. 2.61b Non-Cruciform Inlets Require Bank-to-Turn Maneuvering (cont)

2.17 Roll Angle and Control Surface Sign Convention

Fig. 2.62 Typical Sign Convention for Cruciform Missile Roll Angle and Flight Control Surface Deflection
Fig. 2.63 X Roll Orientation Flight Is Usually Better Than + Roll Orientation Flight

2.18 Trim and Static Stability Considerations

- Fig. 2.64 Trimmed Normal Force Is Defined at Zero Pitching Moment
- Fig. 2.65 Relaxed Static Stability Margin Allows Higher Trim Angle of Attack and Higher Normal Force
- Fig. 2.66 Relaxed Static Stability Margin Reduces Drag
- Fig. 2.67 Missile Static Margin Is Driven by Tail Area and Static Margin Prediction Has Large Uncertainty

2.19 Stability and Control Conceptual Design Criteria

- Fig. 2.68 Stability & Control Cross Coupling Requires Higher Flight Control Effectiveness
- Fig. 2.69 Stability & Control Cross Coupling of a Lifting Body Requires Higher Flight Control Effectiveness

Chapter 3 Propulsion

3.1 Introduction

- Fig. 3.1 Chapter 3 Propulsion—What You Will Learn
- Fig. 3.2 Propulsion Thrust Impacts Flight Range

3.2 Propulsion Alternatives Assessment

- Fig. 3.3 Each Type of Air Breathing Propulsion Has an Optimum Mach Number for Propulsion Efficiency
- Fig. 3.4 Cruise Range Is Driven by Lift-to-Drag Ratio, Specific Impulse, Velocity, and Propellant or Fuel Weight Fraction
- Fig. 3.5 Solid Propellant Rocket Propulsion Has Higher Thrust-to-Weight Capability Than Air-Breathing Propulsion
- Fig. 3.6 Turbine-Based Missile Propulsion Is Capable of Subsonic to Supersonic Cruise

3.3 Turbojet Flow Path, Components, and Nomenclature

- Fig. 3.7 Schematic of Turbojet Flow Path, Components, and Nomenclature
- Fig. 3.8 High Temperature Compressors Are Required to Achieve High Pressure Ratio at High Mach Number
- Fig. 3.9 A High Temperature Turbine Is Required for a High-Speed Turbojet Missile
- Fig. 3.10 Compressor Pressure Ratio and Turbine Temperature Limit Turbojet Maximum Thrust
- Fig. 3.11a Approaches for Mach 3+ Missile Turbojet Propulsion
- Fig. 3.11b Approaches for Mach 3+ Missile Turbojet Propulsion (cont)
- Fig. 3.12 Compressor Alternatives Are Multi-Stage Axial, Single Centrifugal, Multi-Stage Axial + Centrifugal
- Fig. 3.13 A Small Diameter Turbojet Requires a High-Speed Compressor
- Fig. 3.14 Rotors of a High-Pressure Axial Compressor Have High Local Mach Number
- Fig. 3.15 Multi-Stage Compressors Provide Higher Compressor Pressure Ratio

3.4 Turbojet Thrust Prediction

- Fig. 3.16 Turbojet Thrust Is Limited by Inlet Flow and Turbine Maximum Allowable Temperature
- Fig. 3.17 Assumption Specific Heat Ratio = 1.4 Has Sufficient Accuracy for Turbojet Conceptual Design Thrust

3.5 Turbojet Specific Impulse Prediction

- **Fig. 3.18** Turbojet Specific Impulse Decreases with Increased Mach Number and Combustion Temperature
- **Fig. 3.19** Turbojet Has Relatively High Thrust @ Relatively Low Compressor Pressure Ratio but High Specific Impulse Requires High Pressure Ratio

3.6 Subsonic Turbojet Propulsion Efficiency

- **Fig. 3.20** Typical Values of Subsonic Turbojet Propulsion Efficiency
- **Fig. 3.21** Higher Thrust Turbojet and Turbofan Expendable Engines Usually Have More Efficient Performance
- **Fig. 3.22** Turbojet Missile Inlet Best Location Is Driven by Launch Platform Fitment, Radar Cross Section, and Pressure Recovery
- **Fig. 3.23** During Subsonic Cruise, Free Stream Flow into Inlet is Approximately Equal to Inlet Capture Geometry

3.7 Ramjet Flow Path, Components, and Nomenclature

- **Fig. 3.24** Examples of Missile Ramjet Propulsion
- **Fig. 3.25** Schematic of Liquid Fuel Ramjet Engine Flow Path, Components, and Nomenclature
- **Fig. 3.26** Example of Flow Path Design Considerations and Limitations for a Typical Missile Ramjet Engine

3.8 Ramjet Temperature and Specific Impulse Prediction

- **Fig. 3.27** Ramjet Combustion Temperature Increases with Free Stream Mach Number and Fuel Flow
- **Fig. 3.28** High Specific Impulse for a Ramjet Occurs @ Mach 3 to 4 Flight

3.9 Ramjet Thrust Prediction

- **Fig. 3.29** High Ramjet Thrust Occurs at Mach 3 to 5, High Combustion Temperature, and High Inlet Flow
- **Fig. 3.30** High Ramjet Thrust Occurs at High Mach Number, High Fuel Flow, and High Inlet Flow

3.10 Ramjet Inlet Design Considerations

- **Fig. 3.31** A Ramjet with Low Combustion Temperature and High Mach Number Flight Requires a Large Inlet
- **Fig. 3.32** Ramjet Inlet Throat Area Is a Driver for Combustion Mach Number, Mass Flow, Specific Impulse, and Thrust

3.11 Ramjet Combustor Design Considerations

- **Fig. 3.33** A Ramjet in Low Supersonic Flight with High Temperature Combustion May Be Susceptible to Thermal Choking
- **Fig. 3.34** A Relatively Low Mach Number into Combustor Is Desirable for High Ramjet Combustion Efficiency
- **Fig. 3.35** Required Length and Weight of the Ramjet Combustor Is a Function of Combustor Velocity

3.12 Ramjet Booster Integration

- **Fig. 3.36** Ramjet Engine and Booster Integration Options
- **Fig. 3.37** Ramjet Engine, Booster, Inlet Integration Trades
- **Fig. 3.38** Ramjets with Internal Boosters and No Wings Have Lower Drag

3.13 Ramjet Inlet Options

- Fig. 3.39 Ramjet Inlet Options
- Fig. 3.40 Ramjet Inlet Concept Trades
- Fig. 3.41 Examples of Inlets for Supersonic Air-Breathing Missiles

3.14 Supersonic Inlet/Airframe Integration

- Fig. 3.42 Supersonic Inlet-Airframe Integration Tradeoffs Include Drag, Pressure Oscillation, and Inlet Start
- Fig. 3.43 A Supersonic External Compression Inlet with Shock Wave(s) on the Cowl Lip Prevents Spillage
- Fig. 3.44 Nose Shock Wave Angle Is Driven by Type of Shock Wave, Angle of Attack, Surface Deflection, and Mach Number
- Fig. 3.45 A Conical Nose Inlet Typically Has Higher Capture Efficiency Than a Two-Dimensional Wedge Nose Inlet
- Fig. 3.46 Inlet Throat Area Is a Driver for Supersonic Inlet Start Mach Number, Combustion Efficiency, and Thrust
- Fig. 3.47 Optimum Forebody Deflection Angle(s) for Best Pressure Recovery Increases with Mach Number
- Fig. 3.48 Example of Near-Isentropic Supersonic Compression Inlet—3M80 (SS-N-22 Sunburn)
- Fig. 3.49 Oblique Shocks Prior to the Inlet Normal Shock Are Required to Satisfy MIL-E-5007D

3.15 Fuel Alternatives

- Fig. 3.50 High Density Fuel Has Higher Volumetric Performance, but Also Has Higher Observables

3.16 Solid Propellant Rocket Motor Flow Path, Components, and Nomenclature

- Fig. 3.51 Solid Propellant Rocket Motor Design Components, Considerations, and Tradeoffs
- Fig. 3.52 A Rocket Generates Thrust by Converting High Pressure Combustion into High Velocity Exhaust

3.17 Rocket Motor Performance Prediction

- Fig. 3.53 High Propellant Weight Fraction and High Specific Impulse Increase Rocket Burnout Velocity
- Fig. 3.54 Rocket Baseline Pareto Shows Thrust Drivers Are Chamber Pressure and Nozzle Throat Area
- Fig. 3.55 A High Specific Impulse Rocket Requires High Chamber Pressure and Optimum Nozzle Expansion

3.18 Rocket Motor Sizing Process

- Fig. 3.56 High Propellant Weight Flow Rate Is Driven by High Chamber Pressure and Nozzle Throat Area
- Fig. 3.57 Rocket Motor Chamber Pressure Is Driven by Propellant Burn Area and Nozzle Throat Area
- Fig. 3.58 Conceptual Design Sizing of a Solid Propellant Rocket Motor Is an Iterative Process
- Fig. 3.59 Rocket Motor Pressure, Weight Flow Rate, and Thrust Are Proportional to Propellant Burn Area
- Fig. 3.60 Missile Thrust—Time Requirements Drive Solid Propellant Rocket Grain Cross Section Geometry

3.19 Solid Propellant Rocket Motor Production Alternatives

Fig. 3.61a   Solid Propellant Motor Production Alternatives—Case Bonded Propellant vs Extruded Propellant

Fig. 3.61b   Solid Propellant Motor Production Alternatives—Case Bonded Propellant vs Extruded Propellant (cont)

3.20 Solid Propellant Rocket Thrust Magnitude Control

Fig. 3.62   Thrust Magnitude Control Provides Flexibility of Solid Propellant Propulsion Energy Management

Fig. 3.63   Examples of Solid Propellant Rocket Missiles with Pulse Motor Thrust Magnitude Control

Fig. 3.64   Missile Rocket Motor Thrust Magnitude Control Alternatives and Tradeoffs

3.21 Solid Propellant Alternatives

Fig. 3.65   Solid Propellant Primary Tradeoffs Are Performance, Explosive Safety, Toxicity, and Observables

Fig. 3.66   Efficient Loading of Solid Propellant Particles Prevents Crack Propagation through the Binder

Fig. 3.67   Missile Solid Propellant Binder Tradeoffs Include Elasticity, Aging, Energy Added, and Burn Rate

3.22 Solid Propellant Aging

Fig. 3.68   Solid Propellant Rocket Motor Lifetime May Be Driven by a High Temperature Environment

Fig. 3.69   Solid Propellant Rocket Motor Lifetime May Be Driven by Shock and Strain in a Low Temperature Environment

3.23 Solid Propellant Rocket Combustion Stability

Fig. 3.70a   A Driver of Solid Propellant Rocket Motor Combustion Stability Is Motor Geometry

Fig. 3.70b   A Driver of Solid Propellant Rocket Motor Combustion Stability Is Motor Geometry (cont)

Fig. 3.71   Other Drivers of Rocket Combustion Stability Include Chamber Pressure, Oxidizer Size, and Exhaust Particles

3.24 Rocket Motor Case and Nozzle Material Alternatives

Fig. 3.72   Steel and Aluminum Motor Cases Are Low Cost but a Composite Motor Case Is Light Weight

Fig. 3.73   Heat Transfer Drives Solid Propellant Rocket Nozzle Materials, Weight, and Cost

3.25 Ducted Rocket Design Considerations

Fig. 3.74   Ducted Rocket Propulsion Often Provides Longer Standoff Range Than Rocket Propulsion

Fig. 3.75   Ducted Rocket Design Tradeoffs Include Type of Propellant Fuel and Oxidizer

Fig. 3.76   Ducted Rocket Combustion Temperature Increases with Mach Number and Fuel Flow

Fig. 3.77   Ducted Rocket High Thrust Occurs for High Mach Number, High Combustion Temperature, and High Inlet Flow

Fig. 3.78   Ducted Rocket High Specific Impulse Occurs with a High Heating Value Fuel @ Mach 3 to 4 Flight

Chapter 4 Weight

4.1 Introduction

Fig. 4.1 Chapter 4 Weight -What You Will Learn
Fig. 4.2 A Balanced Missile Design Requires Harmonized Mission Requirements and Measures of Merit
Fig. 4.3 Weight Impacts Missile Range and Maneuverability
Fig. 4.4 A Lightweight Missile Has Payoff
Fig. 4.5 Weights of Missile Subsystems Often Drive Flight Performance

4.2 Missile Weight Prediction

Fig. 4.6 Ballistic Missile Launch Weight Is Driven by Range, Payload Weight, Propellant Weight, and Specific Impulse
Fig. 4.7 Staging Provides Range and Weight Payoff for Long Range Ballistic Missiles
Fig. 4.8 A First-Order Estimate of Missile Weight Can Be Derived from Body Geometry Dimensions
Fig. 4.9 Most Subsystems for Missiles Have a Weight Density of about 0.05 lbm per in^3

4.3 Center-of-Gravity and Moment-of-Inertia Prediction

Fig. 4.10 Modeling Missile Weight, Balance, and Moment-of-Inertia Is Based on a Build-up of Subsystems
Fig. 4.11 Missile Moment-of-Inertia Is Driven by Length, Diameter, and Weight

4.4 Missile Airframe Structure Manufacturing Processes

Fig. 4.12a Examples of Missile Structure Manufacturing Processes
Fig. 4.12b Examples of Missile Structure Manufacturing Processes (cont)
Fig. 4.12c Examples of Missile Structure Manufacturing Processes (cont)
Fig. 4.12d Examples of Missile Structure Manufacturing Processes (cont)
Fig. 4.12e Examples of Missile Structure Manufacturing Processes (cont)
Fig. 4.13 Mechanical Fastener Versus Adhesive Bonding Is a Tradeoff for Structural Joint Attachment
Fig. 4.14 Increasing Metal Hardness Increases Strength, but Machining Is More Difficult and Expensive
Fig. 4.15 Missile Structure Parts Count Is Driven by the Manufacturing Process
Fig. 4.16 Missile Turbojet Engine Parts Are Often Castings (Lower Parts Count and Lower Cost)

4.5 Missile Airframe Material Alternatives

Fig. 4.17 Missile Airframe Material Alternatives Include Aluminum, Steel, Titanium, and Composite
Fig. 4.18 Missile Structure Drivers Include Strength and Toughness
Fig. 4.19 Example of Strength—Elasticity Comparison of Missile Structure Material Alternatives
Fig. 4.20 Laminate Graphite Composite Provides a High Strength-to-Weight Structure
Fig. 4.21 A Hypersonic Missile without External Insulation Requires High Temperature and Heavy Structure

4.6 Missile Structure/Insulation Trades

Fig. 4.22 Missile Structure Insulation Concepts and Trades for Short Duration Flight
Fig. 4.23 External Structure Insulation Has High Payoff for Short Duration Flight at High Mach Number

4.7 High Temperature Insulation Materials

Fig. 4.24   There Are Many Considerations for High Temperature Missile Insulation
Fig. 4.25   Maximum Temperature and Insulation Efficiency Are Drivers for High Temperature Insulation
Fig. 4.26   Required Insulation Thickness Is a Consideration for Integrating External Insulation with Airframe

4.8 Missile Aerodynamic Heating/Thermal Response Prediction

Fig. 4.27   A Thermally Thin Surface Has High Heat Transfer, A Thermally Thick Surface Has Low Heat Transfer
Fig. 4.28a   A "Thermally Thin" Surface in Aero Heating Has Rapid Temperature Rise
Fig. 4.28b   A "Thermally Thin" Surface in Aero Heating Has Rapid Temperature Rise (cont)
Fig. 4.29a   A "Thermally Thick" Surface in Aero Heating Has Large Internal Temperature Gradient
Fig. 4.29b   A "Thermally Thick" Surface in Aero Heating Has Large Internal Temperature Gradient (cont)
Fig. 4.30   Airframe Internal Insulation Temperature and Required Thickness Can Be Predicted Assuming Constant Flux Conduction
Fig. 4.31   External Insulation Greatly Reduces Airframe Structure Temperature in Short Duration Flight
Fig. 4.32   A Sharp Nose Tip or Sharp Leading Edge Has High Aerodynamic Heating in Hypersonic Flight
Fig. 4.33   Missiles Have Relatively Low Radiation Heat Loss at Low Altitude, Moderate Temperature, and Moderate Mach Number

4.9 Localized Aerodynamic Heating and Thermal Stress

Fig. 4.34a   A Missile Design Concern Is Localized Aerodynamic Heating and Thermal Stress
Fig. 4.34b   A Missile Design Concern Is Localized Aerodynamic Heating and Thermal Stress (cont)
Fig. 4.35   Examples of Aerodynamic Hot Spots

4.10 Missile Structure Design

Fig. 4.36   Missile Metal Body Structure Weight Is about 22% of the Missile Launch Weight
Fig. 4.37   Missile Structure Is Based on Considering the Cradle-to-Grave Environment
Fig. 4.38   Missile Body Structure Required Thickness Is Based on Considering Many Design Conditions
Fig. 4.39   Higher Structure Design Factor of Safety Is Required for Hazardous Subsystems and Hazardous Flight Conditions
Fig. 4.40   Localized Buckling May Be a Concern for a Thin Wall Structure
Fig. 4.41   MIL-A-8591 Provides a Conceptual Design Procedure to Estimate Captive Carriage Maximum Flight Load
Fig. 4.42   Maximum Body Bending Moment Depends Upon Load Distribution
Fig. 4.43   Body Bending Moment May Drive Body Structure Thickness and Weight
Fig. 4.44   For a Typical Solid Propellant Rocket Motor, about 71% of the Motor Weight Is Propellant
Fig. 4.45   Maximum Expected Operating Pressure (MEOP) Increases with Propellant Grain Initial Temperature
Fig. 4.46a   Solid Propellant Rocket Motor Case Required Thickness and Weight Is Usually Driven by Internal Pressure
Fig. 4.46b   Solid Propellant Rocket Motor Case Required Thickness and Weight Is Usually Driven by Internal Pressure (cont)
Fig. 4.46c   Solid Propellant Rocket Motor Case Required Thickness and Weight Is Usually Driven by Internal Pressure (cont)

Fig. 4.47	Laminate Graphite Composite Rocket Motor Case for Rocket Baseline Missile Is Lighter Weight	
Fig. 4.48	A Low Aspect Ratio Delta Surface Planform Has Lighter Weight Structure	

4.11 Seeker Dome Alternatives

Fig. 4.49	Multi-mode Seeker Dome Material Is Driven by Transmission and Flight Environment
Fig. 4.50	Infrared Seeker Dome Material Is Driven by IR Transmission and Flight Environment
Fig. 4.51	Radar Seeker Radome Material Is Driven by RF Transmission and Flight Environment
Fig. 4.52	A Driver for Radome Weight Is the Optimum Thickness Required for Efficient Transmission

4.12 Missile Power Supply and Flight Control Actuators

Fig. 4.53	Missile Electrical Power Supply Drivers Include Weight, Environment, and Safety
Fig. 4.54	A Thermal Battery Has Lighter Weight for Short Time of Flight—Generator Has Lighter Weight for Long Flight Time
Fig. 4.55	Missile Power Supply Is Usually a Thermal Battery and Most Batteries Are Located Near Electronics
Fig. 4.56	Most Missiles Use Electromechanical Flight Control Actuators

Chapter 5 Flight Performance

5.1 Introduction

Fig. 5.1	Chapter 5: Introduction -What You Will Learn
Fig. 5.2	A Balanced Missile Design Requires Harmonized Mission Requirements and Measures of Merit
Fig. 5.3	Missile Conceptual Design and System Engineering Requires Broad, Creative, Rapid, and Iterative Evaluations

5.2 Missile Flight Performance Envelope

Fig. 5.4	Flight Performance and Trajectory Are Driven by Forces (Aerodynamics, Propulsion, Weight) on the Missile
Fig. 5.5	Flight Trajectory Is Driven by Aerodynamic, Propulsion, and Gravity Forces
Fig. 5.6	Missile Flight Envelope May Be Characterized by Maximum Range, Minimum Range, and Off Boresight
Fig. 5.7a	Missile Flight Envelope Is Defined by the Missile, Target, Launch Platform, and Fire Control System
Fig. 5.7b	Missile Flight Envelope Is Defined by the Missile, Target, Launch Platform, and Fire Control System (cont)
Fig. 5.7c	Missile Flight Envelope Is Defined by the Missile, Target, Launch Platform, and Fire Control System (cont)
Fig. 5.7d	Missile Flight Envelope Is Defined by the Missile, Target, Launch Platform, and Fire Control System (cont)
Fig. 5.7e	Missile Flight Envelope Is Defined by the Missile, Target, Launch Platform, and Fire Control System (cont)
Fig. 5.7f	Missile Flight Envelope Is Defined by the Missile, Target, Launch Platform, and Fire Control System (cont)
Fig. 5.8	Example Targets That Require Extreme Terminal Flight Trajectory Shaping

5.3 Equations of Motion Modeling

Fig. 5.9   Conceptual Design vs Preliminary Design Models of Flight Trajectory Degrees of Freedom (DOF)

Fig. 5.10   Degree of Freedom (1-DOF) Equation of Motion Has Good Accuracy if the Fly-out Is at Low Angle of Attack

5.4 Driving Parameters for Missile Flight Performance

Fig. 5.11a   3 Degrees of Freedom (3-DOF) Equations of Motion Show Drivers for Missile Configuration Sizing

Fig. 5.11b   3 Degrees of Freedom (3-DOF) Equations of Motion Show Drivers for Missile Configuration Sizing (cont)

5.5 Steady-State Flight and Constant Bearing Intercept

Fig. 5.12   Steady-State Flight Range Is Enhanced by High Lift-to-Drag Ratio

Fig. 5.13   Drivers for Cruise Range Are Velocity, Specific Impulse, Lift-to-Drag Ratio, and Weight Fraction of Fuel and Propellant

Fig. 5.14   Steady-State Glide Range Is Driven by Initial Glide Altitude and Lift-to-Drag Ratio

Fig. 5.15   A Constant Bearing Intercept Has Constant Lead Angle

5.6 Boost, Glide, Coast, Ballistic, and Divert Flight

Fig. 5.16   Incremental Boost Velocity Is Driven by Propellant Weight, Launch Weight, Specific Impulse, and Drag

Fig. 5.17   Lofted Boost-Glide Range Is Driven by Initial Glide Velocity, Initial Glide Altitude, and Lift-to-Drag Ratio

Fig. 5.18   Lofted Boost-Glide Velocity Decay Is Driven by Initial Glide Velocity, Initial Glide Altitude, Range, and Lift-to-Drag Ratio

Fig. 5.19   Flight Trajectory Shaping Provides Extended Range for High Performance Missiles

Fig. 5.20   Boost-Glide Trajectory May Have Larger Footprint Capability and Less Exposure to Threat Radar

Fig. 5.21   1 Degree of Freedom Coast Range Is Driven by Initial Velocity, Altitude, Drag, and Weight

Fig. 5.22   2 Degrees of Freedom Coast Prediction Is More Accurate than 1-DOF, but Requires Numerical Solution

Fig. 5.23a   Short Range Ballistic Flight Drivers Include Burnout Velocity, Burnout Angle, Burnout Altitude, and Drag

Fig. 5.23b   Short Range Ballistic Flight Drivers Include Burnout Velocity, Burnout Angle, Burnout Altitude, and Drag (cont)

Fig. 5.24   Drivers of Long-Range Ballistic Flight Are Burnout Velocity and Burnout Attitude Angle

Fig. 5.25   Interceptor Divert with Thrusters is Driven by Propellant Weight, Interceptor Weight, Axial Velocity, and Seeker Lock-on Range

5.7 Turn Performance

Fig. 5.26   Small Turn Radius with Aero Control Requires High Normal Force Coefficient, Light Weight, and Low Altitude

Fig. 5.27   High Turn Rate with Aero Control Requires High Normal Force, High Velocity, Low Altitude, and Light Weight

Fig. 5.28   Small Turn Radius with TVC Requires High Thrust-to-Weight Ratio, High Angle of Attack, and Low Velocity

Fig. 5.29   High Turn Rate with TVC Requires High Thrust-to-Weight Ratio, High Angle of Attack, and Low Velocity

Fig. 5.30   Large or Rapid Heading Change with TVC Requires High Thrust-to-Weight, High Angle of Attack, and Low Velocity

Chapter 6 Other Measures of Merit

6.1 Introduction

Fig. 6.1   Chapter 6 Other Measures of Merit—What You Will Learn

Fig. 6.2   Missile Conceptual Design and System Engineering Requires Broad, Creative, Rapid, and Iterative Evaluations

6.2 Robustness

Fig. 6.3   A Balanced Missile Design Requires Harmonized Mission Requirements and Measures of Merit

Fig. 6.4   Missiles Should Have Robust Capability

Fig. 6.5   Seeker Robustness Considerations Include Weather, Autonomy, Range, Observables, and Countermeasures

Fig. 6.6   Radar Seekers Have Longer Range in Adverse Weather, Infrared Seekers Have Finer Resolution

Fig. 6.7   Cloud Cover Is Pervasive

Fig. 6.8   A Radar Seeker and Sensor Has an Advantage of More Robust Operation Through Clouds

Fig. 6.9   A Consideration in Missile Seeker and Sensor Selection Is the Probability of Precipitation

Fig. 6.10   Aerosols Drive Seeker and Sensor Atmospheric Attenuation if Aerosol Dimension Is Comparable to Wavelength

Fig. 6.11   Threshold Contrast Range of a Passive Electro-Optical (EO) Seeker Is Typically Comparable to Human Sight Range

Fig. 6.12   An Imaging Seeker Has Enhanced Accuracy, Target Acquisition, and Target Discrimination

Fig. 6.13   An Infrared Seeker Focuses Infrared Energy on a Detector

Fig. 6.14   Imaging Infrared Seeker Trades Include Staring Focal Plane Array vs. Scanning Array vs. Pseudo Image

Fig. 6.15   A Large Diameter Infrared Seeker Has Longer Detection Range and Better Resolution

Fig. 6.16   Exo-atmospheric Hit-to-Kill Missiles Use High Resolution Imaging Infrared Seekers

Fig. 6.17   A Long Wave Infrared Seeker Has Longer Detection Range Against a Cold Target

Fig. 6.18   Most Cold Target Energy Is at Long Wavelength—Most Warm Target Energy at Shorter Wavelength

Fig. 6.19   ATR with an Imaging Seeker Requires Signal-to-Noise, Signal-to-Clutter, and Pixels on Target

Fig. 6.20   Semi-active Laser Seeker with Quadrant Detector Has High Accuracy for a Small Laser Spot

Fig. 6.21   A Semi-active Laser Seeker with Large Aperture Has Longer Detection Range

Fig. 6.22   Radar Seekers Are Typically Either Pulse Doppler or Continuous Wave

Fig. 6.23   A Pulse Active Radar Seeker Transmits and Receives Reflected Energy from the Target

Fig. 6.24   An Active Radar Seeker with Large Diameter Has Longer Detection Range and Better Resolution

Fig. 6.25   A Radar Seeker with Polarization Can Provide Automatic Target Recognition (ATR)

Fig. 6.26   A Semi-active Radar Seeker with Large Diameter Has Longer Detection Range

Fig. 6.27   Typical Active Radar Seeker Benefit Is Launch Platform Survivability, Semi-Active Radar Seeker Is Lower Cost

Fig. 6.28	Directional Antenna Advantages Are Gain, Range, Resolution, and Less Sensitive to Countermeasures
Fig. 6.29	Millimeter Wave Seeker Has Shorter Wavelength
Fig. 6.30	A mmW Seeker Has Better Resolution, but a cmW Seeker Typically Has Higher Power and Longer Range
Fig. 6.31	Examples of Millimeter Wave Seekers
Fig. 6.32	Active RF Seeker Transmitter Options Are Traveling Wave Tube Versus Solid State Power Amplifier
Fig. 6.33a	Most Seekers Have a Gimbaled Platform
Fig. 6.33b	Most Seekers Have a Gimbaled Platform (cont)
Fig. 6.33c	Most Seekers Have a Gimbaled Platform (cont)
Fig. 6.34	Examples of Strapdown Seekers
Fig. 6.35	An Imaging Infrared (IIR) Seeker Tradeoff Is Gimbaled Stabilization vs Strapdown Electronic Stabilization
Fig. 6.36	Field of Regard of a Gimbaled Imaging IR Seeker Is Usually Much Larger than Field of View of a Strapdown Seeker
Fig. 6.37	Seekers Are Improved by Global Positioning System (GPS), Inertial Navigation System (INS), and Data Link Sensors
Fig. 6.38	Tactical Inertial Navigation System Drivers Are Cost & Weight, Strategic INS Drivers Are Accuracy & Autonomy
Fig. 6.39	Most US Missile Midcourse Guidance Is Provided by Either Inertial Navigation System (INS) or GPS-INS
Fig. 6.40	Waypoint Guidance Can Provide Midcourse Guidance Updates from Stored Terrain Features
Fig. 6.41	A GPS-INS Based on Strapdown MEMS Gyros Is Low Cost and Light Weight, with Good Accuracy
Fig. 6.42	Drivers for Strapdown INS Accuracy without GPS Are Accelerometer Misalignment, Gyro Bias, and Time
Fig. 6.43	GPS-INS Allows Robust Seeker Lock-on in Adverse Weather and Clutter
Fig. 6.44	Global Positioning System with Inertial Navigation System Allows Precision, Extended Range, and Vertical Impact
Fig. 6.45	A Global Positioning System Guided Short Range Weapon Requires Fast (~ 5s) GPS Acquisition
Fig. 6.46	Pseudolites Provide Resistance to Global Positioning System (GPS) Jammers
Fig. 6.47	USA, Russia, European Union, China, Japan, and India Have Navigation Satellite Systems
Fig. 6.48	A High Bandwidth Data Link Reduces Moving Target Location Error
Fig. 6.49	Military Radar Frequency (RF) Spectrum Requirements Must Compete with Expanding Commercial Requirements
Fig. 6.50	Modern Precision Strike Missiles Often Combine Seeker, Inertial Navigation, Global Positioning System, and Data Link Guidance
Fig. 6.51	Optimum Cruise and Glide Are a Function of Mach Number, Type Propulsion, Altitude, and Planform
Fig. 6.52	Missile Guidance and Control Must Be Robust for Changing Events, Environment, and Uncertainty
Fig. 6.53	Design Robustness Requires Consideration of Large Variation of Atmospheric Properties with Altitude
Fig. 6.54	Design Robustness Requires Consideration of Standard Atmospheric Modeling Differences
Fig. 6.55	Design Robustness Requires Consideration of Uncertainty
Fig. 6.56	Missile Robustness Requires Consideration of Countermeasures and Counter-Countermeasures
Fig. 6.57	Examples of Countermeasure Resistant Seekers
Fig. 6.58	Electromagnetic Compatibility (EMC) Considers EMI, ESD, and EMP Hardening

6.3 Lethality

Fig. 6.59	Target Size, Hardness, and Collateral Drive Missile Warhead Design and Technology
Fig. 6.60	76% of Baghdad Targets Struck First Night of Desert Storm were Time Critical Targets
Fig. 6.61	Type of Target Drives Precision Strike Missile Size, Weight, Speed, Cost, Seeker, and Warhead
Fig. 6.62	Lightweight Multi-Purpose Precision Strike Weapons Are Based on Many Tradeoffs
Fig. 6.63	Unmanned Combat Air Vehicles (UCAVs) Require Lightweight Weapons and Sophisticated Avionics
Fig. 6.64	A Typical Probability of Kill Criteria for a Blast Fragmentation Warhead Is Lethal Radius Greater Than 3 Sigma Miss Distance
Fig. 6.65	Warhead Blast Kill Requires Small Miss Distance
Fig. 6.66	Excess Fuel in Thermobaric (Metal Augmented Charge) Warhead Provides Secondary Blast Combustion
Fig. 6.67	The Primary Kill Mechanism of a Typical Blast Frag Cylindrical Warhead Is the Warhead Fragments
Fig. 6.68	A High Warhead Total Kinetic Energy Requires Charge Weight Be Approximately Equal to Fragments Weight
Fig. 6.69	Warhead High Fragment Velocity Requires High Charge-to-Metal Ratio
Fig. 6.70	Warhead Fragment Kinetic Energy Usually Has a Larger Lethal Radius Than Blast Overpressure
Fig. 6.71	A Blast Fragment Warhead with a Small Spray Angle Has a Larger Lethal Radius
Fig. 6.72	Accurate Guidance Provides Higher Lethality, Lighter Warhead Weight, and Lower Collateral Damage
Fig. 6.73	Small Miss Distance Improves the Number of Warhead Fragment Hits
Fig. 6.74	Small Miss Distance Improves Warhead Fragment Penetration
Fig. 6.75	Blast Frag Warhead Lethality Drivers—Total Fragment Impacts, Target Vulnerable Area, and Target Size
Fig. 6.76	A Directional Warhead Has High Kinetic Energy Density but It Also Has Fuzing and Guidance & Control Risk
Fig. 6.77	Desired Kinetic Energy Directional Distribution Drives the Type of Warhead Design
Fig. 6.78	Hypersonic Hit-to-Kill Enhances Energy on Target, Especially for a Missile with a Small Warhead
Fig. 6.79	Defeating a Deeply Buried Target Requires High Speed Impact, Kinetic Energy Warhead, and Smart Fuzing
Fig. 6.80	Kinetic Energy Penetrator Weight, Density, Length, and Velocity Increase Target Penetration
Fig. 6.81	Kinetic-Kill Targets Include Ballistic Missiles, Aircraft, Projectiles, Buried Targets, and Armored Vehicles
Fig. 6.82	A Kinetic Kill Missile Requires Hit-to-Kill Accuracy
Fig. 6.83	Shaped Charge Warhead Jet Penetration Is Driven by Diameter, Penetrator Density, and Target Density
Fig. 6.84	Shaped Charge Warhead Liner Design Geometry Has Many Trade-offs
Fig. 6.85	Shaped Charge Warhead Liner Material Has Many Trade-offs
Fig. 6.86	Shaped Charge Warhead with High Ductility Requires a Pure Metallic Crystal Lattice Liner without Voids
Fig. 6.87	Optimum Detonation Standoff Distance for a Shaped Charge Warhead is Typically About 4 Charge Diameters from the Target
Fig. 6.88	There Are Multiple Approaches to Minimize Collateral Damage
Fig. 6.89	Low Density Warhead Case Material Has Rapid Deceleration of Fragments, Resulting in Lower Collateral Damage
Fig. 6.90	A Lightweight Warhead with Precision Guidance Accuracy Has Lower Collateral Damage
Fig. 6.91	A Proximity Fuze Selects the Standoff Distance at Warhead Detonation
Fig. 6.92	A Proximity Fuze for a Blast-Frag Warhead Against an Air Target Requires Selecting Fuze Angle and Time Delay

Fig. 6.93a   Air Target Proximity Fuze Minimum Angle Is Driven by Maximum Closing Velocity and Warhead Fragments Velocity

Fig. 6.93b   Air Target Proximity Fuze Minimum Angle Is Driven by Maximum Closing Velocity and Warhead Fragments Velocity (cont)

Fig. 6.94a   Air Target Proximity Fuze Time Delay Is Driven by Fuze Angle, Miss Distance, Closing Velocity, and Target Length

Fig. 6.94b   Air Target Proximity Fuze Time Delay Is Driven by Fuze Angle, Miss Distance, Closing Velocity, and Target Length (cont)

6.4 Miss Distance

Fig. 6.95   CEP Is Approximately Equal to Miss Distance

Fig. 6.96   Guided Advantages - Accuracy and Low Collateral Damage, Unguided Advantages - Low Cost, CCM, R&M

Fig. 6.97   Seeker Proportional Homing Is the Most Common Terminal Guidance Law Against High-Speed Targets

Fig. 6.98   Missile Guidance & Control Fundamentals Can Be Illustrated with a G&C Block Diagram

Fig. 6.99   Example of Gimbaled Seeker Simplified Block Diagram

Fig. 6.100   Proportional Guidance Provides a Constant Bearing Flight Path

Fig. 6.101   A Maneuvering Target and An Initial Heading Error Result in Missile Miss Distance

Fig. 6.102   Effective Navigation Ratio Is a Function of Missile Velocity, Target Velocity, and Engagement Geometry

Fig. 6.103   Proportional Guidance with High Navigation Ratio Quickly Approaches Constant Bearing Trajectory

Fig. 6.104   Terminal Homing Guidance Law Tradeoffs Include Robustness, Required Sensors, Required Target Data, and Accuracy

Fig. 6.105a   Missile Guidance Time Constant Is an Indication of Maneuver Response Time

Fig. 6.105b   Missile Guidance Time Constant Is an Indication of Maneuver Response Time (cont)

Fig. 6.106   Missile Maneuver Time Constant Is Often Driven by Flight Control Effectiveness

Fig. 6.107   Missile Time Constant Contributor from Flight Control Dynamics Is Usually Driven by Actuator Dynamics

Fig. 6.108   Drivers for Radome Time Constant Include Radome Error Slope and Closing Velocity

Fig. 6.109   High Initial Maneuverability Is Required to Eliminate Heading Error

Fig. 6.110a   Maximum Miss Distance from Heading Error Is Driven by Time to Intercept, Navigation Ratio, and Time Constant

Fig. 6.110b   Maximum Miss Distance from Heading Error Driven by Time to Intercept, Navigation Ratio, and Time Constant (cont)

Fig. 6.111   Required Missile Maneuverability for Maneuvering Target Is about 3x the Target Maneuverability

Fig. 6.112   A Target Step Maneuver Requires 6 to 10 Time Constants to Settle Out Miss Distance

Fig. 6.113   A Small Time Constant Is Required for Small Miss Distance Against a High Maneuvering Target

Fig. 6.114   An Aero Control Missile Has Smaller Miss Distance at Low Altitude and High Dynamic Pressure

Fig. 6.115   Weaving, Jinking, and Cork-Screw Maneuvering Targets Require Large Navigation Ratio for Small Miss Distance

Fig. 6.116   Target Flight Trajectory Dynamics Result in Missile Miss Distance

Fig. 6.117   Radar Glint Error Occurs from the Angular Flashes of Target Scatter Centers

Fig. 6.118   Glint Miss Distance Is Driven by Seeker Resolution, Missile Time Constant, Navigation Ratio, and Target Size

Fig. 6.119   Minimum Miss Distance Requires Optimum Time Constant and Optimum Navigation Ratio

6.5 Carriage and Launch Observables

Fig. 6.120	Internal Weapon Carriage Is Required for a Low Radar Cross Section Aircraft Launch Platform
Fig. 6.121	Aircraft Center Bay Is Best for Missile Ejection—Aircraft Side Bay Is Best for Missile Rail Launch
Fig. 6.122	Minimum Smoke Propellant Reduces Launch Observables

6.6 Missile Survivability and Safety

Fig. 6.123	Options for Survivability Include Stealth, Altitude, Speed, Threat Avoidance, Terrain Masking, and Maneuverability
Fig. 6.124	Long Range Strike Missiles Use Speed, Altitude, Maneuverability, and RCS for Survivability
Fig. 6.125	High Altitude Flight and Low Radar Cross Section Enhance Survivability by Reducing Detection Range
Fig. 6.126	There Are Many Contributors to Frontal Radar Cross Section
Fig. 6.127	There Are Many Geometry Contributors to Radar Cross Section
Fig. 6.128	Conceptual Design Radar Cross Section May Be Computed from Simple Shapes Scattering
Fig. 6.129	Typical Assumptions for Conceptual Design Prediction of Radar Cross Section
Fig. 6.130	Pareto of Ramjet Baseline Missile Frontal RCS Shows Most of the Frontal RCS Is from Seeker Antenna
Fig. 6.131	Evaluate RCS Reduction from Shape and Orientation before Considering Radar Absorbing Material
Fig. 6.132	Missile Radar Absorbing Material (RAM) Is Usually Sized for a High Frequency Radar Threat
Fig. 6.133	Examples of Radar Absorbing Material (RAM) and Radar Absorbing Structure (RAS) Alternatives
Fig. 6.134	Example of Radar Absorbing Structure (RAS)
Fig. 6.135	Dallenbach Magnetic RAM Is Popular for Tactical Missiles Because It Has Smaller Thickness
Fig. 6.136	Dallenbach Radar Absorbing Material Thickness Is Driven by Radar Frequency, Dielectric Constant, and Permeability
Fig. 6.137	Flat Disk Seeker Antenna Has Large RCS at Near Normal Incidence and Low RCS at Off Boresight
Fig. 6.138	A Rectangular Flat Surface Has Large RCS at a Normal Incidence Angle and Low RCS at an Inclination Angle
Fig. 6.139	A High Fineness Metallic Nose or a High Fineness Conformal Antenna Has a Low Frontal RCS
Fig. 6.140	Example of Radar Cross Section Reduction for Ramjet Baseline Missile
Fig. 6.141	Reduced RCS Reduces Radar Detection Range
Fig. 6.142	There Are Many Contributors to Infrared Signature
Fig. 6.143	A High-Speed Missile Has High Infrared Signature
Fig. 6.144	Infrared (IR) Detection Range Is Reduced by Atmospheric Attenuation, Especially at Low Altitude
Fig. 6.145	IR Detection Range of Mach 4 Ramjet Baseline Missile Can Be Larger Than RCS Detection Range
Fig. 6.146a	Signature Tests Requirements Are Driven by Effectiveness and Survivability Requirements
Fig. 6.146b	Signature Tests Requirements Are Driven by Effectiveness and Survivability Requirements (cont)
Fig. 6.147	Short Detection Range and High Speed Reduce Threat Exposure Time, Providing Enhanced Survivability
Fig. 6.148	Mission Planning with Threat Avoidance and Reduced Observables Provide Enhanced Survivability

Fig. 6.149   Mission Planning with Threat Avoidance and High Speed Provide Enhanced Survivability
Fig. 6.150   Low Altitude Flight "Under the Radar" Reduces Detection Range
Fig. 6.151   Insensitive Munitions Improve Launch Platform Safety and Survivability

6.7 Reliability

Fig. 6.152   Reliability Is Provided by Few Events, Reliable Parts, Few Parts, Short Flight Time, and Benign Flight
Fig. 6.153   Few Parts and Reliable Parts Provide Higher System Reliability
Fig. 6.154   Example of a Multi-Event Weapon That Requires High Event Reliability for Good System Reliability
Fig. 6.155   Missile Failures Can Be Characterized by Infant Mortality, Environment Overload, and Reliability
Fig. 6.156   Example of Infant Mortality Failure Design Error—Trident II D-5 TVC Failure from Water Plume Load
Fig. 6.157   An Example of Higher Reliability and Safety Technology Is Electronic Safe and Arm

6.8 Cost

Fig. 6.158   Missile Cost Has a Life Cycle
Fig. 6.159   Missile Engineering and Manufacturing Development (EMD) Cost Is Driven by Many Factors
Fig. 6.160   Missile Production Cost Is Driven by Many Factors
Fig. 6.161   Learning Curve and Total Production Are Drivers for Reducing Missile Unit Production Cost
Fig. 6.162   Sensors, Electronics, and Propulsion Subsystems Typically Drive Missile Unit Production Cost
Fig. 6.163   Missiles with a Small Weight Fraction of Sensors and Electronics Are Usually Less Expensive
Fig. 6.164   Missile Production Culture Is Driven by Rate Production of Sensors and Electronics
Fig. 6.165   Missile Electronics Are Usually Based Upon Commercial-Off-The-Shelf (COTS) Electronics
Fig. 6.166   A Low Parts Count Missile Has a Lower Unit Production Cost
Fig. 6.167   Missile Logistics Emphasis Is Different in War and Peace
Fig. 6.168   Storage in a Protected Container Reduces Aging
Fig. 6.169   Logistics Cost Is Lower for a Simple Missile System
Fig. 6.170   Logistics Is Usually Simpler for a Lightweight Missile
Fig. 6.171   Small MEMS Sensors Provide Logistics Health Monitoring, Reduce Cost, and Reduce Weight

6.9 Launch Platform Integration

Fig. 6.172   Missile Carriage Size, Shape, and Weight Limits Are Often Driven by Launch Platform Compatibility
Fig. 6.173   Lightweight Missiles Provide Enhanced Firepower
Fig. 6.174   Example of Lightweight, Small Size, and High Firepower Missile—Precision Strike Brimstone
Fig. 6.175   Missile Electromagnetic Compatibility (EMC) Is a Requirement for Launch Platform Integration
Fig. 6.176   Missile - Launch Platform Problems Include Limited Off Boresight, Minimum Range, and Safety
Fig. 6.177a   Examples of Missile—Launch Platform Integration Problems
Fig. 6.177b   Examples of Missile—Launch Platform Integration Problems (cont)
Fig. 6.177c   Examples of Missile—Launch Platform Integration Problems (cont)

Fig. 6.177d	Example of Missile—Launch Platform Integration Problems (cont)
Fig. 6.177e	Examples of Missile—Launch Platform Integration Problems (cont)
Fig. 6.177f	Examples of Missile—Launch Platform Integration Problems (cont)
Fig. 6.177g	Example of Missile—Launch Platform Integration Problems (cont)
Fig. 6.177h	Example of Missile—Launch Platform Integration Problems (cont)
Fig. 6.177i	Examples of Missile—Launch Platform Integration Problems (cont)
Fig. 6.177j	Example of Missile—Launch Platform Integration Problems (cont)
Fig. 6.177k	Example Missile—Launch Platform Integration Problems (cont)
Fig. 6.177l	Examples of Missile—Launch Platform Integration Problems (cont)
Fig. 6.177m	Examples of Missile—Launch Platform Integration Problems (cont)
Fig. 6.177n	Examples of Missile—Launch Platform Integration Problems (cont)
Fig. 6.177o	Examples of Missile—Launch Platform Integration Problems (cont)
Fig. 6.177p	Examples of Missile—Launch Platform Integration Problems (cont)
Fig. 6.177q	Examples of Missile—Launch Platform Integration Problems (cont)
Fig. 6.177r	Example of Missile—Launch Platform Integration Problems (cont)
Fig. 6.177s	Examples of Missile—Launch Platform Integration Problems (cont)
Fig. 6.177t	Examples of Missile—Launch Platform Integration Problems (cont)
Fig. 6.178	Examples of Aircraft Rail Launched and Ejection Launched Missiles
Fig. 6.179	MIL-STD-8591 Imposes Aircraft Store Suspension and Ejection Launcher Requirements
Fig. 6.180	Examples of Safe Store Separation
Fig. 6.181	Smart Launchers May Replace Pyrotechnic Ejection Launcher for Aircraft with Light-to-Medium Stores
Fig. 6.182	MIL-STD-8591 Addresses Aircraft Store Rail Launchers
Fig. 6.183	There Are Many Types of Store Separation Wind Tunnel Tests
Fig. 6.184	Compressed Carriage Missiles Provide Higher Firepower
Fig. 6.185a	Examples of Missile Carriage and Fire Control Interface
Fig. 6.185b	Examples of Missile Carriage and Fire Control Interface (cont)
Fig. 6.185c	Examples of Missile Carriage and Fire Control Interface (cont)
Fig. 6.186	Network Centric Warfare Display Provides Combat Situational Awareness
Fig. 6.187a	Fire Control System Provides Weapon Selection, Release, and Firing
Fig. 6.187b	Fire Control System Provides Weapon Selection, Release, Firing (cont)
Fig. 6.188	An Aircraft Control Stick and Cockpit Display Provide Weapon Selection, Release, and Firing
Fig. 6.189	A Helmet Mounted Sight Provides a Head-up Display of Missile Parameters
Fig. 6.190	Missile Terminal Guidance Alternatives Are Autonomous, Semi-Active, and Command Guidance
Fig. 6.191	Missile Terminal Guidance Drivers Include Launch Platform Survivability, Missile Sensor Cost, and Accuracy
Fig. 6.192	A Command Guidance Fire Control System Tracks the Missile, Tracks Target, and Command Guides Missile
Fig. 6.193	A Semi-Active Fire Control System Tracks the Target and Illuminates the Target for the Missile Seeker
Fig. 6.194	Multi-mode Guidance Example—An Active Radar Seeker Augmented with an Inertial Navigation System and a Data Link
Fig. 6.195	A Phased Array Radar Fire Control System Has Higher Search Rate, Greater Coverage, and Higher Reliability
Fig. 6.196	An Example of Ship Carriage and Launcher Is the US Mk41 Vertical Launch System (VLS)
Fig. 6.197	Submarine Carriage and Launchers Have Vertical and Horizontal Launch Tubes
Fig. 6.198	Missile and Launch Platform Climatic Environment Is Typically Based on the 1% Probability Extreme
Fig. 6.199	Missile Environment Cooling Time Drivers Include Temperature, Dimensions, Geometry, and Diffusivity
Fig. 6.200	Missile Exposure to Solar Input Requires Design for Higher Than Ambient Temperature

Chapter 7 Missile Sizing Examples and Sizing Tools

7.1 Introduction

Fig. 7.1	Chapter 7 Missile Sizing Examples and Sizing Tools - What You Will Learn
Fig. 7.2	Missile Sizing Requires Harmonizing Aero, Propulsion, Weight, and Trajectory to Meet Flight Performance

7.2 Rocket Baseline Missile

Fig. 7.3	Rocket Baseline Missile Is AIM-7 Sparrow
Fig. 7.4	Rocket Baseline Missile Provides Beyond-Visual-Range (BVR) Intercept
Fig. 7.5	Target Contrast and Size Drive Visual Detection and Recognition Range
Fig. 7.6	Target Contrast, Low Texture, Color, Scintillation, and Motion Increase Visual Detection in Clutter
Fig. 7.7	High Missile Velocity Improves Air Intercept Standoff Range
Fig. 7.8	The Requirement for Missile Flight Range Is Greater for a Tail Chase Intercept
Fig. 7.9	Drawing of Rocket Baseline Missile Configuration
Fig. 7.10	Mass Properties of Rocket Baseline Missile
Fig. 7.11a	Rocket Baseline Missile Definition
Fig. 7.11b	Rocket Baseline Missile Definition (cont)
Fig. 7.11c	Rocket Baseline Missile Definition (cont)
Fig. 7.11d	Rocket Baseline Missile Definition (cont)
Fig. 7.12	Rocket Baseline Missile Has a Boost—Sustain Thrust Time History
Fig. 7.13a	Rocket Baseline Missile Aerodynamic Characteristics
Fig. 7.13b	Rocket Baseline Missile Aerodynamic Characteristics (cont)
Fig. 7.14	High Altitude Launch Enhances the Rocket Baseline Missile Flight Range
Fig. 7.15	Low Altitude Maneuvers Have Smaller Turn Radius for Rocket Baseline Missile
Fig. 7.16	Rocket Baseline Missile Range Is Driven by Specific Impulse, Propellant Weight, Drag, and Static Margin
Fig. 7.17	Rocket Baseline Missile Flight Has Boost, Sustain, and Coast Axial Acceleration
Fig. 7.18	The Rocket Baseline Missile Has a Boost—Sustain—Coast Velocity Profile
Fig. 7.19	Flight Range and Time-to-Target of the Rocket Baseline Missile Satisfy Assumed Requirements
Fig. 7.20	Maneuverability Measures of Merit for Rocket Baseline Missile Against a Maneuvering Target
Fig. 7.21	Rocket Baseline Missile Wing Area Is a Driver for Normal Acceleration Maneuverability
Fig. 7.22	Rocket Baseline Missile Wing Area Is a Driver for Turn Radius Maneuverability
Fig. 7.23	Rocket Baseline Missile Wing Area Is a Driver for Turn Rate Maneuverability
Fig. 7.24	A Lofted Boost—Glide Provides Extended Range for the Rocket Baseline Missile

7.3 Ramjet Baseline Missile

Fig. 7.25	The Ramjet Baseline Missile Is a Chin Inlet Integral Rocket Ramjet (IRR)
Fig. 7.26	Ramjet Baseline Missile Has an Integral Rocket Booster - Ramjet Combustor
Fig. 7.27	Mass Properties of Ramjet Baseline Missile
Fig. 7.28a	Ramjet Baseline Missile Definition
Fig. 7.28b	Ramjet Baseline Missile Definition (cont)
Fig. 7.29	Engine Nomenclature and Flow Path Geometry for Ramjet Baseline
Fig. 7.30a	Aerodynamic Characteristics of Ramjet Baseline Missile
Fig. 7.30b	Aerodynamic Characteristics of Ramjet Baseline Missile (cont)
Fig. 7.30c	Aerodynamic Characteristics of Ramjet Baseline Missile (cont)
Fig. 7.31	Maximum Efficiency of the Ramjet Baseline Missile Inlet Occurs at About Mach 3.5

Fig. 7.32	Maximum Thrust of the Ramjet Baseline Missile Occurs at About Mach 3.5
Fig. 7.33	Maximum Specific Impulse of the Ramjet Baseline Missile Occurs at About Mach 3.5
Fig. 7.34	Rocket Booster of Ramjet Baseline Missile Provides a Boost Greater Than Mach 2.2
Fig. 7.35	Ramjet Baseline Missile Has Best Performance at High Altitude Cruise
Fig. 7.36	Pareto Analysis Shows Ramjet Baseline Missile Range Is Driven by Specific Impulse, Fuel Weight, Thrust, and Drag
Fig. 7.37	Ramjet Baseline Missile Has Ground, System Integration, System Engineering, and Flight Test Demos
Fig. 7.38	Ramjet Baseline Missile Flight Range Uncertainty Is Approximately 7%, 1 Sigma
Fig. 7.39	High Density Fuel and Efficient Packaging Provide Extended Range for Ramjet Baseline Missile
Fig. 7.40	Ramjet Baseline Missile Achieves Velocity Control Through Fuel Flow Rate Control

7.4 Turbojet Baseline Missile

Fig. 7.41	Turbojet Baseline Missile Has Ship, Submarine, Aircraft, and Ground Vehicle Launch Platforms
Fig. 7.42	Most Anti-Ship Missiles Have Air-Breathing Propulsion, Radar Seeker, Inertial Navigation System, and Satellite Navigation
Fig. 7.43	Example of a Potential Conflict Area that Drives Anti-Ship Missile Requirements is the South China Sea
Fig. 7.44	Drawing of Turbojet Baseline Missile Configuration and Subsystems
Fig. 7.45	Turbojet Baseline Missile Inlet Has Geometry Constraints
Fig. 7.46	Mass Properties of Turbojet Baseline Missile
Fig. 7.47a	Turbojet Baseline Missile Geometry Definition
Fig. 7.47b	Turbojet Baseline Missile Geometry Definition (cont)
Fig. 7.48a	Aerodynamic Characteristics of Turbojet Baseline Missile
Fig. 7.48b	Aerodynamic Characteristics of Turbojet Baseline Missile (cont)
Fig. 7.48c	Aerodynamic Characteristics of Turbojet Baseline Missile (cont)
Fig. 7.48d	Aerodynamic Characteristics of Turbojet Baseline Missile (cont)
Fig. 7.49	Turbojet Baseline Missile Engine Is the Teledyne J402-CA-400
Fig. 7.50	Turbojet Baseline Missile Engine Performance—Teledyne J402-CA-400
Fig. 7.51	Turbojet Baseline Missile Cruise Performance
Fig. 7.52	Turbojet Baseline Missile Range Drivers Are Specific Impulse, Fuel Weight, and Zero-Lift Drag Coefficient
Fig. 7.53	Turbojet Baseline Missile Thrust Is a Function of Mach Number, Altitude, and Combustion Temperature
Fig. 7.54	Turbojet Baseline Missile Specific Impulse Is a Function of Mach Number, Altitude, and Combustion Temperature
Fig. 7.55a	Process for Turbojet Baseline Missile Cruise Range Prediction
Fig. 7.55b	Process for Turbojet Baseline Missile Cruise Range Prediction (cont)
Fig. 7.55c	Process for Turbojet Baseline Missile Cruise Range Prediction (cont)
Fig. 7.55d	Process for Turbojet Baseline Missile Cruise Range Prediction (cont)
Fig. 7.56a	Process for Turbojet Baseline Missile Maximum Range Prediction
Fig. 7.56b	Process for Turbojet Baseline Missile Maximum Range Prediction (cont)
Fig. 7.56c	Process for Turbojet Baseline Missile Maximum Range Prediction (cont)
Fig. 7.56d	Process for Turbojet Baseline Missile Maximum Range Prediction (cont)
Fig. 7.57	Turbojet Baseline Missile Centrifugal Compressor Has Nearly Constant Rotational Speed
Fig. 7.58	For Surface Launch, Turbojet Baseline Missile Has a Drop-Off Booster
Fig. 7.59	Booster for Turbojet Baseline Missile Provides Approximately 745 fps Boost Velocity

7.5 Baseline Guided Bomb

Fig. 7.60	Baseline Guided Bomb (GBU-31 JDAM) Has Many Fighter and Bomber Launch Platforms
Fig. 7.61	Drawing of Baseline Guided Bomb Configuration
Fig. 7.62	Mass Properties of Baseline Guided Bomb
Fig. 7.63	Baseline Guided Bomb Definition
Fig. 7.64a	Aerodynamic Characteristics of Baseline Guided Bomb
Fig. 7.64b	Aerodynamic Characteristics of Baseline Guided Bomb (cont)
Fig. 7.64c	Aerodynamic Characteristics of Baseline Guided Bomb (cont)
Fig. 7.64d	Aerodynamic Characteristics of Baseline Guided Bomb (cont)
Fig. 7.64e	Aerodynamic Characteristics of Baseline Guided Bomb (cont)
Fig. 7.64f	Aerodynamic Characteristics of Baseline Guided Bomb (cont)
Fig. 7.64g	Aerodynamic Characteristics of Baseline Guided Bomb (cont)
Fig. 7.65	Baseline Guided Bomb Ballistic Flight Range Is Driven by Launch (Initial) Velocity and Altitude
Fig. 7.66	Baseline Guided Bomb Ballistic Flight Range Is Enhanced by High-Speed and High-Altitude Release
Fig. 7.67	Baseline Guided Bomb Glide Range Is Driven by Lift-to-Drag Ratio and Initial Glide Altitude

7.6 Computer Aided Conceptual Design Sizing Tools

Fig. 7.68	ADAM Is a Missile Conceptual Design Sizing Code
Fig. 7.69a	Tactical Missile Design (TMD) Spreadsheet Is a Missile Conceptual Design Sizing Code
Fig. 7.69b	Tactical Missile Design (TMD) Spreadsheet Is a Missile Conceptual Design Sizing Code (cont)
Fig. 7.70	TMD Spreadsheet Verification Is Based on Comparing with Alternative Prediction and Test Data

7.7 Soda Straw Rocket (DBF, Pareto, Uncertainty Analysis, HOQ, DOE)

Fig. 7.71a	Soda Straw Rocket Customer Requirements for Design, Build, and Fly
Fig. 7.71b	Soda Straw Rocket Customer Requirements for Design, Build, and Fly (cont)
Fig. 7.71c	Soda Straw Rocket Customer Requirements for Design, Build, and Fly (cont)
Fig. 7.72	Soda Straw Rocket Has an Air Pressure Launcher
Fig. 7.73	Soda Straw Rocket Initial Baseline Configuration
Fig. 7.74	Soda Straw Rocket Initial Baseline Cost, Weight, and Balance
Fig. 7.75a	Soda Straw Rocket Initial Baseline Definition
Fig. 7.75b	Soda Straw Rocket Initial Baseline Definition (cont)
Fig. 7.76	Soda Straw Rocket Initial Baseline Has Excess Static Margin
Fig. 7.77	Soda Straw Rocket Has High Acceleration Boost Performance
Fig. 7.78	Most of the Soda Straw Rocket Drag Coefficient Is from Skin Friction
Fig. 7.79a	Soda Straw Rocket Initial Baseline Deterministic Flight Range Meets 90 Feet Range Requirement
Fig. 7.79b	Soda Straw Rocket Initial Baseline Deterministic Flight Range Meets 90 Feet Range Requirement (cont)
Fig. 7.80	Work Instructions for Soda Straw Rocket
Fig. 7.81	Range Pareto Analysis Shows Fixed Diameter Soda Straw Rocket Baseline Range Is Driven by Length and Launch Angle
Fig. 7.82	Maturity Assessment of Soda Straw Rocket Is Based on a Comparison of Predictions and Tests
Fig. 7.83	Because of Uncertainty, Soda Straw Rocket Initial Baseline Does Not Meet Range Requirement
Fig. 7.84	Revised Soda Straw Rocket with Smaller Tails (Lower Static Margin, Lighter Weight, Lower Drag) Meets Range Requirement

List of Figures   451

Fig. 7.85a	House of Quality Translates Customer Requirements into Engineering Emphasis
Fig. 7.85b	House of Quality Translates Customer Requirements into Engineering Emphasis (cont)
Fig. 7.85c	House of Quality Translates Customer Requirements into Engineering Emphasis (cont)
Fig. 7.86	Design of Experiments (DOE) Approaches Are a Tradeoff of Time Required vs. Confidence in Result
Fig. 7.87	A DOE Parametric Study Changes One Factor at a Time, with All Others Held at Base Level
Fig. 7.88	A DOE Adaptive OFAT Sequentially Selects the Values of Parameters, Based on Results
Fig. 7.89	A DOE Full Factorial Evaluates All Combinations of Parameters, but Requires More Time
Fig. 7.90	A DOE Study Should Use Engineering Intuition in Searching the Broad Possible Design Space
Fig. 7.91	A Design of Experiments (DOE) Study Should Search the Broad Possible Design Space
Fig. 7.92	Engineering Experience Should Guide Design of Experiments (DOE) Study
Fig. 7.93	A DOE Parametric OFAT Study Provides a Quick Evaluation but Has Lowest Confidence
Fig. 7.94	DOE Adaptive OFAT Provides a Quick Evaluation with Higher Confidence Than Parametric OFAT
Fig. 7.95	A Full Factorial Design of Experiments (DOE) Study Provides Highest Confidence but Requires More Time

Chapter 8 Development Process

8.1 Missile Technology and System Development Process

Fig. 8.1	Chapter 8 Development Process—What You Will Learn
Fig. 8.2	The US Has a Systematic Research, Technology Development, and Acquisition Process for Missile Development
Fig. 8.3	Engineering and Manufacturing Development (EMD) Requires Subsystem Tests
Fig. 8.4	Technology Readiness Level (TRL) Indicates the Maturity of Technology
Fig. 8.5	Conceptual Design Requires Broad Range of Alternatives—Production Requires Configuration Management of Single Design
Fig. 8.6	Missile Development, Integration, Tests, Design Verification, and Validation Is an Integrated Process
Fig. 8.7a	Examples of Missile Development Tests and Facilities
Fig. 8.7b	Examples of Missile Development Tests and Facilities (cont)
Fig. 8.7c	Examples of Missile Development Tests and Facilities (cont)
Fig. 8.7d	Examples of Missile Development Tests and Facilities (cont)
Fig. 8.7e	Examples of Missile Development Tests and Facilities (cont)
Fig. 8.8	Missile Development Flight Test Should Cover the Extremes and Corners of the Flight Envelope
Fig. 8.9	A Relatively Large Number of Test Flights Are Required for a High Confidence Missile System
Fig. 8.10	Missile Block Upgrades Often Provide Improved Flight Performance, Guidance, Flight Control, and Propulsion
Fig. 8.11a	Technology Roadmap Is a Plan for the Development Process
Fig. 8.11b	Technology Roadmap Is a Plan for the Development Process (cont)
Fig. 8.12	Conceptual Design and System Engineering Is Required to Explore Alternative Approaches
Fig. 8.13a	Some Attributes of a Good Technology Development and Demonstration Program
Fig. 8.13b	Some Attributes of a Good Technology Development and Demonstration Program (cont)
Fig. 8.14a	Example of a Typical Technology Development Process—Missile Aerodynamics
Fig. 8.14b	Example of a Typical Technology Development Process—Missile Aerodynamics (cont)
Fig. 8.14c	Example of a Typical Technology Development Process—Missile Aerodynamics (cont)
Fig. 8.14d	Example of a Typical Technology Development Process—Missile Aerodynamics (cont)
Fig. 8.15	There Is Currently Relatively Low US Emphasis on Funding Defense Programs
Fig. 8.16a	There Is Currently Relatively Low Competition for US Missile Development
Fig. 8.16b	There Is Currently Relatively Low Competition for US Missile Development (cont)

8.2 Examples of State-of-the-Art Advancement

Fig. 8.17　Missile Technologies Have Transformed Warfare
Fig. 8.18　US Tactical Missile Follow-On Programs Occur about Every 24 Years
Fig. 8.19　US Strategic Missiles and Their Follow-on Programs Occurred During the Cold War
Fig. 8.20　Example of Missile Technology State-of-the-Art Advancement—Air-to-Air Missile Maneuverability
Fig. 8.21　Example of Missile Technology State-of-the-Art Advancement—Ramjet Propulsion Mach Number

8.3 Enabling Technologies for Missiles

Fig. 8.22　There Are Many Enabling Technologies for Missiles

Chapter 9 Some Lessons Learned

Fig. 9.1　Lessons Learned Base on Career in Missile Design, Development, and System Engineering
Fig. 9.2　Quickly Switch Back-and-Forth Focus from the Big Picture to a Detail Picture (See the Forest As Well As the Trees)
Fig. 9.3　Use Standards As a Guide, to Avoid Going Off Course
Fig. 9.4　Find the Driving Parameters for Most Important Measures of Merit
Fig. 9.5　Select the Best Value for the Driving Measures of Merit
Fig. 9.6　Start with a Good Baseline Design
Fig. 9.7　Confirm Accuracy of Computer Input Data and Model (Bad Input or Bad Model Leads to Bad Output)
Fig. 9.8　Conduct Balanced, Unbiased Tradeoffs
Fig. 9.9　Include System Integration
Fig. 9.10a　Evaluate Many Alternatives
Fig. 9.10b　Evaluate Many Alternatives (cont)
Fig. 9.11　Search a Broad Design Solution Space (Global Optimization vs. Local Optimization)
Fig. 9.12　Evaluate and Refine as Often as Possible for Conceptual Design—Maintain Strict Configuration Management for Production Design
Fig. 9.13　Be Faster and More Accurate Than the Competition
Fig. 9.14　Provide Balanced Emphasis of Analytical vs. Experimental
Fig. 9.15　Consider Law of Unintended Consequences-Decisions May Have Bad Unforeseen Consequences
Fig. 9.16　Keep Track of Assumptions and Develop Real-Time Documents
Fig. 9.17a　Conduct Broad, Unbiased, Creative, Iterative, and Rapid Design Evaluations
Fig. 9.17b　Conduct Broad, Unbiased, Creative, Iterative, and Rapid Design Evaluations (cont)
Fig. 9.18　Establish a Diverse Team for a Balanced Design
Fig. 9.19　Use Creative Skills for Program Planning
Fig. 9.20　Use Input from the Entire Organization
Fig. 9.21　Provide Balanced Investments for Near, Mid, and Long-Term Payoff
Fig. 9.22　Use the Design, Build, and Fly Process for Broader Knowledge and Understanding
Fig. 9.23　Provide Broad Documentation

Chapter 10 Summary

10.1 Missile Design Guidelines

Fig. 10.1　Missile Conceptual Design and System Engineering Sizing Guidelines for Aerodynamics
Fig. 10.2　Missile Conceptual Design and System Engineering Sizing Guidelines for Flight Control
Fig. 10.3　Missile Conceptual Design and System Engineering Sizing Guidelines for Propulsion
Fig. 10.4　Missile Conceptual Design and System Engineering Sizing Guidelines for Weight

Fig. 10.5   Missile Conceptual Design and System Engineering Sizing Guidelines for Flight Performance
Fig. 10.7   Missile Conceptual Design and System Engineering Sizing Guidelines for Guidance
Fig. 10.8   Missile Conceptual Design and System Engineering Sizing Guidelines for Lethality

10.2 Wrap Up

Fig. 10.9a Wrap Up (Part 1 of 3)
Fig. 10.9b Wrap Up (Part 2 of 3)
Fig. 10.9c Wrap Up (Part 3 of 3)

Follow-up Communication

I would appreciate receiving any questions, comments, and corrections that you may have on this material, as well as any data, photographs, drawings, videos, examples, or references that you may offer.

Thank you,

Gene Fleeman
Missile Design, Development, and System Engineering
E-mail: GeneFleeman@msn.com
Web Site: https://sites.google.com/site/eugenefleeman/

Index

1-DOF modeling, 136, 142
2-DOF modeling, 390
3-DOF pitch simulation modeling, 366
3-DOF point mas modeling, 136, 137
4-DOF modeling, 390
5-DOF modeling, 390
6-DOF modeling, 390

A

Abbreviations, 390
Acceleration capability advantage of rocket, 372
Accelerometer, 170
Active guidance, 416
Actuator, 371, 386
Actuator bandwidth, 388, 404
Active radar, 159–161, 259, 425
Adaptive one factor at a time (OFAT), 320, 323
Adhesive bonding, 105
Adjoint miss distance, 414
Advanced Concept Technology Demonstration (ACTD), 378
Advanced Design of Aerodynamic Missiles (ADAM), 307
Advanced Medium Range Air-to-Air Missile (AMRAAM), 211, 212, 416
Advanced Precision Kill Weapon System (APKWS), 194
Advanced Strategic Air Launched Missile (ASALM), 374, 425
Advanced technology demonstration (ATD), 378
Adverse weather, 151, 170
Aerodynamic center, 27, 28, 33, 34, 405
Aerodynamic configuration buildup, 18, 425
Aerodynamic configuration sizing and system engineering, 18
Aerodynamic efficiency, 35
Aerodynamic heating, 29, 112–116, 409
Aerodynamic surface geometry, 32
Aerodynamic trim, 50–52
Aeroelastic instability, 405
Aerosol, 153
AGM-130 missile, 154, 419
AIM-7 Sparrow missile, 184, 264, 425
AIM-120 AMRAAM, 368
Aim bias, 387
Air-breathing propulsion, 56, 57, 287
Aircraft carriage and fire control interfaces, 416
Aircraft carrier strike group, 8
Aircraft control stick, 256
Aircraft integration, 76
Aircraft weapon carriage alternatives, 211

Airframe castings, 102, 106, 375
Airframe machining, 105, 408
Airframe manufacturing, 102–109, 375, 379, 408
Airframe materials, 365
Airframe maximum Mach number, 376
Airframe structure, 102–109, 115, 379
Airframe weight, 97–128
Airframe welding, 103, 365
Air Launched Cruise Missile (ALCM), 243
Air-to-air (ATA), 12
Air-to-surface (ATS), 12
Air turbo rocket (ATR), 158
Aluminum, 92, 107, 370
Ammonium perchlorate (AP), 370
Angle of attack (AoA), 33–35, 38, 39, 41, 43–45, 51, 77, 136, 146, 147, 379
Angular acceleration, 388
Angular resolution, 388
Antenna, 162, 216, 219, 220
Anti-armor missile, 391
Anti-radiation homing (ARH), 177, 412
Anti-radiation missiles (ARM), 390
Anti-ship missiles, 286, 287, 425
Apache helicopter, 240
Apogee, 366
Armor targets, 189
Army Tactical Missile System (ATACMS), 11
ASALM missile, 374, 425
Aspect ratio, 35, 124
Atmosphere, 175, 411
Atmospheric attenuation, 153, 223, 411
Automatic target acquisition (ATA), 391
Automatic target recognition (ATR), 158, 160
Autopilot design, 373
AWACS, 391
Axial compressor, 62, 370, 406
Axial force coefficient, 381

B

Ballistic flight, 143, 144, 306
Ballistic missile time of flight, 127, 371
Bandwidth, 173, 375
Bank-To-Turn (BTT) maneuvering, 391
Base drag, 363
Baseline guided bomb, 301–307
Baseline ramjet missile, 277–286
Baseline rocket missile, 264–276
Baseline turbojet missile, 286–300
Battery

Battle damage assessment (BDA), 391
Battle damage indication (BDI), 391
Beam rider
Beam width, 372
Bending moment, 120, 121
Beyond visual range (BVR), 264
BILL missile, 185
Binder, 88, 89
Blast/fragmentation warheads, 181, 182, 184, 186, 195, 366, 413
Blast overpressure, 184
Blast wave theory, 357
Block upgrade, 331
Boattail, 23–24, 363
Body bending, 20, 120, 121, 404
Body drag, 22–23
Body fineness ratio, 404
Body normal force, 24–26, 363
Boeing, xv
Bonding, 105
Booster, 73–74, 277, 283, 300, 372, 374, 379
Boost flight performance, 139–147
Boost-glide, 139–147
Boost velocity, 19, 300, 426
Boron fuel, 370
Brahmos missile, 46
Breguet range equation, 371, 373
Brilliant Anti-armor Technology (BAT), 391
Brimstone missile, 238, 250
Buckling, 119
Buried targets, 188, 189
Burn area, 83, 84, 370, 371
Burn rate, 89, 371
Burn rate exponent, 371, 407
Burnout velocity, 81, 143, 144, 366
BVR standoff missiles, 264

C

C^3 targets, 376
C4ISR, 13, 363, 375, 376, 379
Campaign model, 404
Canard control alternatives, 87
Canard control induced roll, 41, 43
Canister Launch System (CLS), 391
Captive flight carriage, 120
Captive trajectory simulation (CTS), 392
Capture efficiency, 77, 364
Carbon based fuels, 365
Carbon composite structure, 108, 124
Carriage, 120, 211, 237, 240, 253, 254, 260, 261, 367, 416
Case bonded propellant, 85, 86
Casting, 102, 106, 375
Catastrophic kill (K-kill), 394
CBU-57 dispenser, 391, 416
Center of gravity (cg), 101–104, 367, 379
Center of pressure (cp), 382, 387, 392, 425
Centimeter wave radar (cmW radar), 392
Centimeter wave seeker (cmW seeker), 163
Centrifugal compressor, 299, 368, 370

Chamber pressure, 82, 83, 91, 365, 426
Characteristic velocity, 371, 407
Chemical vapor deposition (CVD), 392
Circular error probable (CEP), 373, 375
Cloud cover, 152, 411
Clutter, 158, 170
Coast flight, 139–144
Collateral damage, 185, 193, 194, 198, 372, 413
Combined effects bomblets (CEBs), 391
Combustion, 64, 68–72, 78, 81, 90–91, 94, 182, 294, 295
Combustor, 71–72, 364, 368, 370
Command and control (C&C and C2), 376
Command, control, and communication (C3), 391
Command, control, communication, and intelligence (C3I), 391
Command, control, communication, computers, intelligence, surveillance, reconnaissance (C4ISR), 391
Command guidance, 81, 257, 258, 375, 416
Commercial off the shelf (COTS) electronics, 234
Composite motor case, 92, 408, 410
Compressed carriage, 253, 416
Compression molding, 104
Compressor pressure ratio, 59, 62, 64, 364, 370, 376, 377
Compressor rotation rate, 370
Compressor stages, 418
Compressor total pressure loss, 364, 368
Compressor type, 370
Computational fluid dynamics (CFD), 391
Computer-aided sizing, 307–309
Conceptual design, 4–8, 16, 18, 19, 52–53, 63, 84, 120, 130, 135
Conductivity, 365, 383
Configuration sizing parameters, 4, 18, 136, 137
Contrast, 154, 158, 265, 373
Control surfaces, 37, 363, 364, 405, 425
Convection heat transfer, 383
Conventional Air Launched Cruise Missile (CALCM), 391
Cool-down time, 262
Cooling, 262
Cost, 230–237
Counter-countermeasures (CCM), 177, 198
Countermeasures (CM), 151, 162, 177
Critical design review (CDR), 391
Cross coupling, 52, 53
Cross-linked double base propellants (XLDB propellants), 398
Cruciform, 48, 49
Cruise flight performance, 293
Cruise missiles (CM), 30, 31
Crystal alloys, 192

D

Data link, 167, 173, 174, 259, 366, 375, 377–379, 411, 412
Data Compendium (DATCOM), 358, 359
Defense Support Program (DSP), 392
Degrees of freedom (DOF), 135, 136, 142
Dallenbach magnetic film RAM, 218, 219

Demonstration and Validation (Demval), 377
Density, 3, 80, 101, 187, 189, 190, 193, 285
Design, build, and fly (DBF), 309
Design case studies, 374, 421
Design criteria, 52–54
Design of Experiments (DOE), 322, 323
Design maturity, 368
Detection range, 156, 157, 159–161, 213, 221, 225, 226
Detector, 155, 158
Development process, 325–340
Development testing and evaluation (DT&E), 392
Diameter tradeoff, 19–20
Digital Scene Area Matching Correlation (DSMAC), 392
Direct Attack Guided Rocket (DAGR), 42
Discrimination, 10, 154
Divert flight, 139–144
Divert and attitude control system (DACS), 392
Documentation, 352, 377, 378, 402
Dome bandpass, 410
Dome error slope, 385
Dome materials, 125, 126, 366
Dome weight, 126
Double base propellant, 398
Drag, 20–26
Drawings, 267, 268, 301, 374, 379
Dual Combustor Ramjet-scramjet (DCR), 392
Ducted rocket combustion temperature, 94
Ducted rocket fuel, 93, 95
Ducted rocket gas generator, 365
Ducted rocket maximum Mach number, 94
Ducted rocket oxidizer, 93
Ducted rocket range, 93
Ducted rocket specific impulse, 95, 365
Ducted rocket thrust, 94
Ductility, 192
Dwell time, 375
Dynamic pressure, 25, 46, 208, 366, 369, 374

E
Ejection launcher, 250, 251
Elasticity, 89, 108, 241
Electrical discharge machining (EDM), 103
Electromagnetic compatibility (EMC), 178, 239
Electromagnetic interference (EMI), 178
Electromagnetic pulse (EMP), 178
Electromechanical (EM) actuators, 128
Electronics, 128, 232–234
Electro-optical (EO), 154
Electro-optical countermeasures (EOCM), 392
Electrostatic discharge (ESD), 178
Emissivity, 387
Enabling technologies, 340, 378, 379
Engagement model, 201
Engineering and manufacturing development (EMD), 231, 326
Environment, 89
Equations of motion modeling, 135–136, 373
Ethylene propylene dimethyl monomer (EPDM), 392
Evolved Sea Sparrow Missile (ESSM), 48

Excalibur guided projectile, 171
Exit velocity of nozzle, 386
Exo-atmospheric Kill Vehicle (EKV), 156
Exocet missile, 46, 226
Expansion ratio, 370
Experimental, 347, 426
Exploratory development, 368, 378, 379
Explosive, 88, 412
Explosively formed penetrator, projectile (EFP), 392
Extruded propellant, 85, 86

F
F-16, 240
F-18, 256
F-22 weapons bay, 211
F-35, 256
Faceted dome, 21
Factor of safety (FOS), 119, 365, 371, 408
Failure, 229, 278
False alarm rate (FAR), 392
Fastener, 105
Fiber Optic Gyro (FOG), 393
Field of regard (FOR), 167
Field of view (FOV), 167, 372
Filament winding, 104
Filter, 426
FIM-92 Stinger missile, 403
Fineness ratio, 20, 404
Finite difference time domain (FDTD), 393
Finite element modeling (FEM), 393
Fire control system (FCS), 132–134, 237, 240, 255, 256
Fire power, 238, 253
Fire power kill (F-kill), 393
Flame holders, 381
Flap control, 40
Flare stabilizer, 28–29
Flight control, 47–48
Flight control actuators, 127–128
Flight control alternatives, 37–46
Flight control deflection, 364
Flight control effectiveness, 52, 53, 203, 367
Flight envelope, 131–134, 330, 366, 374
Flight loads, 408
Flight path angle, 371, 372, 379
Flight performance envelope, 130
Flight performance measures of merit, 6, 18, 97, 99, 129–130, 136, 149, 150, 263, 331, 354, 363, 366, 371–372, 401–402, 410, 412
Flight test, 53, 257, 284, 330, 374
Flight trajectory, 131, 135, 141, 209, 366
Flow path, 58–62, 67–68, 80–81, 279
Flush inlet, 364
Fly-out, 136, 372
Focal plane array (FPA), 155, 374, 411
Forrestal aircraft carrier, 227
Forming, 365
Forward air control (FAC), 392
Forward looking infrared (FLIR), 393
F-pole range, 383, 385, 417

Fracture, 373
Fragment velocity, 183
Free-to-roll tails, 42
Frequency modulation continuous waveform mmW (FMCW mmW), 393
Fuel, 56, 67, 68, 70, 80
Fuel-to-air ratio, 364
Fuel weight, 56, 284, 368
Full factorial DOE evaluation, 321, 323
Fuze, 194–197, 228
Fuzing, 187, 188

G
Gas generator, 365, 370
GBU-31 JDAM guided bomb, 301
GBU-39 small diameter bomb, 31
Gel propellant, 408
Generator, 127
Geometric theory of diffraction (GTD), 393
Gimbaled seeker, 167, 199, 366, 411
Glide flight performance, 366
Glint, 209, 210, 367, 372
Global positioning system (GPS), 167, 171, 172, 174
Government Furnished Material (GFM), 393
Government Furnished Property (GFP), 393
GPS/INS, 375
GPS/INS/data link, 411
Grain configuration, 85
Graphite composite structure, 108, 124
Greenhouse gases, 411
Griffin missile, 180
Ground Based Interceptor (GBI), 393
Ground launch platform, 369
Guidance & control, 187, 199, 368
Guidance law, 198, 202
Guided bomb aerodynamic characteristics, 302–305
Guided bomb ballistic range, 306
Guided bomb baseline, 307
Guided bomb glide range, 307
Guided Multiple Launch Rocket System (GMLRS), 185, 326
Guided weapon advantages, 413
Gun-launched XM982 Excalibur guided projectile GPS, 67
Gyro, 169, 170

H
Hang-fire, 243, 367
Hardware-in-loop (HWL), 394
Harop, 177
Harpoon missile, 243, 286, 300, 425
HCl contrail, 212
Heading error, 200, 205, 206, 367, 414
Heat capacity, 399
Heating value of fuel, 95, 364, 370
Hellfire missile, 180, 185, 212, 233, 241, 250
Helmet mounted sight, 257
Her Majesty's Explosive (HMX), 393, 413
High density polyethylene (HDPE), 393
High firepower missile, 238

High-g Boost Experiment (HiBEX), 393
High Mobility Artillery Rocket System (HIMARS), 11
High power microwave (HPM), 394
High-speed Anti-radiation Missile (HARM), 393
High speed cruise missile (HSCM), 394
Hinge moment (HM), 33–34, 43, 363, 405
Hit-to-kill, 156, 188, 190
Homing guidance law, 202
Hot isostatic press (HIP), 393
House of quality (HOQ), 318, 319, 368, 373, 374, 377–379, 418
HWL simulation, 394
Hydrocarbon fuel, 370, 407
Hydroxyl-terminated polybutadiene binder (HTPB), 394
Hypersonic missile (HM), 109, 404

I
Identification-friend-foe (IFF), 394
Ignition, 379
Imaging infrared seeker alternatives, 125–126
Imaging infrared seeker dome, 125, 126
Imaging infrared seeker target acquisition, 154
Impact angle, 25
Induced roll, 41, 43
Inert weight, 404, 418
Internal carriage, 211, 249
Inertial navigation system (INS), 167, 168, 171, 174, 259, 287
Infrared (IR) dome, 125
Infrared seeker, 125, 151, 155, 156
Infrared signature, 222, 378
Initial operational capability (IOC), 377
Inlet capture efficiency, 364
Inlet integration, 73
Inlet sideslip, 379
Inlet start, 76, 78, 370
Insensitive munition (IM), 227
Isentropic flow, 79, 407
Instantaneous field of view (IFOV), 372
Insulation efficiency, 111, 409
Insulation material, 110–111
Insulation maximum temperature, 111
Insulation thermal diffusivity, 387
Insulation thickness, 111
Integral rocket-ramjet (IRR), 277
Integrated circuits (ICs), 394
Integrated product and process development (IPPD), 377
Intercontinental ballistic missile (ICBM), 394
Inertial navigation system (INS), 167, 168, 171, 174, 259, 287
Inter-Tropical Convergence Zone (ITCZ), 394
Investment casting, 102
Isentropic flow, 370

J
Jamming, 172
Javelin missile, 212
JDAM guided bomb, 158, 301, 375, 425
Jet interaction (JI), 394

Jet penetrator, 190
Jet tabs, 406
Jet vanes, 405
Joint Air-to-Ground Missile (JAGM), 394
Joint Air-to-Surface Strike Missile (JASSM), 394
Joint attachment, 105
Joint Helmet Mounted Cueing System (JHMCS), 394
Joint Standoff Weapon (JSOW), 394
Joint Tactical Information Display System (JTIDS), 394
JP-10 fuel, 370

K
Kill probability, 181
Kinetic energy, 183, 184, 187–189, 367
Kinetic kill missiles, 190
Kinetic energy warhead, 188, 385

L
LADAR seeker, 394
Laminate graphite composite, 108, 124
Laser cross section (LCS), 395
Laser detection and ranging sensor (LADAR sensor), 394
Laser seeker, 158, 159
Latency time, 372
Lattice fin, 40, 41, 364
LAU-142 launcher, 250
Launch platform compressed carriage, 253, 416
Launch platform integration, 239–244
Launch platform safety, 227
Launch platform stores, 251
Launch separation, 246
Launch velocity, 386
Launcher, 241, 250, 251
Lead angle, 139
Learning curve, 232
Lessons learned, 341–352
Lethality, 178–197
Lethal radius, 181, 184, 366, 373
Lifting body, 24–26, 53, 363
Lift-to-drag ratio, 25, 56, 137, 138, 140, 307
Light weight missiles, 92, 145
Linear wing theory, 425
Line-of-sight (LOS), 367
Line-of-Sight Anti-Tank missile (LOSAT missile), 395
Liner, 191, 192
Liquid hydrocarbon fuel, 370, 407
Lithium battery, 410
Load distributions, 120
Load-out configurations, 8
Lock-on-after-launch (LOAL), 395
Lock-on-before launch (LOBL), 395
Loft-ballistic trajectory, 366
Loft-glide trajectory, 141
Long-Range Anti-Ship Missile (LRASM), 144
Long wave infrared (LWIR), 187
Logistics, 235–237, 379
Low Cost Autonomous Attack Submunition (LOCAAS)

Low observables, 367
Low temperature environment, 90

M
Machining, 105, 373, 408
Mach number range propulsion types, 376
Maneuverability, 17, 38, 42, 43, 46, 98, 205, 206, 212, 213, 274, 276, 339
Maneuver acceleration, 384, 414
Maneuver law alternatives, 47–49
Manufacturing, 102–109, 365, 373, 375, 379
Mass properties, 267, 278, 289, 302
Maximum effective operating pressure (MEOP), 122
Maximum range trajectory, 131, 297–299
Maximum thrust, 59, 282, 370, 371
Mean aerodynamic chord (mac), 382, 386, 387, 395, 405
Measures of merit, 149–262
Medium range missiles, 395, 425
Medium wave infrared (MWIR), 395
Metal Augmented Charge (MAC) warhead, 182
Metal casting process, 106, 375
Metal fuels, 365
Meteor missile, 93
Method of moments (MoM), 395
MGM-140 ATACMS, 403
Micro-electromechanical systems (MEMS), 169, 237, 415
Midcourse guidance, 168, 169
MIL-A-8591, 120
Military Standard (MIL STD), 250, 252
Military Technology Control Regime (MCTR), 395
Millimeter wave, 162, 163
MIL-STD-8591, 250, 252
Minimum range, 131, 239
Minimum smoke, 212, 408
Minuteman III ICBM, 175, 371
Miss distance contributors, 414
Miss distance reduction, 210, 372
Missile aerodynamics, 334, 335
Missile ageing, 89
Missile carriage, 237, 254
Missile cost, 230
Missile countermeasures, 151, 162, 177
Missile counter-countermeasures, 177
Missile cruise, 293, 295–297
Missile diameter tradeoff, 19–20
Missile defense system, 9
Missile development process, 325–340
Missile dome, 21
Missile electromagnetic compatibility (EMC), 178, 239
Missile electronics, 234
Missile failures, 229
Missile fineness ratio, 20
Missile flight control, 38
Missile flight envelope, 131–134, 330
Missile flight test, 83, 247, 284, 330
Missile funding, 336
Missile GPS/INS, 375, 411
Missile guidance, 175, 199, 202, 203

Missile hardware in loop simulation, 394
Missile high temperature environment, 89
Missile infrared signature, 222, 378
Missile integration with fire control system, 132–134, 240, 255, 256, 258, 260, 367
Missile integration with launch platform, 132–134
Missile kill criteria, 181
Missile launch, 8, 99, 117, 239–249, 365, 380
Missile lethality, 178–197
Missile logistics, 235, 376
Missile low temperature environment, 90
Missile manufacturing, 102–106
Missile maneuverability, 17, 206, 339, 367
Missile maximum Mach number, 376
Missile miss distance, 181, 185, 186
Missile modeling, 7, 101, 135–136
Missile moment of inertia, 101–102
Missile navigation, 167, 168, 171, 174
Missile observables, 211
Missile parts count, 106, 234
Missile power supply, 127–128
Missile production rate, 379
Missile propulsion, 57
Missile radar cross section (RCS), 21, 66, 211
Missile reliability, 227
Missile robustness, 177
Missile seeker, 153, 259, 366
Missile signature tests, 224
Missile sizing, 263, 425
Missile storage, 235
Missile subsystems, 3, 99, 365
Missile survivability, 212–227
Missile Systems Technical Committee (MSTC), 375
Missile targets, 132–134
Missile technology assessment, 403
Missile technology development, 325, 333–335, 378
Missile terminal guidance, 198, 257, 258, 367
Missile test facilities, 330
Missile time constant, 204, 210, 367, 414
Missile types, 12
Missile warhead, 178
Missile weight, 99–102, 365
Mission planning, 225, 226
Mission requirements, 6, 13–14, 18
Mixed compression inlet, 370
Mk41 Vertical Launch System (VLS), 260
MK 84 warhead, 425
Mobility kill (M-kill), 395
Modeling, 7, 101, 135–136
Modulus of elasticity, 382, 409
Moment-of-inertia, 101–102
Mono-static radar, 216
Most important requirements (MIRs), 373
Motor case, 92–93, 123
Moving targets, 173, 412
Multifunctional Information Distribution System (MIDS), 395
Multi-Mission Cruise Missile (MMCM), 375–379

Multi-mode guidance, 259
Multi-mode propulsion, 177
Multi-mode seeker, 125
Multiple independent target reentry vehicles (MIRVs), 395
Multiple Launch Rocket System (MLRS), 11, 395
Multi-purpose aircraft, 180
Multi-stage compressors, 62
MWIR seeker, 372

N

Navigation, 167, 168, 171, 174
Navigation ratio, 201, 205, 206, 209, 211, 367
Navy Theater Wide (NTW), 396
NEAR MISL3 computer code, 359
Network centric warfare, 255
Neutral static stability, 369, 374
Nickel-based super alloys, 368
Nielsen Engineering and Research (NEAR), 357
Noise, sensor, 19, 158
Nomogram, 357
Non-cooperative target identification (NCTID), 396
Normal distribution function, 413
Normal force coefficient, 32, 148, 373, 405
Normal shock wave, 79, 370
Northrop Grumman, 194
Nose drag, 22, 23
Nose fineness, 21, 22
Nose geometry, 21
Nose tip bluntness, 23
Nose inlet, 77
Nose wave drag, 22
Nozzle expansion ratio, 82, 365, 370
Nozzle materials, 92
Nusselt number, 384

O

Oblique shock wave, 79
Observables, 80, 88, 151, 211–212, 225, 365, 367
Observe, orient, decide, act (OODA), 396
Off boresight, 131, 219, 239, 366
One-factor-at-a-time (OFAT) DOE, 320, 322, 323
Operational tests and evaluation (OT&E), 396
Optimum cruise, 174, 412
Oxidizer, 91, 93, 370

P

PAC-2 missile, 396
Pareto analysis and sensitivity, 6, 309–323
Parts count, 106, 234, 365
Passive imaging IR seekers, 154
Patriot Advanced Capability (PAC-3) missile, 396
Payload weight, 99, 365, 371
Penetration, 186, 189, 190, 367
Phased array radar, 260
Physical theory of diffraction (PTD), 396
Pintle motor, 365
Pitch control, 382
Pitching moment coefficient, 50, 382

Pixel detector bandwidth, 388
Plume, 81, 229
Podded drop-off booster, 407
Polarization, 160
Polyetheretherketone (PEEK), 396
Polyetherketone (PEK), 396
Polyimide, 409, 410
Polyphenylene (PPE), 396
Power supply, 378, 379, 396
Precipitation, 153
Precision guided munition, 357, 396, 425
Precision strike accuracy, 174, 178–180, 238, 359, 367, 376, 421
Precision strike standoff missiles, 174, 178
Predator, 11, 180, 236
Preliminary design, 135
Preliminary design review (PDR), 396
Probability density function, 396
Probability of kill (P_K), 181, 373
Processor, 392
Production cost, 231, 232, 234, 367
Production drawings, 267, 288
Program Definition and Risk Reduction (PDRR), 377
Propellant aging, 89–90
Propellant burn rate, 371
Propellant characteristic velocity, 371, 407
Propellant density, 371
Propellant grain, 122, 371
Propellant properties, 175
Propellant safety, 212
Propellant specific impulse, 56
Propellant weight, 83, 99, 139, 144, 272, 365
Propellant weight fraction, 81
Proportional guidance, 200, 201, 411
Propulsion technology, 359, 360
Prototype missile, 377, 378
Proximity fuzing, 194–196
Pseudo imaging seeker, 155
Pseudolites, 172
Pulse motor, 87, 365
Pulse radar, 159
Pultrusion manufacturing, 365
Pursuit guidance, 360
Pyrotechnic and non-pyrotechnic launchers, 251
Python missile, 405

R
RBS-15 missile, 406
Radar beam width, 372
Radar absorbing material (RAM), 216, 217, 219
Radar absorbing structure (RAS), 217, 218
Radar atmospheric transmission, 388
Radar bi-static, 387
Radar countermeasures, 151, 162, 177
Radar cross section (RCS), 21, 66, 211, 213–215, 221, 363, 372, 373, 378, 414
Radar frequency (RF), 173, 219, 367
Radar frequency counter measures (RFCM), 397

Radar glint, 209, 367
Radar resolution, 156
Radar seeker and sensors, 126, 151, 152
Radar signal processing, 391
Radar sites, 375
Radiant intensity, 372, 383
Radiation heat transfer, 116
Radio frequency (RF), 397
Radome error slope, 204
Radome weight, 126
Rail launchers, 252, 367
Ramjet baseline missile, 277–285
Ramjet booster, 73–74
Ramjet combustion, 68, 72
Ramjet combustor, 71–72
Ramjet cruise drag, 407
Ramjet fuel, 397
Ramjet flow path, 67–68
Ramjet inlet, 74–75
Ramjet maximum Mach number, 376
Ramjet thermal choking, 71
Ramjet thrust, 69–70
Raytheon, 194, 260
RDX (Royal Demolition Explosive), 396
Reaction jet control (RJC), 397
Real gas effects, 370
Recognition range, 265, 373
Recovery pressure, 66, 78
Reduced smoke propellant, 212
Reference area, 369, 423
Reliability, 227–230
Request for proposal (RFP), 373–380
Research, technology, and acquisition (RT&A), 385
Resin transfer molding (RTM), 365
Resolution, 151, 156, 160
Resonance, 20
Rhenium, 406
Ring laser gyro (RLG), 397
Rise time, 415
Risk, 187, 368
Robustness, 150–178
Rocket baseline missile, 264–276
Rocket motor case bonded propellant, 85, 86
Rocket motor case material, 193
Rocket motor combustion chamber, 366
Rocket motor end-of-boost velocity, 371
Rocket motor internal insulation, 114
Rocket motor nozzle, 92
Rocket motor performance prediction, 81–82
Rocket motor propellant weight flow rate, 83, 84, 370
Rocket motor throat, 82, 83
Rocket motor thrust, 87
Rocket motor thrust magnitude control, 87
Rocket motor weight, 121, 410
Rocket nozzle, 92
Rocket plume, 81
Rocket propellant
Roland missile, 185, 258, 274

Roll control, 37, 363
Roll cross coupling, 52, 53
Rolling airframe (RA), 48, 364, 406
Rolling Airframe Missile (RAM), 396
Rolling moment, 382
Root mean square (RMS), 397
Root sum of squares (RSS), 397
Rotational speed, 299
Russia 3M80 (SS-N-22 Sunburn) missile, 79

S
Safe and arm (S&A), 230
Safety, 212–226
Satellite communication (SATCOM), 397
Satellites, 13
Scanning array, 155
Scenario definition, 363
Scramjet engine demonstrator (SED), 397
Scramjet propulsion, 406
Seeker alternatives, 14
Seeker antenna, 216, 219
Seeker countermeasures, 177
Seeker detector, 372
Seeker dome, 125–126
Seeker field of regard (FOR), 167
Seeker field of view (FOV), 167
Seeker focal plane array, 374
Seeker gimbals, 199, 366
Seeker/GPS/INS/data link synergy, 411
Seeker range, 404
Seeker resolution, 210
Self-Defense Launcher (SDL), 397
Semi-active guidance, 416
Semi-active laser (SAL) seeker, 158, 159
Semi-active radar, 161, 425
Sensor fuzed weapon (SFW), 228
Shaped charge warhead, 190–192
Shock, 79, 90
Shock wave, 76, 77
Short period dynamics, 388
Short range missiles, 14
Sideslip, 379
Sign convention, 26–27, 49–50
Signal-to-noise (S/N) ratio, 19, 158
Signature tests required, 224
Simulation, 7, 366, 368, 378
Sintering, 365
Sizing tools, 263–324
Sizing verification and validation, 327
Skid-To-Turn (STT) maneuvering, 47, 48
Skin friction drag, 34
Stall, 41, 43, 45
Standoff Land Attack Missile (SLAM), 397
Slender body theory, 425
Slender wing theory, 405, 425
Slurry fuel, 370
SM-3 kinetic kill warhead, 403
SM-3 Standard missile, 403

Small Diameter Bomb (SDB), 397
Smoke, 212, 408
Soda straw rocket design, build, and fly (DBF), 309–323
Soda straw rocket Design of Experiments (DOE), 309–323
Soda straw rocket Pareto sensitivity, 309–323
Soda straw rocket uncertainty analysis, 309–323
Software, 359
Solid propellant aging, 89–90
Solid propellant alternatives, 88–89
Solid propellant binders, 89
Solid propellant characteristic velocity, 371
Solid propellant fuel, 93
Solid propellant grain, 122
Solid propellant oxidizer, 93
Solid propellant smoke, 212
Solid propellant rocket motor chamber pressure, 83
Solid propellant rocket motor exit velocity, 386
Solid propellant rocket motor lifetime, 89
Solid propellant rocket motor pintle, 365
Solid propellant rocket motor sizing, 83–85
Solid propellant rocket thrust magnitude control, 86–87
Sound pressure level, 381, 397
Sparrow missile, 48, 184, 212, 264
Specific detectivity, 382
Specific fuel consumption (sfc), 397
Specific heat at constant pressure, 382
Specific heat ratio, 63, 370
Specific impulse, 56, 64, 68
Spectral radiance, 384
Split canards, 364
Spreadsheet, 308, 309, 422–425
SRAM missile, 397
SS-N-22 Sunburn, 79
Stability & control, 52, 53
Stabilization, 46, 166
Stabilizers, 36
Staging, 100
Stand-off Missile (SOM), 14
Starstreak missile, 258
Statement of work (SOW), 375, 379, 402
State-of-the-art (SOTA), 12, 337–339
Statically unstable missile, 364
Static margin, 27, 52, 272, 313, 366, 373
Steady-state flight, 137–139
Steel, 92, 107
Storage, 235, 375
Store separation, 246, 251, 252
Storm Shadow/Scalp missile, 403, 406
Strapdown seeker, 166, 167, 366, 372, 411
Strategic missiles, 2, 338
Strength, 105, 107, 108, 365
Structure buckling, 119
Structure design, 117–124
Structure thickness, 121
Structure weight, 121
Stunner missile, 405
Submarine carriage and integration, 261

Subsonic cruise missile (SCM), 30, 31
Subsystems packaging, 3
Supersonic inlet, 76–79
Surface Launched AMRAAM (SLAMRAAM), 397
Surface planform geometry and integration, 36
Surface-to-air (STA), 397
Surface-to-air missile (SAM), 363, 375
Surface-to-surface (STS) missile, 403
Surface targets, 375, 376
Sustain thrust, 270
Survivability, 212–227
Synthetic aperture radar (SAR), 374, 411
System analysis, 363
System engineering, 4–8
System integration, 7, 284, 344
System-of-Systems, 8–12
System requirements flow-down, 363
System requirements review (SRR), 397

T

Tactical Air-Launched Decoy (TALD), 398
Tactical Missile Design (TMD) Spreadsheet, 308, 423–424
Tactical missiles, 2, 218, 364, 405, 419
Tail area sizing, 52
Tail control, 405
Target detection, 398, 414
Target error (TE), 398
Target glint, 372
Target kill, 404
Target latency, 372
Target location error (TLE), 173
Target maneuverability, 206
Target recognition, 158, 160
Target signal attenuation, 366
Target span resolution, 381
Target weave, 367
Taylor method, 357
Technical requirements document (TRD), 398
Technologies, 337, 340
Technology availability date (TAD), 378
Technology roadmap, 332, 377
Technology Readiness Level (TRL), 326, 377, 378
Teledyne Model 370, J402-CA-400 turbojet engine, 292, 293
Telemetry (TM), 398
Television (TV), 398
Tensile stress, 388
Terminal guidance, 198, 257, 258, 367
Terrain contour matching (TERCOM), 398
Terrain masking, 212
Tests and test facilities, 328–330
Theater ballistic missiles (TBMs), 375
Theater High Altitude Area Defense (THAAD) missile, 398
Thermal battery, 127, 128, 366, 410
Thermal choking, 71, 370
Thermal diffusivity, 387
Thermally thick, 112–114

Thermally thin, 112, 113
Thermal response prediction, 112–115
Thermal stress, 116–117
Thermobaric warhead, 182
Third stage rocket motor (TSRM), 398
Threat radar site destruction, 141
Thrust magnitude control (TMC), 86–87
Thrust-to-weight, 57, 146, 147
Thrust vector control (TVC), 364, 366
Time constant, 202–207, 210
Time critical targets (TCTs), 13, 179, 375, 376
Time of flight, 127, 371, 385
Titanium, 107
Tomahawk missile, 211, 243, 375
Total impulse, 383, 426
Total kinetic energy, 183
Tow missile, 258
Toxicity, 88
Track-via-missile (TVM), 398
Trajectory shaping, 135, 141
Transmit, 159
Transmitter, 164, 414
Transportation, 375
Traveling wave tube (TWT), 164
Trident missile, 21, 229
Trim, 50–52
Tungsten, 408
Turbine, 57, 59, 63
Turbojet baseline missile, 286–300
Turbojet compressor, 406
Turbojet engine rotational speed, 299
Turbojet exit velocity, 386
Turbojet flow path, 58–62
Turbojet maximum Mach number, 376
Turbojet nozzle, 82
Turbojet propulsion efficiency, 65–66
Turbojet range prediction, 295–299
Turbojet specific impulse, 64
Turbojet thrust, 63
Turbojet turbine materials, 377
Turbojet weight and balance, 312, 379, 402
Turbulent flow, 365
Turn radius, 145, 146, 272, 275, 366, 372
Turn rate, 145, 146, 276, 372
TVC, 46, 146, 147, 229, 364, 405, 410

U

Ultimate strength, 372
Ultraviolet (UV), 398
Uncertainty analysis, 309–323
Unmanned air vehicle (UAV), 13, 372, 375, 377
Unmanned combat air vehicles (UCAVs), 180, 369

V

Vacuum assisted resin transfer molding (RTM), 365
Validation, 327, 377, 378
Verification, 309, 327, 377, 378, 425
Vertical Launch System (VLS), 260, 376, 377, 379, 416
Visual detection range, 265, 373

Visual recognition range, 265, 373
Volumetric efficiency, 26, 371
Vortex shedding, 44
Vulnerable area, 186

W

Warhead blast overpressure, 184, 372
Warhead charge energy, 372
Warhead collateral damage, 185
Warhead design considerations, 178, 187
Warhead detonation, 192, 194
Warhead directed energy, 187
Warhead ductility, 192
Warhead fragmentation spray angle, 184
Warhead fragments kinetic energy, 183, 184, 187
Warhead fragments penetration, 186
Warhead fragments velocity, 183
Warhead fuzing, 188
Warhead jet penetration, 190
Warhead kinetic energy, 183, 184, 187, 188
Warhead lethal radius, 186
Warhead liner, 191
Warhead probability of kill, 181
Warhead safe and arm (S&A), 230
Warhead weight, 387
Wave drag, 22
Wavelength, 153, 157, 162
Weather, 151, 170, 411
Weaving target miss distance, 367, 414
Weight of actuator, 127–128
Weight of airframe, 102–109
Weight of baseline guided bomb, 301–307, 418
Weight of dome, 125–126
Weight of fuel, 387
Weight of insulation, 109–111
Weight of missile, 97, 99–101
Weight of power supply, 127–128
Weight of propellant, 99
Weight of ramjet baseline missile, 72
Weight of rocket motor, 121–124
Weight of rocket baseline missile, 124
Weight of structure, 109–110, 117–124
Weight of warhead, 387
Welding, 103, 365, 373
Wide Area GPS Enhancement (WAGE), 398
Window dome, 21
Wind, 247
Wind tunnel, 76, 252, 373
Wind tunnel store carriage loads, 247
Wind tunnel store separation, 247
Wing aerodynamic center, 405
Wing aerodynamic chord, 405
Wing maximum angle of attack, 364
Wing-body-tail configuration, 367
Wing flight control, 43, 44
Wing friction drag, 34
Wingless missiles, 42
Wing normal force, 363, 369, 405
Wing sweep, 35
Wing vortex shedding, 44
Wooden round storage, 377, 415

Y

Yaw control, 363
Yawing moment, 382
Yield strength, 393

Z

Zero lift drag, 22, 25, 294
Zero lift drag coefficient, 22, 25, 294, 368, 369, 373, 374

Supplemental Materials

Supplemental materials for this book are available on its website. Supplemental materials include the following:

- A presentation of the Missile Design and System Engineering short course that is the basis of this textbook. The slides are in full color. Embedded with the slides are over 100 videos illustrating missile design considerations, development testing, manufacturing, and technologies.
- A Tactical Missile Design spreadsheet in Microsoft Excel format. The spreadsheet models the configuration sizing methods for rocket-powered, turbojet-powered, and ramjet-powered missiles as well as guided bombs.
- Missile design case studies. These were conducted by Georgia Tech graduate students.
- A presentation of Soda Straw Rocket Science projects. It is an aerospace engineering outreach program for students.